Optical Channels

Fibers, Clouds, Water, and the Atmosphere

Applications of Communications Theory
Series Editor: R. W. Lucky, *AT & T Bell Laboratories*

Recent volumes in the series:

COMPUTER COMMUNICATIONS AND NETWORKS
John R. Freer

COMPUTER NETWORK ARCHITECTURES AND PROTOCOLS
Edited by Paul E. Green, Jr.

DATA TRANSPORTATION AND PROTECTION
John E. Hershey and R. K. Rao Yarlagadda

DEEP SPACE TELECOMMUNICATIONS SYSTEMS ENGINEERING
Edited by Joseph H. Yuen

DIGITAL PHASE MODULATION
John B. Anderson, Tor Aulin, and Carl-Erik Sundberg

DIGITAL PICTURES: Representation and Compression
Arun N. Netravali and Barry G. Haskell

ERROR-CORRECTION CODING FOR DIGITAL COMMUNICATIONS
George C. Clark, Jr., and J. Bibb Cain

FIBER OPTICS: Technology and Applications
Stewart D. Personick

FUNDAMENTALS OF DIGITAL SWITCHING
Edited by John C. McDonald

MODELING AND ANALYSIS OF COMPUTER COMMUNICATIONS NETWORKS
Jeremiah F. Hayes

MODERN TELECOMMUNICATIONS
E. Bryan Carne

OPTICAL CHANNELS: Fibers, Clouds, Water, and the Atmosphere
Sherman Karp, Robert M. Gagliardi, Steven E. Moran, and Larry B. Stotts

OPTICAL FIBER TRANSMISSION SYSTEMS
Stewart D. Personick

PRACTICAL COMPUTER DATA COMMUNICATIONS
William J. Barksdale

Optical Channels

Fibers, Clouds, Water, and the Atmosphere

Sherman Karp
Lutronix, Inc.
San Diego, California

Robert M. Gagliardi
University of Southern California
Los Angeles, California

Steven E. Moran
SAIC
San Diego, California

Larry B. Stotts
DARPA
Arlington, Virginia

Springer Science+Business Media, LLC

Library of Congress Cataloging in Publication Data

Optical channels: fibers, clouds, water, and the atmosphere / Sherman Karp...[et al.].
p. cm.—(Applications of communications theory)
Includes bibliographies and index.

DOI 10.1007/978-1-4899-0806-3
1. Optical communications. I. Karp, Sherman. II. Series.
TK5103.59.066 1988 88-15063
621.38′0414—dc19 CIP

Originally published by Plenum Press, New York in 1988
MyCopy version of the original edition 1988

Preface

When we were first approached by Dr. Lucky to write this book we were very enthusiastic about the prospect, since we had contemplated a similar project for quite some time. The difficulty lay in how best to digest the vast amount of data on optical propagation, reduce it to a book of manageable size, and simultaneously form the transition from the physics of propagation to the engineering of optical channels. This is the intent of *Optical Channels*. In accomplishing our goal it was necessary to condense the material on optical propagation and, in so doing, we have left a large amount to be handled via references. We have tried to make these decisions in a consistent manner so that the book will be uniform in its treatment of this topic.

We identify four channels for consideration: the free-space channel, which is characteristic of a tranquil atmosphere or a space-to-space link; the turbulent channel, which is characteristic of the atmospheric channel; the scatter channel in two forms, clouds and water; and the fiber optic channel. For each of these channels we have tried to reduce the applicable propagation theory to a level that can be used for engineering design. This has been done by example, but here again decisions had to be made on which examples to present. We have not tried to present any material on optical components and consequently other references on engineering would be necessary to supplement this book.

We have presented the material on optical propagation from two parallel points of view, radiative transport and Maxwell's equations, and have tried to show the relationship between the two. Since the relationship is complex, most of this discussion has been relegated to an appendix and the references. In Chapter 1 we introduce the concept of sources, both point and distributed, by using the traditional concepts of radiance. The propagation of radiance in the free-space environment is discussed, as are methods for considering distributed and noise sources. In Chapter 2 we introduce the concept of the mutual coherence function and show how it is propagated between source and receiver aperture and then between aperture and detector. In the process we introduce the concept of a linear channel and discuss how this affects the propagation of mutual coherence. In so doing, we discuss the issues of signal narrowbandedness and time-space separability. When the channel is separable we see that all spatial effects can be analyzed by passing the mutual coherence function through a channel transfer function, a process much like that covered by conventional filter theory. This concept is used in Chapter 5 for the turbulence channel, and in parts of Chapter 7 for the scatter channel. In Chapter 3 we present an overview on how to make signal-to-noise calculations when the mutual coherence is also included, for both coherent and incoherent systems. Each type of system is described as necessary. We also discuss how to make bit error calculations, with special focus on the avalanche detector. In Chapter 4 we use results from radiative transport theory to describe the behavior of the fiber optic channel. Generic results are presented wherever possible. In Chapter 5 we present a thorough review of the known results in propagation through turbulent media. These results are then applied to both incoherent and coherent communication systems. In Chapter 6 we present a review of the known physical constituents of the optical scatter channels, e.g., clouds, haze, fog, and water. This chapter brings this material together for the first time in a manner and form that permit us to consider the optical scatter channel in its general systems context. Finally, in Chapter 7 we again present a summary review of the known transmission results for optical scatter channels. Topics discussed include pulse stretching in clouds and blue-green propagation into and through water. Whenever possible, general examples are chosen from which insight into the basic problems can be presented.

This book was written to be used primarily as a reference book for engineers working on optical communication systems. Consequently it uses specific individual optical channels for examples. For those interested in fiber optics, Chapter 1 will give some insight into the behavior and characterization of optical sources. To understand this topic there is no need for a discussion of mutual coherence, so one would then go to Chapter 3 for a discussion of signal-to-noise ratios and bit error probabilities. In Chapter 4 we cover the effects of scattering, absorption, and pulse stretching as required to size communication links. We have not included information on couplers and other components that would also be needed to design fiber optic networks.

Those interested in designing transatmospheric systems would again start in Chapter 1. However, they would next proceed to Chapter 2 for the required information on the mutual coherence function and channel transfer functions. From there one would proceed to Chapter 3 to learn about calculating communication system signal-to-noise ratios and bit errors.

Finally, one would end at Chapter 5, in which the turbulent channel is discussed with design examples.

Those interested in underwater communications would also proceed through Chapters 1-3, skip to Chapter 6 for information on the scattering properties of water, and go on from there to Chapter 7 for models and examples of the underwater channel.

For those engineers interested in the cloud channel or other similar channels in which use of mutual coherence breaks down, the route would be Chapters 1, 3, 6, and 7. We point out here that Chapter 7 considers models for both the forward-angle scattering channel and the general scattering channel. For the latter, Monte Carlo techniques are also presented.

In summary, we have tried to make this book as useful as possible as a resource for engineers interested in designing communication systems for a variety of optical channels. In the process of doing so, we feel that we have also written a state-of-the-art reference book on optical channels that will be useful for those in other disciplines, such as optical radar systems and lidars. Finally, we feel that because of the breadth of material covered, the volume will also serve as a handy reference book for researchers in the field of optical propagation. We would like to thank Drs. G. M. Lee, P. A. Bello and R. F. Lutomirski for their technical insights and advice. We also thank Mrs. Gilda Corley for typing the manuscript.

Sherman Karp
Robert M. Gagliardi
Steven E. Moran
Larry B. Stotts

Contents

Chapter 1. Introduction

Chapter 2. Coherence Theory and Random Channels

Chapter 3. Optical Receivers

Chapter 4. The Fiber Optic Channel

Chapter 5. The Turbulence Channel

Chapter 6. The Optical Scatter Channel and Its Properties

Chapter 7. Mathematical Models for Energy Propagation in the Optical Scatter Channel

Optical Channels

1

Introduction

In recent years a great deal of interest has centered on using optical frequencies as carriers of information. This interest has been motivated by improved optical sources, notably the laser. The advantages with respect to optical frequencies lie in the large potential bandwidth available, as a percentage of the carrier frequency, and the extremely short wavelengths with the attendant small component size. Although there was an initial flurry of activity in optical communications after the advent of the laser, scientists and engineers were soon faced with the realities of providing reliable communications under adverse conditions. Clouds, rain, and turbulence dispelled much of the initial enthusiasm, and it was not until the development of virtually lossless glass fibers that optical communications became a commercial reality. Nevertheless, just as there are still obstacles in the deployment of fiber optic communications systems, so too have strides been taken in the understanding and utilization of other means of optical communications. Other books have focused either on the theory of optical communications or on the device technology that either exists or is being developed. In this book we will focus on the physical channels over which optical transmissions might pass and try to develop a common framework by which they can all be considered. We will consider component characterization only insofar as it aids our understanding of optical transmission.

1.1. The Optical Transmission System

A diagram of the generic optical communication system is shown in Figure 1.1a. A source of light is modulated with either analog or digital information and transmitted to the receiver. The various modulation formats that can be used with an optical carrier will be discussed in Chapter 3. The modulated optical beam contains a temporally modulated carrier centered

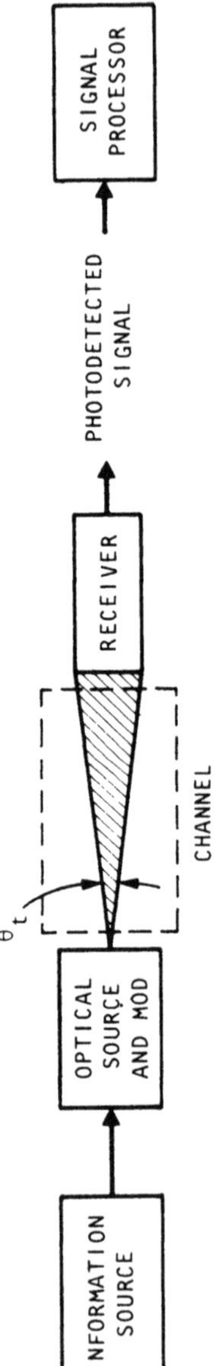

Figure 1.1a. The generic optical communication system.

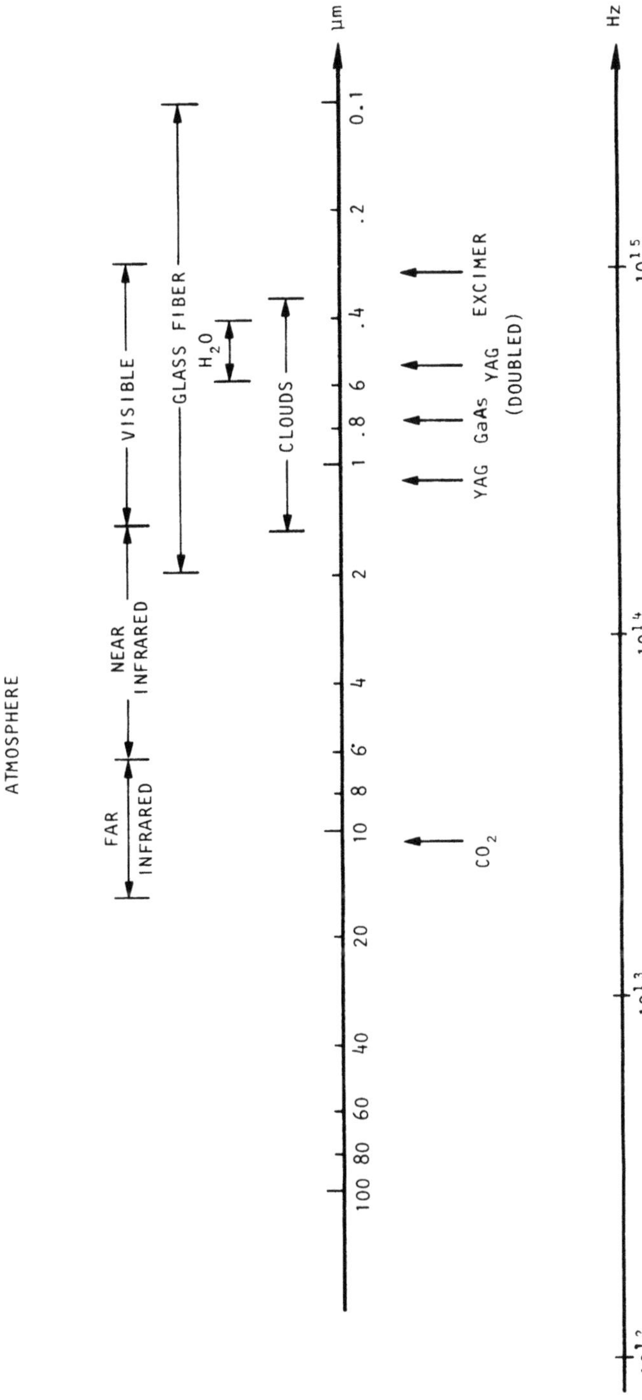

Figure 1.1b. The electromagnetic spectrum.

at a frequency f_c, corresponding to the optical wavelength $\lambda_c = c/f_c$, where c is the speed of light ($c = 3 \times 10^8$ m/s). For optical sources, the wavelength λ_c will lie somewhere in the optical region of the electromagnetic spectrum shown in Figure 1.1b. The corresponding carrier frequency in hertz is also shown. Notice that optical frequencies fall in the range of 10^{13} to 10^{15} Hz, with wavelengths in the range of 0.1 to 10 μm. The diagram also indicates the spectral location of some well-known laser sources.

The optical beam can be considered as a tight bundle of rays emanating from the source in the direction of propagation. Thus a source with a transmitted solid angle of Ω_s in steradians transmits rays in all directions within this solid angle. The power radiated by the source is finite, and consequently at any distance from the source there is a corresponding power density (watts/area) defining the strength of the field. The beam pattern that describes the radiation is usually symmetric and is most often characterized by its planar angular cross section θ_s, in radians. For a *diffraction-limited* optical transmitting system (which will be defined later in the book) having a characteristic diameter d, the planar angular beamwidth is approximately given by

$$\theta_s = \frac{\lambda_c}{d} \tag{1.1.1}$$

Thus, for example, a transmitting source with a 6-inch diameter operating at a wavelength of 1 μm will produce a beam with a planar angle of approximately 6.5 μrad and a solid angle on the order of 4×10^{-11} sr. Since one milliradian equals 0.057 degrees, the ability to produce pencil-thin electromagnetic beams with relatively small transmitting components at optical frequencies is immediately apparent.

The source field propagates over a *transmission channel* toward the receiver. The channel defines the electromagnetic path that carries the source field to the receiver. The channel can be *guided*, as in optical fibers, or can be *unguided*, as in space wave systems. In guided channels, the emitted field from the source is collected in an optical waveguide and confined to a specific path. In an unguided channel, the source field is permitted to propagate without confinement toward the receiver. The important unguided channels are the *free-space* channel, the *atmospheric* channel, and the *cloud and underwater* channel. The free-space channel represents a vacuous medium, and the only effect on field transmission is the inherent geometric spreading loss predicted by Maxwell's equation. The atmospheric channel corresponds to field propagation through the earth's atmosphere, the latter containing air turbulences and scattering particulates, such as gases, vapors, and water droplets, all of which impact directly on any impinging electromagnetic field. The cloud and water channels correspond to strongly scattering media, with significantly more severe effects on propagating fields. At optical frequencies, where the field wavelengths become commensurate with the physical size of the atmospheric particulates, raindrops, and water molecules, the radiation interaction is extremely severe, requiring levels of analyses beyond typical microwave channel modeling. This book attempts to present this optical channel modeling in a cohesive manner.

The *receiver* collects the impinging optical field, from the channel, converting the detected field to an electronic signal via photodetection. This electronic signal therefore represents the available waveform for all subsequent signal processing, either for demodulation in analog modulation formats, for decoding in digital data formats, or for signal processing in imaging, synchronization, and estimation type operation. Thus, the entire communication processing must rely on the photodetected signal. The latter, in turn, depends on the characteristics of the transmitted signal field, the effect of the channel on all propagating fields, and the manner in which the receiving optics collects and detects this field. A secondary objective of this book is to present a formal study of each of these considerations.

In addition to the transmitted source field, the receiver collects light transmissions from unwanted sources as well, such as the background sky, sunlit clouds, stars, and planets. These sources present an interfering field (in some cases they themselves may represent the desired source field) that propagates through the same media and combines with the desired source field to represent the combined "signal and noise" field that the receiver detects. It is therefore necessary to generate field models for both the desired and undesired light waves, and to incorporate the response of the photodetecting receiver to the combined fields. In the remaining part of this first chapter, we begin by characterizing the various properties of both the optical source and background noise fields encountered in optical transmissions.

1.2. Source Fields

An optical source (laser, flashlamp, etc.) is generally designated as either a *point source* or an *extended source.* For a point source, the optical field appears to emanate from a single point in space (i.e., the transmitter location), and all subsequent field analyses are based on treating the emanating light field in this way. For an extended source, the optical field appears to be generated over a spatial area that extends over some definite domain that can no longer be considered a single point. In particular, light being emitted from all points on the surface area must be properly accounted for, and a key point in extended source modeling is to understand, or properly model, the interaction of the radiated fields from all such points. (This subject will be developed shortly.)

1.2.1. Point Sources

The most elementary source field is that produced by a point source. This is a spherically symmetric field function, producing at any point **r** away from the source the complex field

$$f(t, \mathbf{r}) = \frac{A(t)}{|\mathbf{r}|} e^{j(k|\mathbf{r}| - \omega_c t)} \tag{1.2.1}$$

where $A(t)$ represents the complex modulation, $\omega_c = 2\pi f_c$ is the optical carrier frequency in radians per second, and $k = 2\pi/\lambda_c$. If $A(t)$ is a constant (not dependent on t), the source is *monochromatic* at frequency f_c. If $A(t)$ is real, it represents pure amplitude modulation centered at f_c, and if $A(t)$ is complex, it represents both amplitude and phase modulation at f_c. Since we can always rewrite $f(t, \mathbf{r})$ as

$$f(t, \mathbf{r}) = \frac{A(t)}{|\mathbf{r}|} e^{-j\omega_c(t - [|\mathbf{r}|/c]} \tag{1.2.2}$$

we see that the point source field propagates out along a ray line at the speed c if no channel effects are involved (i.e., free-space propagation). The *intensity* of the field at point **r** is defined as

$$\begin{aligned} I(t, \mathbf{r}) &= |f(t, \mathbf{r})|^2 \\ &= \frac{|A(t)|^2}{|\mathbf{r}|^2} \end{aligned} \tag{1.2.3}$$

Thus, the point source field intensity decreases as the distance squared out along a propagating ray line, even with free-space transmission. Since intensity is proportional to $|A(t)|^2$, time variations in the square of the field amplitude correspond to *intensity modulations* of the optical carrier.

An *isotropic* point source radiates equally in all ray directions. Its source power is obtained by integrating the field intensity over all solid angles (4π steradians), or

$$\begin{aligned} P_s(t) &= \int_{4\pi} I(t, \mathbf{r})|\mathbf{r}|^2 \, d\Omega \\ &= 4\pi |A(t)|^2 \end{aligned} \tag{1.2.4}$$

Thus, the field intensity at distance $R = |\mathbf{r}|$ in Equation (1.2.2) for an isotropic source is equivalently

$$I(t, \mathbf{r}) = \frac{P_s(t)}{4\pi R^2} \tag{1.2.5}$$

Thus, the isotropic field intensity decreases as R^2, and temporally varies proportionally to the transmitter power variation.

If the point source is not an isotropic radiator but instead radiates all its power $P_s(t)$ into a solid angle Ω_s (as in Figure 1.2) which is less than 4π steradians, the intensity within this solid angle will be $4\pi/\Omega_s$ times greater than if we radiated the power isotropically. Looked at another way, an isotropic source would have to be increased by the factor $4\pi/\Omega_s$ to have the same intensity within this solid angle. Consequently, we consider such a source to have a *transmitting gain* equal to $4\pi/\Omega_s$ over the isotropic source for the same radiated power. Thus, the transmitting system in Figure 1.2 has gain

$$G_s = \frac{4\pi}{\Omega_s} \tag{1.2.6}$$

It is common to refer to the product P_sG_s of a source as the *effective isotropic radiated power* (*EIRP*), to reflect this gain. Notice that the gain increases as the beamwidth θ_s (Equation 1.1.1) decreases, so that optical sources have the capability of extremely high EIRP. Of course, with this gain comes a pointing requirement of a fraction of Ω_s to maintain the radiated power at the correct direction over the distance r to where the intended receiver will be.

For a source with gain G_s, Equation (1.2.5) becomes

$$I(t, \mathbf{r}) = \frac{P_s(t)G_s}{4\pi R^2} \tag{1.2.7}$$

Equation (1.2.7) reflects the power per unit area that is radiated by a point source to a distance R units away. At this location the receiver would reside with a collecting aperture of area A_r. If the receiver area is aligned at an angle θ with respect to the normal to the arriving ray line direction from

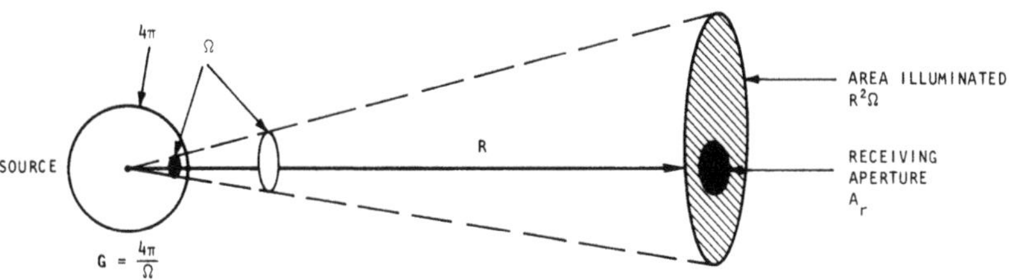

Figure 1.2. Graphical representation of gain and the range equation.

the transmitter (Figure 1.3), the amount of average power collected over the free-space channel becomes

$$P_r = \frac{P_s G_s A_r}{4\pi R^2} \cos\theta \tag{1.2.8}$$

where $P_s = \overline{P_s(t)}$, with the overbar denoting time averaging.

Notice that for the point source we maximize our collected power by setting $\theta = 0$ (by aligning with the source). Equation (1.2.8) is the range equation in frequency-independent form, describing the power flow from the source to the receiver. From Figure 1.2, we see that P_r/P_s is actually a propagation loss, measured as the ratio of the collecting aperture to the beamfront area $R^2\Omega_s$ illuminated by the source at the distance R. When factors leading to the quantification of A_r and G_s are considered, frequency becomes a part of the range equation. For example, when we relate the gain of an antenna to its aperture area A by

$$G = \frac{4\pi A}{\lambda^2} \tag{1.2.9}$$

the range equation can take either of the following forms:

$$P_r = \frac{P_s G_s G_r \lambda^2}{(4\pi)^2 R^2} \tag{1.2.10a}$$

$$= \frac{P_s A_s A_r}{\lambda^2 R^2} \tag{1.2.10b}$$

Notice that when we equate Equations (1.2.9) and (1.2.6) we find that

$$\Omega_s = \frac{\lambda^2}{A_s} \quad \text{and} \quad \Omega_r = \frac{\lambda^2}{A_r} \tag{1.2.11}$$

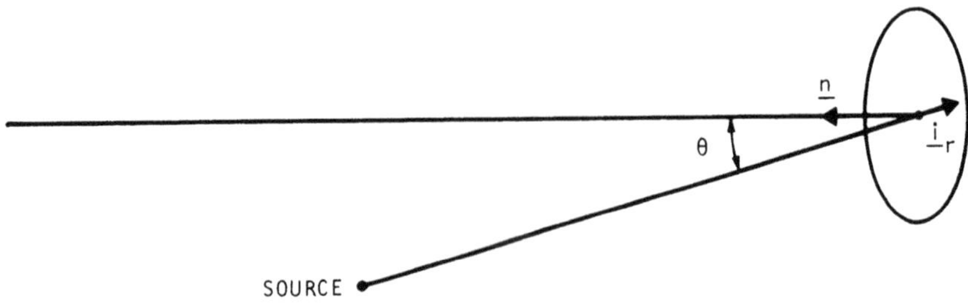

Figure 1.3. Alignment of transmitter and receiver.

Whenever the source and receiver solid angle and area are related as above, we say the transmitter and/or receiver is *diffraction limited*, and designate its beam angle by Ω_{dl}. Equation (1.2.11) can be derived from diffraction theory, where it can be shown that Ω_{dl} is the minimum solid angle that can be associated with an antenna having any aperture area A. By reciprocity, this is true for reception as well as transmission.

1.2.2. Extended Sources

Historically, two major models of energy transport from extended sources have evolved. The principles of traditional radiometry and the theory of radiative transfer which they support are perhaps the oldest, simplest, and intuitively the most appealing. They are based on the notion of optical power propagating along geometrical rays and are derived entirely on the basis of heuristic energy balance considerations which assume that the propagation process is linear in power; coherent phase interference effects within the underlying complex electromagnetic field are not accounted for. Consequently, the traditional radiometry describes, in some approximation, the radiation fields generated by substantially incoherent sources, the blackbody serving as an important example.

In contrast to the traditional radiometry, the wave optics approach is based on Maxwell's equations and the fact that the propagation process is linear in complex field amplitude. It also employs the concepts of optical coherence theory to correctly account for interference effects within the electromagnetic field.

The wave optics and radiometric approaches have in the past been utilized as completely independent approaches to electromagnetic energy propagation with little attention to the possibility of a connection between the two. Indeed, with the development of highly coherent sources such as the laser, serious questions have been raised regarding the applicability of the traditional radiometric laws to sources which exhibit a high degree of coherence. These issues are addressed by a relatively new branch of optical physics known as the generalized radiometry. In the following sections we will discuss the fundamentals of traditional radiometry. In Chapter 2 we will develop the coherence theory and in Appendix A generalized radiometry, which are necessary for an understanding of the channel effects discussed in following chapters.

1.2.2.1. The Traditional Radiometry

We recall that in Equation (1.2.8), the $\cos\theta$ factor merely reflected the fact that the transmitter and receiver were misaligned (off boresight). In fact, if the transmitter is extended, only one point can be aligned. All other points will experience a cos θ loss. It therefore becomes necessary to model the transmission system with a more detailed geometry, shown in Figure 1.4. The source, having an arbitrary shape, is located at the origin of a spherical coordinate system. The receiver is chosen to be aligned with a point on the source at a distance R. We define the projected area of the source (to the receiver) as the integral

$$A_s = \int_{\text{source}} \cos\theta \, dA_s \tag{1.2.12}$$

where the integration is taken over all portions of the source area that are in view of the receiver. The solid angle projected from the source to the receiver is A_s/R^2. It is also necessary to define a viewing angle as the solid angle over which the receiver can collect power. This is called the field of view of the receiver and is designated by Ω_{fov}. For a diffraction-limited receiver, $\Omega_{\text{fov}} = \Omega_{\text{dl}}$. From Equation (1.2.8) we see that the definition of the receiving aperture A_r is consistent with Equation (1.2.12). Thus, the solid angle projected from the receiver to the source is A_r/R^2.

1.2.2.2. *The Radiant Emittance and Irradiance*

As was the case for the point source, the radiant power flowing across a unit area of a real or imaginary surface will often be of interest. Let $E(\mathbf{r}_s)$ describe the amount of radiant power per unit area emanating from the surface Σ_s. We have included a dependence on the variables $\mathbf{r}_s$ to account

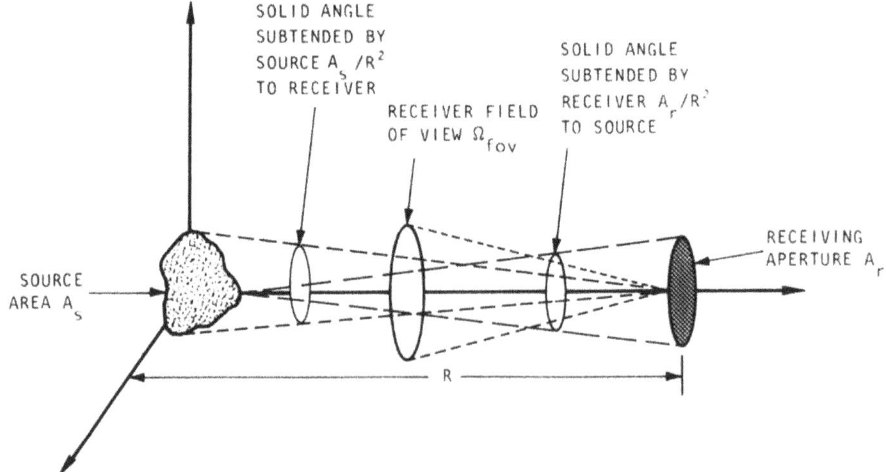

Figure 1.4. Geometric factors in radiance for transmission and reception.

for the fact that, for many sources, the power does not radiate equally from all parts of the surface. The amount of power radiating from the differential source area $d^2\mathbf{r}_s$ is then given by $E(\mathbf{r}_s)\, d^2\mathbf{r}_s$, allowing $E(\mathbf{r}_s)$ to be related to the total radiant power by the integral

$$P_s = \int_{\Sigma_s} E(\mathbf{r}_s)\, d^2\mathbf{r}_s \tag{1.2.13}$$

If Σ_s lies in the plane of a primary source, then $E(\mathbf{r}_s)$ is called the source radiant emittance. If Σ_s does not lie in the plane of the primary source but receives (reflects) radiant power from a source at some other location in space (such as the moon reflecting sunlight), then $E(\mathbf{r}_s)$ is replaced by the irradiance function $H(\mathbf{r}_s)$. In MKS units both are specified in watts per square meter.

1.2.2.3. The Traditional Radiance

Although $E(\mathbf{r}_s)$ does represent all the power radiated from the area $d^2\mathbf{r}_s$, it is not always the case that this power radiates equally in all directions. Hence there is a need for a finer characterization. If we describe the amount of power radiating from the same infinitesimal area into an infinitesimal solid angle $d\Omega$ as $N(\mathbf{r}_s, \theta) \cos\theta$, then we have

$$E(\mathbf{r}_s) = \int_{\Omega} N(\mathbf{r}_s, \theta) \cos\theta\, d\Omega \tag{1.2.14}$$

and

$$P_s = \int_{\Sigma_s} \int_{\Omega} N(\mathbf{r}_s, \theta) \cos\theta\, d^2\mathbf{r}_s\, d\Omega \tag{1.2.15}$$

$N(\mathbf{r}_s, \theta)$ is the traditional source radiance (sometimes called the source brightness) and has the dimensions of power per unit area per unit solid angle. In Equation (1.2.14), θ is the angle measured at the surface relative to the surface normal, and $\cos\theta\, d^2\mathbf{r}_s$ represents the projected surface area in the direction θ. Thus, for example, a source cannot radiate in a direction where it projects zero area; i.e., at $\theta = 90°$. Similarly, a source cannot contribute power from a surface element not in view of the receiver, and consequently we are interested, usually, in the power radiated into a hemisphere. As a final refinement, it is necessary to recognize that sources do not radiate equally at all frequencies (such a source would be called white). Rather sources have spectral emission characteristics which are quite varied.

Thus, if we consider the radiance of a source in the spectral range df to be $N(\mathbf{r}_s, \theta, f)\, df$, then it follows that

$$N(\mathbf{r}_s, \theta) = \int_{\text{all frequencies}} N(\mathbf{r}_s, \theta, f)\, df$$

$$E(\mathbf{r}_s) = \int_{\text{all frequencies}} \int_{\text{all angles}} N(\mathbf{r}_s, \theta, f) \cos\theta\, d\Omega\, df \tag{1.2.16}$$

$$P_s = \int_{\text{surface area}} \int_{\text{all angles}} \int_{\text{all frequencies}} N(\mathbf{r}_s, \theta, f) \cos\theta\, d\Omega\, d^2\mathbf{r}_s\, df$$

$N(\mathbf{r}_s, \theta, f)$ is called the traditional spectral radiance of a source and has the units of power per unit area per unit solid angle per unit frequency. In this book we will primarily consider frequency in units of cycles per second, or hertz. In practice, however, it is often common to measure these quantities in wavelength units λ, and bandwidths also in wavelength units $\Delta\lambda$. Since the bandwidth in hertz is related to the bandwidth in wavelength by

$$B = df = \frac{c\Delta\lambda}{\lambda^2} \tag{1.2.17}$$

we have that

$$N(\mathbf{r}_s, \theta, f)B = N_\lambda(\mathbf{r}_s, \theta, \lambda)\, \Delta\lambda$$

or

$$N\left(\mathbf{r}_s, \theta, \frac{c}{\lambda}\right) \frac{c}{\lambda^2} = N_\lambda(\mathbf{r}_s, \theta, \lambda) \tag{1.2.18}$$

Many of the curves presented in this chapter were taken from the literature where units of wavelength were commonly used and are reproduced directly.

1.2.2.4. The Traditional Radiant Intensity

In many instances one is concerned only with the amount of power radiated per steradian in a given direction from the entire source plane. The function which describes this dependence is called the traditional radiant intensity and can be expressed in terms of the traditional spectral radiance as

$$J(\theta) = \cos\theta \int_{\Sigma_s} \int N(\mathbf{r}_s, \theta, f)\, d^2\mathbf{r}_s\, df \tag{1.2.19}$$

which has dimensions of watts per unit solid angle. Furthermore, one can define a traditional spectral radiant intensity by suppressing the integration over frequency, or

$$J(\theta,f) = \cos\theta \int_{\Sigma_s} N(\mathbf{r}_s, \theta, f)\, d^2\mathbf{r}_s \tag{1.2.20}$$

$J(\theta, f)$ has dimensions of power per unit solid angle per unit frequency.

1.2.2.5. The Lambertian Source

Two assumptions are often made about sources to ease computation. The first is that there is no dependence on the coordinate position $\mathbf{r}_s$. This is a surface uniformity assumption which suppresses the variable $\mathbf{r}_s$ from all the quantities above. The other common assumption is that the radiance from a source is the same in all directions. Such a pattern is called isotropic and the source bearing this property is generally referred to as a Lambertian source or surface. Notice that when the isotropic assumption is made, Equation (1.2.16) reduces to

$$E(\mathbf{r}_s, f) = \pi N(\mathbf{r}_s, f) \tag{1.2.21}$$

and when the surface uniformity assumption is also made

$$E(f) = \pi N(f) \tag{1.2.22}$$

The radiant intensity is given by

$$J(\theta) = J(0)\cos\theta \tag{1.2.23}$$

with

$$J(0) = \int_{\Sigma_s}\int N(\mathbf{r}_s, f)\, d^2\mathbf{r}_s\, df \tag{1.2.24}$$

1.3. Power Collected by a Receiver

To compute the power collected by a receiver, we have only to refer back to Figure 1.2 and use the expression $N(\mathbf{r}_s, \theta, f)$. We have a projected receiver collecting aperture A_r located a distance R from the source. The field of view of the receiving system is Ω_{fov}. The source projects a solid angle A_s/R^2 to the receiver, and the receiver projects a solid angle A_r/R^2 to the source. The received power is then determined by integrating the spectral radiance $N(\mathbf{r}_s, \theta, f)$ over each of its three variables.

Integration in frequency: In communication systems as well as most other optical systems, we are generally interested in sources that have relatively narrow spectral widths. Over these bandwidths, the sources are generally constant enough for the integration to be performed by merely multiplying the value of spectral radiance by the bandwidth B. It is only in rare cases that this approximation will not be perfectly adequate.

Integration in solid angle: Another simplification arises at distances from the source larger than the dimension of the collecting aperture. This is also almost always going to be true. For these cases, the receiver projects a solid angle of A_r/R^2 to the source and we can merely multiple the spectral radiance by this factor.

Integration in source area: If we assume an average value of the source throughout the receiver field of view, then we can reduce this integration to one of two cases:

1. If the source is contained totally within the receiver field of view, then the integration yields A_s.
2. If the source totally fills the receiver field of view, then the projected area projected at the source is $R^2\Omega_{\text{fov}}$. This then becomes the multiplying factor to account for the integration over the source area.

These two cases are shown in Equation (1.3.1a) for case 1 and Equation (1.3.1b) for case 2:

$$\int_{\Sigma_s} = N(\mathbf{r}_s, \theta, f)\frac{A_s}{R^2}A_s B \qquad (1.3.1a)$$

$$\int_{\Sigma_s} = N(\mathbf{r}_s, \theta, f)A_r\Omega_{\text{fov}}B \qquad (1.3.1b)$$

1.4. Noise Sources

Since the characterization we are presenting here is based upon power flow, the differences between signal sources and noise sources are determined by the application. Equation (1.3.1a) is characteristic of a source that is small compared to the field of view of the receiver, that is $A_s/R^2 < \Omega_{\text{rov}}$. This would be the case for discrete sources such as distant stars or possibly even the emission from, or reflection off, an aircraft or spacecraft. When measuring such sources, we often do not know the physical area A_s. We therefore make the measurement called the spectral irradiance, $H(f)$. It is also common in this case to assume uniformity and isotropy yielding

$$P_n = \frac{N(f)A_r A_s B}{R^2} = N(f)A_r B\Omega_s = H(f)A_r B; \qquad \Omega_s = \frac{A_s}{R^2} \qquad (1.4.1)$$

thus relating $H(f) = N(f)\Omega_s$. Notice that this is an intensity and is determined by dividing the received power by the receiver aperture and the bandwidth. The bandwidth B is generally determined by the characteristic of the signaling component, which is usually narrowband relative to the noise. Equation (1.3.1b) is valid for both large objects as well as diffuse sources, such as the sky, which encompass the receiver field of view. This case can also be pursued further to gain our first insight into system understanding. To do so, let us consider the case where the receiver field of view is equal to its diffraction limit, i.e.,

$$\Omega_{\text{fov}} = \Omega_{\text{dl}} = \frac{\lambda^2}{A_r} \tag{1.4.2}$$

and the source is uniform and isotropic. Then the noise power can be rewritten as

$$P_n = \lambda^2 N(f) B \tag{1.4.3}$$

Notice that there is no dependence on any geometrical or structural parameters for this condition. Recalling that Ω_{dl} is the minimum field of view that can be obtained in a system, we can define the ratio D as

$$D = \frac{\Omega_{\text{fov}}}{\Omega_{\text{dl}}} \tag{1.4.4}$$

which we will call the spatial dimension of the system. The noise power can then be written in general as

$$P_n = \lambda^2 N(f) BD \tag{1.4.5}$$

Finally, since we know that B is the temporal dimension of the signal per second, we see that we can interpret $\lambda^2 N(f)$ as the energy per dimension or the equipartition energy of the source, and we denote this as N_{ob}, the single-sided spectral density. Going a step further, we know that in a time T a function can be represented by $2BT$ Nyquist samples and reconstructed by $2BT$ Nyquist sampling functions. Similarly, it can be shown[(1)] that a spatial function can be represented by D spatial samples and reconstructed by D spatial eigenfunctions. Consequently, just as the relationship $2BT = 1$ defines a temporal degree of freedom with

$$2B = \frac{1}{T} \text{ samples per second} \tag{1.4.6}$$

so does the relationship $\Omega_{\text{dl}}(A/\lambda^2) = 1$ define a spatial degree of freedom with

$$\frac{1}{\Omega_{\text{dl}}} = \frac{A}{\lambda^2} \quad \text{samples per steradian} \tag{1.4.7}$$

1.4.1. Blackbody Radiation[2]

At the close of the nineteenth century, the most exciting area of physics was determining the underlying mechanisms of blackbody radiation. This was the name given to "pure-temperature radiation," or radiation resulting from a system in thermal equilibrium. Since this is the fundamental source of primary (sun, stars, plasmas, etc.) and secondary (moon, planets, atmosphere, etc.) electromagnetic noise and, in addition, was so basic in the development of the quantum theory, we digress somewhat to give the reader a little feeling into the nature of these noise sources.

With the successes of the kinetic theory of gases then behind them, scientists sought the derivation of the blackbody distribution from basic thermodynamics. It was known that in a sealed chamber at a given temperature, the spectral distribution of radiation is independent of the material out of which the chamber is made. In 1884, Boltzmann showed that the energy in this distribution varied as the fourth power of the temperature. The constant of proportionality, $\delta = 0.567 \times 10^{-4}\,\text{erg cm}^{-2}\,\text{deg}^{-4}\,\text{s}^{-1}$, is known as the Stefan–Boltzmann constant and the law as the Stefan–Boltzmann law. In 1893, Wien showed that this functional dependence must be of the form

$$\psi_\lambda = T_0^5 f(\lambda T_0) = \frac{1}{\lambda^5} F(\lambda T_0) \tag{1.4.8}$$

where T_0 is the temperature in degrees Kelvin. Based on an assumption he proposed the formula

$$\psi_\lambda \sim \lambda^{-5}\, e^{b/\lambda T_0} \tag{1.4.9}$$

called Wien's law. En route to this, he developed what is called Wien's displacement law, which states that if the product λT_0 is held constant, ψ_λ / T_0^5 is the same at all temperatures.

In 1900, Rayleigh made a suggestion which was based on a technique used successfully in statistical mechanics. This was followed up with a calculation by Jeans, resulting in the Rayleigh–Jeans formula. The idea relied on the equipartition of energy and went as follows: one can expand the electromagnetic field contained in an enclosure into orthogonal and independent modes (degrees of freedom). By then associating the total energy kT_0 with each degree of freedom (which worked so well in the ideal gas law), an expression for the radiation can be obtained. Jeans calculated the number of modes per cycle to be D_{BB} where

$$D_{BB} = \frac{8\pi f^2}{C^3} \tag{1.4.10}$$

yielding the expression for the energy to be

$$E = kT_0 D_{BB} = \frac{8\pi f^2 kT_0}{C^3} \tag{1.4.11}$$

The obvious flaw in this expression was that it also predicted an infinite energy instead of the fourth-power dependence on T_0. It was also felt that this was unfortunately the best that could be expected from the classical theories as then known. The two expressions, E and ψ_λ, are plotted in Figure 1.5 along with the experimentally derived curve. The results of Lummer and Pringsheim were accurate enough to show that neither the Rayleigh–Jeans law nor Wien's law was correct, but rather than Wien's theory was asymptotically correct at high frequencies and the Rayleigh–Jeans model was asymptotically correct at low frequencies.

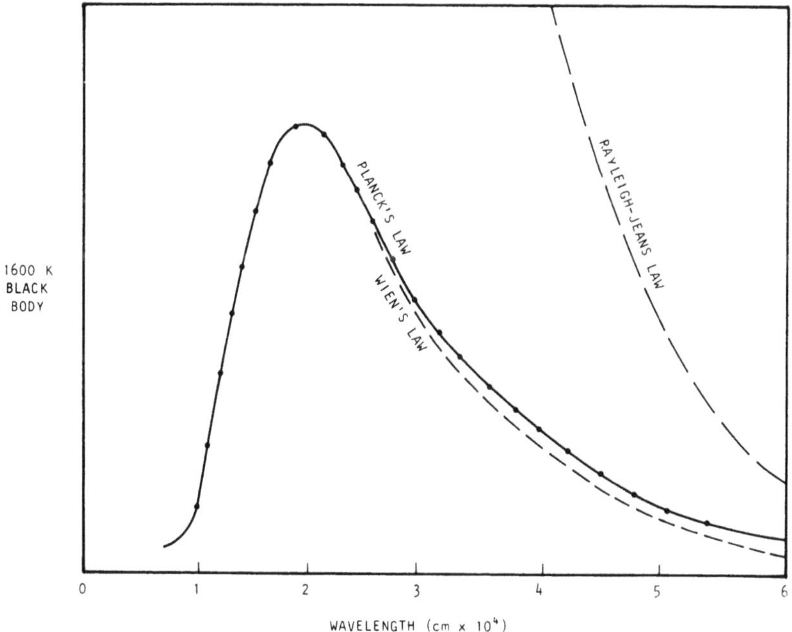

Figure 1.5. Blackbody radiation, Wien's law, and the Rayleigh–Jeans approximation. The experimental curve is fit exactly by Planck's law.

This was the state of the theory as seen by Planck around 1900. Asymptotically correct formulas existed at both the high and low frequencies; the functional dependence on T_0 with λT_0 constant was known; and the total energy as a function of T_0 was known. In addition, an accurate measurement of the function had been made. Planck's derivation of the correct equation for the blackbody law came through his understanding of thermodynamics. He curve-fitted the second derivative of entropy with respect to energy to yield the correct two asymptotic forms and worked this back to obtain the energy equation. The answer was so accurate that it was clearly the correct formula. He then set out to obtain a model whereby he could develop this equation from basic principles. Again he used thermodynamic arguments. The model he adopted consisted of a perfectly reflecting enclosure filled with dipole oscillators, radiating at discrete frequencies. He assumed that the total energy consisted of a finite number of incremental amounts distributed randomly among the oscillators. Furthermore, he assumed that the emission and the absorption of radiation by these oscillators took place not continuously but in jumps of fixed energy (the incremental amounts): this was the first quantum assumption. The remainder of his argument was complicated and has been superseded by more revealing descriptions. The most commonly used derivation multiplies the mode density D_{BB} by the quantum equivalent of kT_0 which is

$$\frac{hf}{e^{hf/kT_0} - 1} \tag{1.4.12}$$

(This was first pointed out by Einstein although the derivation of the energy was developed by Lorentz.) Thus we see that the equation for blackbody radiation can be derived to be

$$\begin{aligned} \psi_f &= \frac{8\pi f^2}{c^3} \frac{hf}{e^{hf/kT_0} - 1} \\ &= \frac{8\pi hcT_0^5}{(\lambda T_0)^5} \frac{1}{e^{hc/k\lambda T_0} - 1} \end{aligned} \tag{1.4.13}$$

which fits the measured curves exactly; h is called Planck's constant, and the incremental energy per oscillator exists in integral amounts of the incremental quantity hf. Notice that the integral of Equation (1.4.13) is

$$\int_0^\infty \psi_f \, df = \left(\frac{2\pi^5 k^4}{15c^2h^3}\right) T_0^4 \tag{1.4.14}$$

the Stefan–Boltzmann law, with δ given in terms of universal constants. By differentiating Equation (1.4.13) with respect to λ, and solving for the maximum value, λ_n, we obtain

$$\lambda_m = \frac{hc}{kT_0 q} \tag{1.4.15}$$

where q is the solution to the transcendental equation

$$\frac{x}{1 - e^{-x}} = 5; \qquad q = 4.965114\ldots \tag{1.4.16}$$

and

$$\lambda_m T_0 = \frac{(hc/k)}{q} = 0.28971 \text{ cm degree} \tag{1.4.17}$$

which is Wien's displacement law. Thus, we see that blackbody radiation can be characterized completely by a single unique number which reflects the temperature of the source at its peak value of emission. It is therefore quite common to specify every source by its effective blackbody temperature, i.e., however misleading, as that blackbody temperature producing a comparable power in the same bandwidth. If the radiator is an isotropic blackbody, then the spectral radiance (in a unit time) for each degree of polarization can be obtained by radiating into 4π steradians at the velocity c and then dividing by 2 or

$$\begin{aligned} N_{\mathrm{BB}}(f) &= c \cdot \frac{1}{4\pi} \cdot \psi_f \cdot \tfrac{1}{2} \\ &= \frac{f^2}{c^2} \frac{hf}{e^{hf/kT_0} - 1} \end{aligned} \tag{1.4.18}$$

with $\lambda^2 N(f)$ becoming

$$\lambda^2 N_{\mathrm{BB}}(f) = \frac{hf}{e^{hf/kT_0} - 1} \tag{1.4.19}$$

the equipartition energy for a blackbody.

The equipartition energy in Equation (1.4.19) has two asymptotic values. At the low frequencies defined by $hf < kT_0$, we have that

$$\frac{hf}{e^{hf/kT_0} - 1} \sim \frac{hf}{1 + (hf/kT_0) - 1} = kT_0 \tag{1.4.20}$$

or an equipartition energy of kT_0. At the high-frequency end, defined by $hf > kT_0$,

$$\frac{hf}{e^{hf/kT_0} - 1} \sim \frac{hf}{e^{hf/kT_0}} = hf e^{-hf/kT_0} \tag{1.4.21}$$

and we have an equipartition energy of $hf \exp(-hf/kT_0)$. This general relationship is shown in Figure 1.6.

Early workers in this field viewed Equation (1.4.13) and Figure 1.6 as an indication that one could transmit huge amounts of information at very low power levels for frequencies $f \gg kT_0/h$, since the thermal noise levels were extremely low. In fact, this breakpoint in frequency more aptly describes the transition in field properties from classical to quantum mechanical. And while it is true that the thermal noise becomes small, a quantum or quantization noise that is proportional to frequency predominates. The "effective" temperature of this quantum noise is hf/k, which is approximately 15,000 °K at $\lambda = 1\ \mu$m. This topic will be discussed further in Chapter 3.

1.4.2. Solar Irradiance

Although the blackbody source has immense theoretical importance in the field of physics, it in fact does not occur naturally in nature. Perhaps the source that most closely resembles a blackbody and exists in nature is the sun. Figure 1.7 shows a plot of the solar irradiance measured outside the atmosphere and a corresponding blackbody of 5900 K. Notice that with the exception of the very short wavelengths, there is a close resemblance.

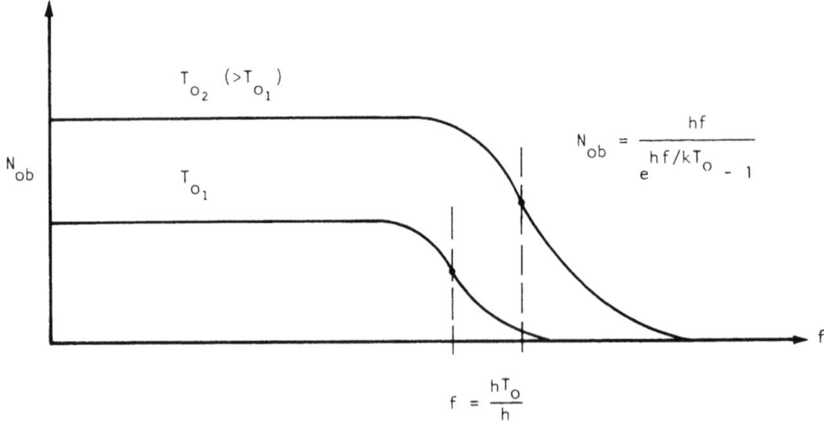

Figure 1.6. Quantum mechanical equipartition energy.

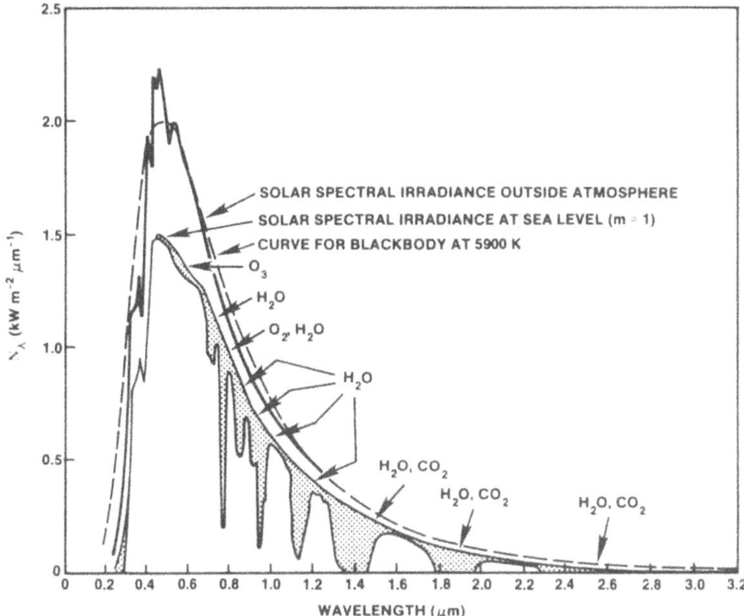

Figure 1.7. Spectral distribution curves related to the sun. The shaded areas indicated absorption at sea level due to the atmospheric constituents shown.[7]

Notice also that as the energy passes through the atmosphere, the constituents of the atmosphere cause a frequency-dependent attenuation which alters the nature of the radiation. Thus we see that optical systems which require transmission through the atmosphere will have better performance at some frequencies than at others. In Figure 1.7, it is assumed that the propagation is vertical through the atmosphere, which is defined as having an air mass of one ($m = 1$, where m is the amount of air in an infinite vertical column). An air mass of one is equivalent to approximately 10 km at sea level. Consequently, a horizontal communication system of 10 km will experience the same sort of attenuation with the exception of an absence of some of the high-altitude constituents such as ozone (O_3). Clearly, if one wanted to transmit 20 km, an even more severe condition would be found. Similarly, if one were to transmit at an angle θ incident to the vertical, a greater air mass would be encountered. As a simple rule of thumb, if we define T_∞ as the fractional transmission through the atmosphere at the vertical, then one can scale this to another angle θ by the factor $(T_\infty)^{\sec\theta}$, where $\sec\theta$ = air mass. In Figure 1.8 we show the extension to several air masses using more detailed models; see also Refs. 3–6.

The solar constant, H_0, is defined as the integral of the solar irradiance curve and is expressed in watts per square meter. For the curves presented,

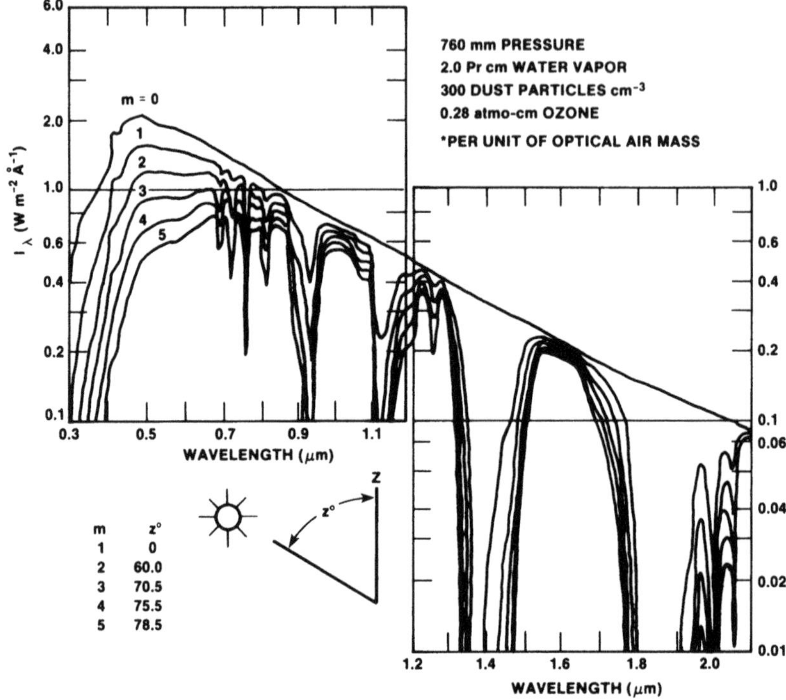

Figure 1.8. Solar spectral irradiance curves at sea level for various optical air masses. The value of the solar constant used in this calculation was 1322 W/m^2.[7]

H_0 is 1322 W/m^2. It should be noted that these irradiance values are really guides and that the true values are greatly dependent upon atmospheric content and location. Table 1.1 compares the solar irradiance at sea level under dry and moist atmospheric conditions with what exists on a mountaintop under the same conditions. The exo-atmospheric solar irradiance is also shown.

Table 1.1. Order of Magnitude Irradiance Levels for Several Light Sources

Light conditions	Irradiance level (W/m^2)
Solar irradiation at sea level	
Clear day	10^3
Lightly overcast day	10^2
Heavily overcast day	10
Well-lighted interiors, walls, ceilings, floors	1
Lunar irradiation, full moon, sea level, horizontal surface	10^{-3}
First magnitude star's irradiance on surface normal to rays	10^{-8}

1.4.3. Daytime Sky Radiance

When discussing the transmission of the solar radiance through the atmosphere, we only considered attentuation as a loss in the spectral irradiance. In fact, this loss can come from either absorption or scattering. Energy that is absorbed is either translated into molecular motion or reradiated at lower frequencies, usually in a continuum. On the other hand, energy that is scattered retains its incident frequency (possibly Doppler shifted) and goes off in a different direction. When viewing the sun directly, this energy would not be collected. However, when looking away from the sun, scattered energy would be encountered. Furthermore, this scattered energy will be frequency dependent as determined by the scattering mechanism. For example, the spectral radiance of the sky region within 30° of the sun is produced mainly by large-particle, or Mie, scattering for the UV/visible/near-infrared wavelength spectrum (Chapter 6). This phenomenon creates spectral radiance values which are inversely proportional to wavelength and explains why the sky is perceived as "white" in that portion. This is also true at 180° from the sun angle. Outside those two regions, small particles begin to contribute significantly to the radiance such that they are the dominant factor at 90° from the sun angle. This phenomenon is referred to as Rayleigh scattering and is characterized by an inverse fourth-power relationship in wavelength. This accentuates the short wavelengths of the solar spectrum and results in the sky being blue in the region of sky away from the sun. In addition, this portion of the sky becomes polarized with its maximum value at 90° from the sun. Figure 1.9 is an example of the spectral sky radiance distribution at sea level for a view angle of 45° under clear sky conditions.(4) This is a multimodal spectrum. The first mode consists of the solar spectrum filtered by the fourth power of the wavelength. This mode lies approximately between 0.3 and 3 μm and falls off four orders of magnitude between about 0.4 μm and 4 μm. The second mode closely resembles a blackbody of approximately 300 °K. This reflects the ambient temperature of the earth exciting the atmosphere. Multimodal spectra are not uncommon and usually occur when a source both emits its own thermal energy and reflects the energy from another source. The moon is a good example. For other references to sky radiance, see Refs. 6–9.

1.4.4. Nighttime Sky Radiance

In optics when we say "as different as night and day," we mean 60 dB. That is the difference between the solar irradiance and the irradiance from a full moon. In Table 1.1 we list some of the irradiance levels for various light sources. Nevertheless, even at these low levels, backgrounds can

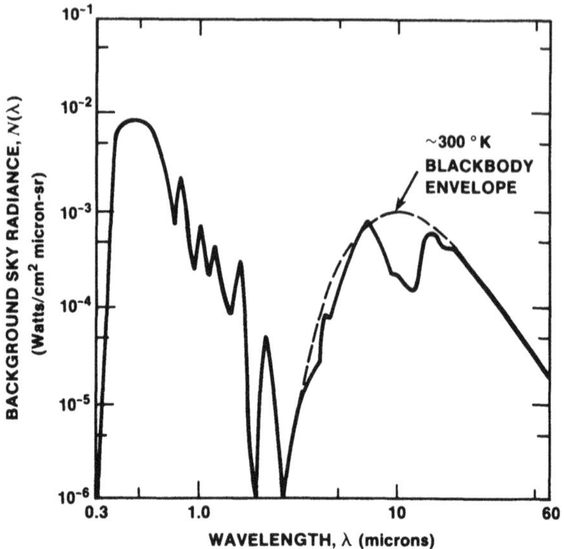

Figure 1.9. Diffuse component of typical background radiance from sea level; zenith angle, 45°; excellent visibility.[4]

seriously impact the performance of an optical system. Under nighttime illumination conditions, the dominant optical noise sources will be the sky radiance created by zodiacal, galactic, and scattered skylight sources. Figure 1.10 shows the typical spectral radiance from these sources.[10] The moon and planets can also contribute to the received noise. Figure 1.11 shows the spectral albedo of the various planets. Spectral albedo is defined as the ratio of total reflected energy to total incident energy. Figure 1.12 illustrates planetary and lunar spectral exo-atmospheric irradiance calculated using this information.[11] Of course, these values are several orders of magnitude lower than the daytime sky radiance, and we would expect better performance at night than in the daytime.

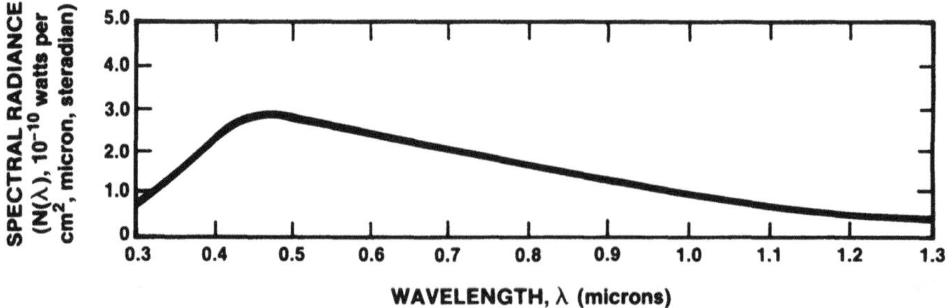

Figure 1.10. Nighttime sky radiance from zenith due to zodiacal light, galactic light, and scattered starlight.[10] (Courtesy of National Bureau of Standards.)

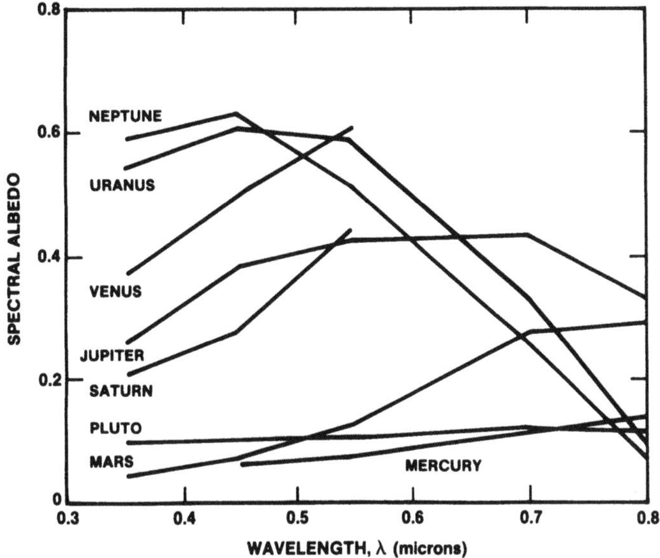

Figure 1.11. Spectral albedo as a function of wavelength.[10,11]

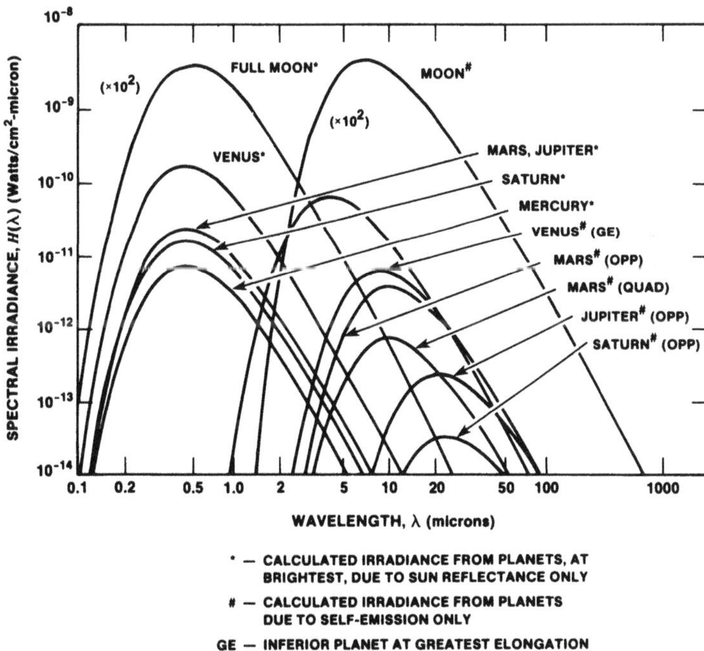

Figure 1.12. Calculated planetary and lunar spectral irradiance outside the terrestrial atmosphere.[10,11]

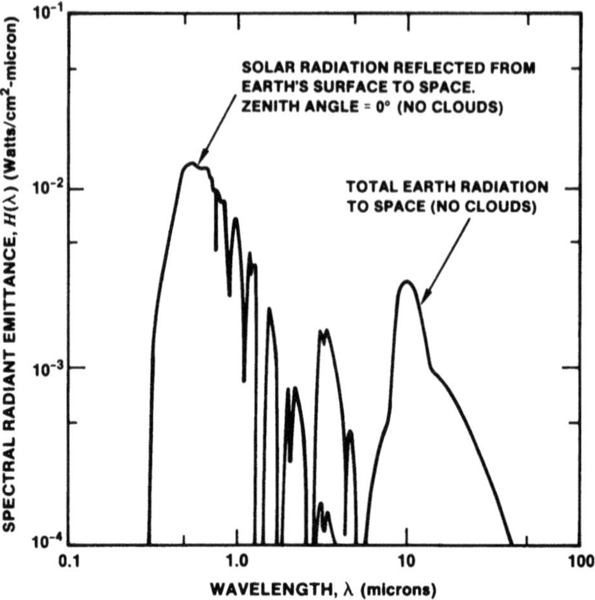

Figure 1.13. Spectral radiant emittance of the earth.[12]

1.4.5. Earthshine Radiance

For earth/satellite optical systems one has to contend with earthshine as the dominant noise source. This is a combination of surface reflectance, thermal surface emission, and atmospheric emission and scatter. Figure 1.13 illustrates typical earthshine radiance for zenith angle viewing through a clear sky.

References

1. R. M. Gagliardi and S. Karp, *Optical Communications*, Wiley-Interscience, New York (1976).
2. Sir Edmund Whittaker, *A History of the Theories of Aether and Electricity*, Vol. I, *The Classical Theories*, Harper & Brothers, New York (1951).
3. N. S. Kopeika and J. Bordogna, Background noise in optical communication systems, *Proc. IEEE* **58**, 1571–1577 (1970).
4. K. Seyrafi, Electro-Optical Systems Analysis, Electro-optical Research Company, Los Angeles (1973).
5. W. J. Smith, *Modern Optical Engineering*: *The Design of Optical Systems*, McGraw-Hill Book Company, New York (1966).
6. M. Ross, *Laser Receivers*, John Wiley and Sons, New York (1966).
7. W. L. Wolfe and G. J. Zissis, eds., *The Infrared Handbook*, The Environmental Institute of Michigan, Ann Arbor, Michigan (1978).

8. J. R. Clark and J. R. Baird, Optical communications in the atmosphere: Scattering and noise considerations, in Optical Communication Systems, Report of an NSF Grantee-User meeting held at the Naval Electronic Laboratory Center, San Diego, November 12–14, 1975, pp. 124–140.
9. A. Ben-Shalom, A. D. Devir, and S. G. Lirson, Infrared spectral radiance of the sky, *Opt. Eng.* **22**, 460–463 (1983).
10. W. K. Pratt, *Laser Communications Systems*, John Wiley and Sons, New York (1969).
11. R. C. Ramsey, Spectral irradiance from stars and planets above the atmosphere from 0.1 to 100 microns, *Appl. Opt.* **1**, 465–471 (1962).
12. I. L. Goldberg, Radiation from Planet Earth, U.S. Army Signal Research and Development Laboratory Report 2231, AD-266-790 (September, 1961).

2

Coherence Theory and Random Channels

In this chapter we will introduce the concept of field coherence, which is necessary to describe fully the propagation of electromagnetic fields in both free space and nonvacuous channels. In particular, the ideas of temporal and spatial coherence will be presented together with the concept of the mutual coherence function. We will derive the equations which govern the propagation of both the field and the mutual coherence function in free space. In doing so, we will quantify the notions of monochromaticity and narrowbandedness. Finally, we will show how the presence of both nonvacuous channels and receiver optics affects the quantities defined above. Since the concepts discussed in this chapter do not relate strongly to problems in fiber transmission, those readers primarily interested in fibers can skip this chapter and proceed to Chapter 3. The concepts in this chapter lay the groundwork for subsequent introduction of generalized radiometry, which describes power flow for partially coherent fields. However, because of the specialized nature of this topic, it will be pursued in Appendix A.

2.1. Field Properties

So far we have not been able to describe the effects of a channel on propagation other than by simple geometric divergence or absorption. In this section we will start to evolve our propagation model to include the concept of temporal and spatial coherence. It is shown[(1)] that for monochromatic fields satisfying the Helmholtz equation (the steady-state or single-frequency solution to the wave equation), we can describe the field at the receiver aperture, $f_{r_a}(\mathbf{r}_a, \omega)$, in terms of the field over the extended source

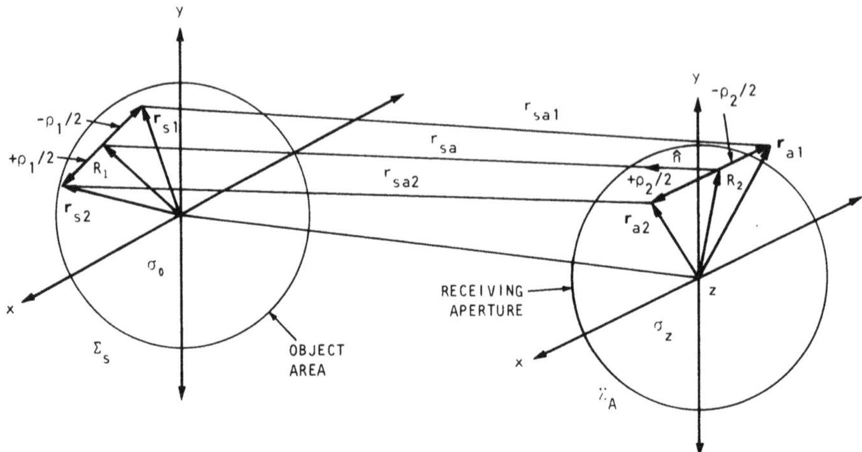

Figure 2.1. Geometry describing the propagation of electromagnetic fields.

plane Σ_s by $f_{r_s}(\mathbf{r}_s, \omega)$,† where

$$f_{r_a}(\mathbf{r}_a, \omega) = \frac{\omega}{j2\pi c} \int_{\Sigma_s} \frac{f_{r_s}(\mathbf{r}_s, \omega)\, e^{-j(\omega/c)r_{sa}}}{r_{sa}} \, d^2\mathbf{r}_s; \qquad \frac{\omega}{c} = k = \frac{2\pi}{\lambda} \quad (2.1.1)$$

The geometry for this equation is depicted in Figure 2.1. Here r_{sa} is the scalar distance from a point in the source plane to a point in the aperture plane. Since $\mathbf{r}_s$ and $\mathbf{r}_a$ are the coordinates of the source plane and the aperture, respectively, we write r_{sa} as

$$r_{sa} = (R^2 + |\mathbf{r}_s - \mathbf{r}_a|^2)^{1/2}; \qquad \mathbf{r}_i = (x_i, y_i) \quad (2.1.2)$$

with R the distance between the source plane and the aperture plane. We are then led readily to the Fresnel approximation

$$r_{sa} = R\left[1 + \frac{(x_s - x_a)^2}{R^2} + \frac{(y_s - y_a)^2}{R^2}\right]^{1/2} \cong R + \frac{(x_s - x_a)^2}{2R} + \frac{(y_s - y_a)^2}{2R};$$

$$R^3 \gg \frac{\pi}{4\lambda}[(x_s - x_a)^2 + (y_s - y_a)^2]^2_{\max} \quad (2.1.3)$$

†We have omitted the obliquity or inclination factor $\cos(\mathbf{n}, r_{sa}) = 1$.[(2)]

for which

$$f_{r_a}(\mathbf{r}_a, \omega) = \frac{\omega}{2\pi c} \frac{e^{-j(\omega/c)R}}{jR} e^{-j(\omega/2cR)|\mathbf{r}_a|^2}$$

$$\int_{\Sigma_s} f_{r_s}(\mathbf{r}_s, \omega)\, e^{-j(\omega/cR)(\mathbf{r}_s \cdot \mathbf{r}_a)}\, e^{j(\omega/2cR)|\mathbf{r}_s|^2}\, d^2\mathbf{r}_s;$$

$$\mathbf{r}_s \cdot \mathbf{r}_a = x_s x_a + y_s y_a \tag{2.1.4}$$

and the Fraunhoffer approximation

$$R \gg \frac{\pi}{\lambda} |\mathbf{r}_s|^2_{\max} \tag{2.1.5}$$

for which

$$f_{r_a}(\mathbf{r}_a, \omega) = \frac{\omega}{2\pi c} \frac{e^{-j(\omega/c)R}}{jR} e^{-j(\omega/2cR)|\mathbf{r}_a|^2} \int_{\Sigma_s} f_{r_s}(\mathbf{r}_s, \omega)\, e^{-j(\omega/cR)(\mathbf{r}_s \cdot \mathbf{r}_a)}\, d^2\mathbf{r}_s \tag{2.1.6}$$

For purposes of determining the relationship between the source field and its influence at the aperture plane, we can disregard the terms outside the integrals since they are not a function of f_{r_s} or $\mathbf{r}_s$. Thus we generally characterize these two approximations only in terms of the integrals. Notice also that the only difference between the two integrals lies in the additional quadratic-phase term in the Fresnel approximation. This is the diffraction term encountered at close distances. In communications the distances are generally large enough so that the Fraunhoffer approximation is valid and Equation (2.1.6) is applicable. It is also true that at these longer distances we can extend the integral over the source surface Σ_s to infinity, yielding the well-known result that the field over the aperture is the two-dimensional Fourier transform of the source field. In fact, it is customary to define the spatial frequency $\mathbf{f} = \mathbf{r}_s/\lambda R$ and rewrite Equation (2.1.6) as

$$f_{r_a}(\mathbf{r}_a, \omega) = \int_{-\infty}^{\infty} f_{r_s}(\mathbf{f}, \omega)\, e^{-j2\pi \mathbf{f} \cdot \mathbf{r}_a}\, d^2\mathbf{f} \tag{2.1.7}$$

So far we have only considered field propagation in the context of a single propagating frequency. In a more general context, we are interested in transmitting signals with a finite bandwidth. When viewed as a Fourier spectrum, Equation (2.1.1) indicated how each frequency propagates. We will, in general, receive a field at the aperture, which we can relate to Equation (2.1.1) by

$$f_{r_a}(\mathbf{r}_a, t)\, e^{j\bar{\omega} t} = \int_{-\infty}^{\infty} f_{r_a}(\mathbf{r}_a, \omega)\, e^{j\omega t} \frac{d\omega}{2\pi} \tag{2.1.8}$$

Although $\bar{\omega}$ is usually referred to as the carrier frequency in a narrowband process, it in fact applies in the general case.[3] When Equation (2.1.1) is inserted into Equation (2.1.8), it takes the form

$$f_{r_a}(\mathbf{r}_a, t)\, e^{j\bar{\omega}t} = \int_{\Sigma_s} \frac{d^2\mathbf{r}_s}{2\pi c r_{sa}} \left[\int_{-\infty}^{\infty} -j\omega' f_{r_s}(\mathbf{r}_s, -\omega')\, e^{j\omega'[t-(r_{sa}/c)]} \frac{d\omega'}{2\pi} \right] \tag{2.1.9}$$

which can be solved to yield

$$f_{r_a}(\mathbf{r}_a, t)\, e^{j\bar{\omega}t} = \int_{\Sigma_s} \frac{d^2\mathbf{r}_s}{2\pi c r_{sa}} \left[\frac{d}{dt} f_{r_s}\left[\mathbf{r}_s; \left(t - \frac{r_{sa}}{c} \right) \right] e^{j\omega[t-(r_{sa}/c)]} \right] \tag{2.1.10}$$

What we have is an integration of all the differential Huygens wavelets at their retarded times, over the surface of the source—one of the oldest postulates in the theory of wave propagation. However, we are more interested in narrowband processes, and for such processes an interesting approximation can be made in Equation (2.1.9). Since the function ω is monotonic and almost constant over a narrow band of frequencies, we can use the mean value theorem for integrals to move it outside the ω integration, yielding

$$\begin{aligned} f_{r_a}(\mathbf{r}_a, t)\, e^{j\bar{\omega}t} &= \int_{\Sigma_s} \frac{j\bar{\omega}\, d^2\mathbf{r}_s}{2\pi c r_{sa}} \left[\int_{-\infty}^{\infty} f_{r_s}(\mathbf{r}_s, \omega)\, e^{j\omega[t-(r_{sa}/c)]} \frac{d\omega}{2\pi} \right] \\ &= \int_{\Sigma_s} \frac{j\bar{\omega}\, d^2\mathbf{r}_s}{2\pi c r_{sa}} \left\{ f_{r_s}\left[\mathbf{r}_s; \left(t - \frac{r_{sa}}{c} \right) \right] e^{j\bar{\omega}[t-(r_{sa}/c)]} \right\} \end{aligned} \tag{2.1.11}$$

This then shows that when the signal is narrowband, we can consider the integral as an integral of elements of the radiated field at the retarded time. This, of course, is not surprising and, in fact, is intuitively used by communication engineers in signal design and analysis.

2.2. The Mutual Coherence Function (MCF)

In a more general context, it is common for the source to generate a random field or for the field to undergo random effects en route to the receiver. In these cases, the description of the field as a deterministic function is inadequate, and it becomes necessary to introduce convenient statistical properties that are confirmed in practice. Since it is also necessary to ensure wave propagation, we start with Equation (2.1.1) and consider the covariance function defined as

$$E[f_{r_a}(\mathbf{r}_a,\omega)f^*_{r_a}(\mathbf{r}'_a,\omega')]$$

$$=\int_{\Sigma_s}\int_{\Sigma_s}\frac{\omega\omega'}{(2\pi c)^2 r_{sa}r'_{sa}}E[f_{r_s}(\mathbf{r}_s,\omega)f^*_{r_s}(\mathbf{r}'_s,\omega')]\,e^{-j[(\omega/c)r_{sa}-(\omega'/c)r'_{sa})]}\,d^2\mathbf{r}_s\,d^2\mathbf{r}'_s$$

$$=\int_{\Sigma_s}\int_{\Sigma_s}\frac{\omega\omega'}{(2\pi c)^2 r_{sa}r'_{sa}}R_{r_s}(\mathbf{r}_s,\mathbf{r}'_s;\omega,\omega')\,e^{-j[(\omega/c)r_{sa}-(\omega'/c)r'_{sa}]}\,d^2\mathbf{r}_s\,d^2\mathbf{r}'_s$$

$$=R_{r_a}(\mathbf{r}_{a'}\mathbf{r}'_a;\omega,\omega') \tag{2.2.1}$$

where $R_{r_s}(\mathbf{r}_s,\mathbf{r}'_s;\omega,\omega')$ is the corresponding covariance function of the source and $E[\cdot]$ denotes ensemble average. What we have defined here is a joint auto-covariance function in frequency and space which then allows us to define the mutual coherence function at the receiver as the Fourier transform of $R_{r_a}(\mathbf{r}_a,\mathbf{r}'_a;\omega,\omega')$. In the literature, $R_{r_a}(\mathbf{r}_a,\mathbf{r}'_a;\omega,\omega')$ is referred to as the mutual power spectrum of the received field with $R_{r_s}(\mathbf{r}_s,\mathbf{r}'_s;\omega,\omega')$ the mutual power spectrum of the source. The two-dimensional Fourier transform of the mutual power spectrum over (ω,ω') takes the form

$$\Gamma_{r_a}(\mathbf{r}_a,\mathbf{r}'_a;t_1,t_2)$$

$$=\frac{1}{(2\pi)^2}\int_{-\infty}^{\infty}\int_{-\infty}^{\infty}R_{r_a}(\mathbf{r}_a,\mathbf{r}'_a;\omega,\omega')\,e^{j(\omega t_1-\omega' t_2)}\,d\omega\,d\omega'$$

$$=\int_{\Sigma_s}\int_{\Sigma_s}\frac{d^2\mathbf{r}_s\,d^2\mathbf{r}'_s}{(2\pi c)^2 r_{sa}r'_{sa}}\left\{\frac{d}{dt}\cdot\frac{d}{dt'}\Gamma_{r_s}\left[\mathbf{r}_s,\mathbf{r}'_s;\left(t_1-\frac{r_{sa}}{c}\right),\left(t_2-\frac{r'_{sa}}{c}\right)\right]\right\}$$

$$=\int_{\Sigma_s}\int_{\Sigma_s}\frac{\bar{\omega}^2\,d^2\mathbf{r}_s\,d^2\mathbf{r}'_s}{(2\pi c)^2 r_{sa}r'_{sa}}\left\{\Gamma_{r_s}\left[\mathbf{r}_s,\mathbf{r}'_s;\left(t_1-\frac{r_{sa}}{c}\right),\left(t_2-\frac{r'_{sa}}{c}\right)\right]\right\};\quad\text{narrowband} \tag{2.2.2}$$

$\Gamma_{r_i}(\mathbf{r}_i,\mathbf{r}'_i;t_1,t_2)$ is called the mutual coherence function.[†]

Equation (2.2.2) is a rather formidable result which is difficult to apply in practice. Fortunately, in the vast majority of applications one is concerned with propagation in the region where Fresnel or Fraunhoffer diffraction is valid and Equation (2.2.2) can, with further approximations, be cast into a much simpler form. Let us focus on the first form in Equation (2.2.2) with

[†] $\Gamma_{r_a}(\mathbf{r}_a,\mathbf{r}'_a;t_1,t_2)$ is also equal to $E[f_{r_a}(\mathbf{r}_a,t_1)f^*_{r_a}(\mathbf{r}'_a,t_2)]$.

Equation (2.2.1) inserted for $R_{r_a}(\mathbf{r}_a, \mathbf{r}'_a; \omega, \omega')$. The cissoid involving time and distance takes the form

$$e^{+j\omega[(t-\mathbf{r}_{sa}/c)-(t'-\mathbf{r}'_{sa}/c)]} \tag{2.2.3}$$

Let us assume that the Fresnel conditions in Equation (2.1.3) apply. It is readily seen that this approximation is equivalent to replacing spherical wavefronts with quadratic surfaces. The applicability of the Fresnel approximation implies that the accuracy of this approximation dictates bounds on the relative sizes of the source, receiver plane, propagation distance R, and radiation bandwidth. In this small angle regime we also have $r_{sa}r'_{sa} \approx R^2$. It is also convenient at this time to introduce the following change of variables

$$\begin{aligned} t_1 &= t + \tau/2 \qquad & t_2 &= t - \tau/2 \\ \omega &= \omega_+ + \omega_-/2 \qquad & \omega' &= \omega_+ - \omega_-/2 \end{aligned} \tag{2.2.4}$$

which can be equivalently expressed as

$$\begin{aligned} t &= (t_1 + t_2)/2 \qquad & \tau &= t_1 - t_2 \\ \omega_+ &= (\omega + \omega')/2 \qquad & \omega_- &= \omega - \omega' \end{aligned} \tag{2.2.5}$$

and

$$\begin{aligned} \mathbf{r}_s &= \mathbf{R}_1 + \boldsymbol{\rho}_1/2 \qquad & \mathbf{r}_a &= \mathbf{R}_2 + \boldsymbol{\rho}_2/2 \\ \mathbf{r}'_s &= \mathbf{R}_1 - \boldsymbol{\rho}_1/2 \qquad & \mathbf{r}'_a &= \mathbf{R}_2 - \boldsymbol{\rho}_2/2 \end{aligned} \tag{2.2.6}$$

which can be expressed as

$$\begin{aligned} \mathbf{R}_1 &= (\mathbf{r}_s + \mathbf{r}'_s)/2 \qquad & \mathbf{R}_2 &= (\mathbf{r}_a + \mathbf{r}'_a)/2 \\ \boldsymbol{\rho}_1 &= \mathbf{r}_s - \mathbf{r}'_s \qquad & \boldsymbol{\rho}_2 &= \mathbf{r}_a - \mathbf{r}'_a \end{aligned} \tag{2.2.7}$$

In terms of these new variables, we can rewrite Equation (2.2.2) as

$$\begin{aligned} &\Gamma_{\mathbf{r}_a}(\mathbf{R}_2, \boldsymbol{\rho}_2; t, \tau) \\ &\quad = \frac{1}{(2\pi Rc)^2} \\ &\quad = \int_{-\infty}^{\infty}\int_{-\infty}^{\infty}\int_{\Sigma_s}\int_{\Sigma_s} \left(\omega_+^2 + \frac{\omega_-^2}{4}\right) R_{\mathbf{r}_s}(\mathbf{R}_1, \boldsymbol{\rho}_1; \omega_+, \omega_-) \\ &\quad \cdot e^{-j\omega_+(\mathbf{r}_1-\mathbf{R}_2)\cdot(\boldsymbol{\rho}_1-\boldsymbol{\rho}_2)/Rc}\, e^{j\{\omega_+\tau+\omega_-[t-(R/c)]\}} \\ &\quad \cdot \exp\left\{j\omega_-\left[\frac{|\mathbf{R}_1 - \mathbf{R}_2|^2 + |\boldsymbol{\rho}_1 - \boldsymbol{\rho}_2|^2/4}{2Rc}\right]\right\} d\omega_+\, d\omega_-\, d^2\mathbf{R}_1\, d^2\boldsymbol{\rho}_1 \end{aligned} \tag{2.2.8}$$

which is the general expression for the mutual coherence function in the aperture plane. Unfortunately, the Fresnel approximation alone does not produce a substantial simplification of the propagation equation. However, when coupled with the narrowband approximations to be introduced below, a more tractable result is obtained.

2.3. The Narrowband Approximation

When we assume that the signal is narrowband, we imply that there is a carrier frequency denoted by $\bar{\omega}$ and a bandwidth denoted by $\Delta\omega$ such that $\Delta\omega/\bar{\omega}$ is much less than unity. We can see that the variables in Equations (2.2.4) and (2.2.5) imply that ω_+ relates to the location of the carrier frequency and ω_- relates to the spectral width of the source. Further, it can be shown that $\omega_+^2 - \omega_-^2/4$ in Equation (2.2.8) can be approximated by

$$\omega_+^2 - \frac{\omega_-^2}{4} \approx \bar{\omega}^2; \qquad \omega_+ \approx \bar{\omega} \tag{2.3.1}$$

for narrowband sources. In addition, we require that

$$\omega_- \ll \left\{\frac{2Rc}{[|\mathbf{R}_1 - \mathbf{R}_2|^2 + |\boldsymbol{\rho}_1 - \boldsymbol{\rho}_2|^2]/4}\right\}_{\max}$$

or

$$R \gg \frac{\omega_-}{2c}\{[|\mathbf{R}_1 - \mathbf{R}_2|^2 + |\boldsymbol{\rho}_1 - \boldsymbol{\rho}_2|^2]/4\}_{\max} \tag{2.3.2}$$

be satisfied. This condition ensures that there will be no differential phase shift across the spectral width of the source defined by the ω_- integration and, hence, the final cissoid in the integrand of Equation (2.2.8) can be replaced by unity:

$$\begin{aligned} &\Gamma_{r_a}(\mathbf{R}_2, \boldsymbol{\rho}_2; t, \tau) \\ &\quad = \frac{\bar{\omega}^2}{(2\pi Rc)^2} \int_{-\infty}^{\infty}\int_{-\infty}^{\infty}\int_{\Sigma_s}\int_{\Sigma_s} R_{r_s}(\mathbf{R}_1, \boldsymbol{\rho}_1; \omega_+, \omega_-)\, e^{j\{\omega_+\tau + \omega_-[t-(R/c)]\}} \\ &\qquad \cdot e^{-j\bar{\omega}(\mathbf{R}_1-\mathbf{R}_2)(\boldsymbol{\rho}_1-\boldsymbol{\rho}_2)/Rc}\, d\omega_+\, d\omega_-\, d^2\mathbf{R}_1\, d^2\mathbf{R}_2 \end{aligned} \tag{2.3.3}$$

Equation (2.3.3) applies only over those regions of space for which Equation (2.3.2) is valid. A further simplication of Equation (2.3.3) is affected by introducing the wavenumber and wavevector

$$k_0 = \frac{\bar{\omega}}{c} = \frac{2\pi}{\bar{\lambda}} \tag{2.3.4}$$

$$\mathbf{k} = k_0(\mathbf{R}_1 - \mathbf{R}_2)/R$$

The differential area $d^2\mathbf{R}_1$ can, in the small-angle approximation associated with the Fresnel and Fraunhoffer regions, be expressed as

$$d^2\mathbf{R}_1 \approx R^2\, d\Omega \approx R^2\, d^2\mathbf{k}/k^2 \tag{2.3.5}$$

with $d\Omega$ the solid angle subtended at $\mathbf{R}_2$ by a source area element at $\mathbf{R}_1$. We note that $\mathbf{k}$ determines the angular location, $\hat{\mathbf{n}}$, of the point $\mathbf{R}_1$ in Σ_s as viewed from $\mathbf{R}_2$ in Σ_a through the relationship

$$\hat{\mathbf{n}} = \hat{\mathbf{n}}_\perp + (1 - |\mathbf{n}_\perp|^2)^{1/2}\mathbf{k} \tag{2.3.6}$$

Here $\hat{\mathbf{n}}$ is a unit vector that points from $\mathbf{R}_2$ to $\mathbf{R}_1$ as shown in Figure 2.2, and $\hat{\mathbf{n}}_\perp$ is the component of $\hat{\mathbf{n}}$ that is orthogonal to the axis of propagation (z-axis). Making these substitutions and performing the integration in ω_+ and ω_-, we obtain

$$\Gamma_{r_a}(\mathbf{R}_2, \boldsymbol{\rho}_2; t, \tau) = \int_{-\infty}^{\infty} e^{-j\mathbf{k}\cdot\boldsymbol{\rho}_2}\, d^2\mathbf{k} \left\{ \int_{\Sigma_s} \Gamma_{r_s}\left[\left(\mathbf{R}_z + \frac{R\mathbf{k}}{k_0} \right), \boldsymbol{\rho}_1; \left(t - \frac{R}{c} \right), \tau \right] e^{j\mathbf{k}\cdot\boldsymbol{\rho}_1}\, d^2\boldsymbol{\rho}_1 \right\} \tag{2.3.7}$$

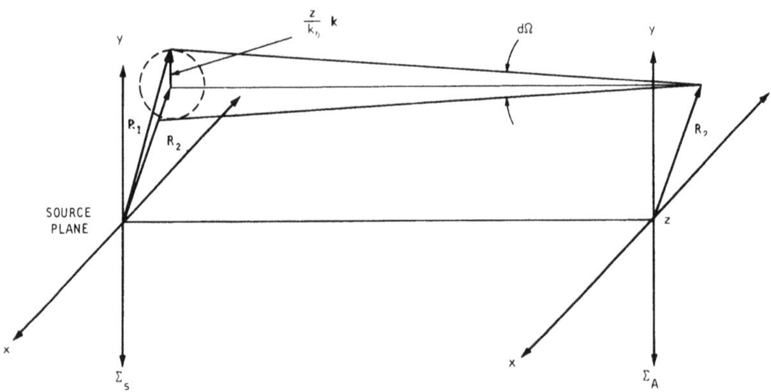

Figure 2.2. Illustrating propagation from the source plane to the detector plane.

This is the narrowband expression for the propagation of the mutual coherence function from the source to the aperture.

While we have been able to reduce the equation for the propagation of mutual coherence to a workable form, there are still many additional approximations that can be made about its properties. We will relegate all but two of these forms to Appendix A. The two conditions that we will assume for the remainder of this chapter are coherence separability and wide-sense spatially incoherent source fields.

2.4. Coherence Separability

In most practical cases of interest, the spectral properties of the source are position independent, and fields for which this representation is valid are said to be coherence separable or cross-spectrally pure. When this can be assumed, the mutual power spectrum can be factored into a product of spatial and spectral components, or

$$R_{\mathbf{r}_s}(\mathbf{R}_1, \boldsymbol{\rho}_1; \omega_+, \omega_-) = J_s(\mathbf{R}_1, \boldsymbol{\rho}_1) R_s(\omega_+, \omega_-) \tag{2.4.1}$$

$J_s(\mathbf{R}_1, \boldsymbol{\rho}_1)$ is a spatial covariance function and is referred to as the mutual intensity function or spatial coherence function. Equations (2.3.3) and (2.3.4) yield

$$\Gamma_{r_a}(\mathbf{R}_2, \boldsymbol{\rho}_2; t, \tau)$$
$$= C_s\left[\left(t - \frac{R}{c}\right), \tau\right] \int_{-\infty}^{\infty} \int_{\Sigma_s} J_s\left[\left(\mathbf{R}_2 + \frac{R\mathbf{k}}{k_0}\right), \boldsymbol{\rho}_1\right] e^{j\mathbf{k}\cdot(\boldsymbol{\rho}_1 - \boldsymbol{\rho}_2)}\, d^2\mathbf{k}\, d^2\boldsymbol{\rho}_1 \tag{2.4.2}$$

where $C_s(t, \tau)$ is the temporal covariance function of the source which is retarded by the propagation time R/c in the variable t:

$$C_s\left[\left(t - \frac{R}{c}\right); \tau\right] = \frac{1}{(2\pi)^2} \int_{-\infty}^{\infty} \int_{-\infty}^{\infty} R_s(\omega_+, \omega_-)\, e^{j\{\omega_+\tau + \omega_-[t-(R/c)]\}}\, d\omega_+\, d\omega_- \tag{2.4.3}$$

Notice also that in the narrowband approximation, coherence separability at the source yields coherence separability at the aperture plane, and the temporal covariance function of the source is directly propagated to the receiver at the speed of light c.

2.5. Partially Coherent and Incoherent Spatial Fields

The degree of spatial coherence of a source is generally determined by the correlation between the fields emanating from different spatial points. To aid in discussing this topic, it is convenient to introduce the concept of quasi-homogeneity. A field is said to be quasi-homogeneous when the spatial coherence function can be separated into two terms, that is, when

$$J_s(\mathbf{R}_1, \boldsymbol{\rho}_1) = I_{r_s}(\mathbf{R}_1)\,\Delta(\boldsymbol{\rho}_1) \tag{2.5.1}$$

This is closely akin to the quasi-stationary assumption used in temporal processes. In the quasi-stationary assumption the second-order statistics are primarily determined by the time differences $(t_1 - t_2)$. However, it is noted that the underlying mechanisms may vary slowly (such as from day to night), causing slowly changing parameters which, while affecting the level or mean of the process, do not affect the statistical behavior. Similarly, in Equation (2.5.1) we see that while the average intensity of the field $I_{r_s}(\mathbf{R}_1)$ may vary slowly with $\mathbf{R}_1$, the statistical mechanisms occur on the scale $\boldsymbol{\rho}_1 = \mathbf{R}_1 - \mathbf{R}_2$ and are not affected by the slow changes in $\mathbf{R}_1$. For such sources, Equation (2.4.2) becomes

$$\begin{aligned}
&\Gamma_{r_a}\left[\mathbf{R}_2, \boldsymbol{\rho}_2; \left(t - \frac{R}{c}\right), \tau\right] \\
&= C_s\left[\left(t - \frac{R}{c}\right), \tau\right] \int_{\Sigma_s} I_{r_s}\left[\left(\mathbf{R}_2 + \frac{R\mathbf{k}}{k_0}\right)\right]\left[\int_{\Sigma_s} \Delta(\boldsymbol{\rho}_1)\, e^{j\mathbf{k}\cdot\boldsymbol{\rho}_1}\, d^2\boldsymbol{\rho}_1\right] e^{-j\mathbf{k}\cdot\boldsymbol{\rho}_2}\, d^2\mathbf{k} \\
&= C_s\left[\left(t + \frac{R}{c}\right), \tau\right] \int_{\Sigma_s} I_{r_s}\left(\mathbf{R}_2 + \frac{R\mathbf{k}}{k_0}\right) \Delta'(\mathbf{k})\, e^{-j\mathbf{k}\cdot\boldsymbol{\rho}_2}\, d^2\mathbf{k}
\end{aligned} \tag{2.5.2}$$

where

$$\Delta'(\mathbf{k}) = \int_{\Sigma_s} \Delta(\boldsymbol{\rho}_1)\, e^{j\mathbf{k}\cdot\boldsymbol{\rho}_1}\, d^2\boldsymbol{\rho}_1$$

Notice that $\Delta'(\mathbf{k})$ has the behavior of a linear filter in the variable $\mathbf{k}$, in that it filters the angular intensity distribution of the source field $I_{r_s}[\mathbf{R}_2 + (R\mathbf{k}/k_0)]$. When a large degree of spatial coherence exists in the field, $\Delta(\boldsymbol{\rho}_1)$ will be constant for large values of $\boldsymbol{\rho}_1$. $\Delta'(\mathbf{k})$ in turn will approach a delta function located at $\mathbf{k} = 0$ and the last integral in Equation (2.5.2) will approach the value $I_{r_s}(\mathbf{R}_2)$. If, at the other extreme, we consider the source to be totally incoherent, then $\Delta(\boldsymbol{\rho}_1) = \delta(\boldsymbol{\rho}_1)$ and $\Delta'(\mathbf{k})$ in turn becomes a constant. From Fourier transform theory we know that as the width of the function $\Delta'(\mathbf{k})$ in $\mathbf{k}$ space becomes larger, then the width of $\Delta(\boldsymbol{\rho}_1)$ in $\boldsymbol{\rho}_1$ space becomes smaller. Conversely, as $\Delta'(\mathbf{k})$ becomes narrower, then $\Delta(\boldsymbol{\rho}_1)$ becomes broader. With reference to Figure 2.2, this implies that as the source spatial coherence increases, the angular filtering action of $\Delta'(\mathbf{k})$ appears to come from smaller regions of the source plane in the superposition integral (2.5.2). Thus, the coherence properties of the field at a particular $\mathbf{R}_2$ in σ_z appear as an increasingly smaller region of σ_0 about the point $\mathbf{R}_1 = \mathbf{R}_2$. Note, however, that generation of the entire coherence function in σ_z for all $\mathbf{R}_2$ requires that the effects of the full extended source be taken into account in the integral (2.5.2) through the argument of the function $I_{r_s}[\mathbf{R}_2 + (R\mathbf{k}/k_0)]$.

Let us now consider the limit of source spatial incoherence in more detail. For this case Equation (2.5.2) reduces to

$$\Gamma_{r_a}\left[\mathbf{R}_2, \boldsymbol{\rho}_2; \left(t - \frac{R}{c}\right), \tau\right] = C_s\left[\left(t - \frac{R}{c}\right); \tau\right] \int_{\Sigma_s} I_{r_s}\left(\mathbf{R}_2 + \frac{R\mathbf{k}}{k_0}\right) e^{-j\mathbf{k}\cdot\boldsymbol{\rho}_2}\, d^2\mathbf{k}$$

and finally $\mathbf{R}_1 = \mathbf{R}_2 + (R\mathbf{k}/k_0)$, $d^2\mathbf{R}_1 = (R^2/k_0^2)d^2\mathbf{k}$ results in

$$\Gamma_{r_a}\left[\mathbf{R}_2, \boldsymbol{\rho}_2; \left(t - \frac{R}{c}\right), \tau\right] = R'_{r_a}[\mathbf{R}_2, \boldsymbol{\rho}_2] C_s\left[\left(t - \frac{R}{c}\right), \tau\right] \tag{2.5.3}$$

where

$$R'_{r_a}[\mathbf{R}_2, \boldsymbol{\rho}_2] = \frac{\bar{\omega}^2}{(Rc)^2}\, e^{j(\bar{\omega}/Rc)\mathbf{R}_2\cdot\boldsymbol{\rho}_2} \int_{\Sigma_s} I_{r_s}(\mathbf{R}_1)\, e^{-j(k_0/R)\boldsymbol{\rho}_2\cdot\mathbf{R}_1}\, d^2\mathbf{R}_1. \tag{2.5.3a}$$

Equation (2.5.3a) is a form of the well-known Von Citert–Zernike theorem.

We observe that if one defines the effective source as

$$I_{r_s}\left(\mathbf{R}_2 + \frac{R\mathbf{k}}{k_0}\right) \Delta'(\mathbf{k}) = I_{r_s}(\mathbf{R}_1)\, \Delta'\left[(\mathbf{R}_1 - \mathbf{R}_2)\frac{k_0}{R}\right] \tag{2.5.4}$$

where we have set $\mathbf{k} = (\mathbf{R}_1 - \mathbf{R}_2)(k_0/R)$, then Equation (2.5.2) retains this Fourier relationship of the incoherent source and readily transitions to the coherent source, strongly displaying the nature of $\Delta'(\mathbf{k})$ as the transform of the source coherence function.

The intensity at the receiver aperture becomes

$$R'_{r_a}(\mathbf{R}_2, 0) C_s\left[\left(t - \frac{R}{c}\right), 0\right] \tag{2.5.5}$$

where we have included the temporal variation in the emittance. For the radiometric systems considered in Chapter 1, all the power collected by an aperture at a receiver of area A_r is usable. In such cases, we can calculate the collected power to be

$$\begin{aligned} P_r(t) &= R_{r_a}(\mathbf{R}_2, 0) C_s\left[\left(t - \frac{R}{c}\right), 0\right] A_r \\ &= \frac{A_r}{\bar{\lambda}^2 R^2}\left\{\int_{\Sigma_s} I_{r_s}(\mathbf{R}_1) C_s\left[\left(t - \frac{R}{c}\right), 0\right] d^2\mathbf{R}_1\right\} \end{aligned} \quad (2.5.6)$$

We see by reference to Equation (1.1.3a) that the bracketed term in Equation (2.5.6) is the source power $P_s(t)$. Since we integrate over the source area A_s, we also see that

$$\frac{d^2\mathbf{R}}{R^2} = d^2\Omega_s \quad (2.5.7)$$

is an integration in solid angle. (Had we included the obliquity factor in our derivation, this result would yield the source area projected onto the line-of-sight direction as stated in Section 1.3.1.) By referring to Equation (1.3.1a) we also see that we can relate the source radiance or brightness to

$$N(\mathbf{R}_1, \theta) = I_{r_s}(\mathbf{R}_1) C_s\left[\left(t - \frac{R}{c}\right), 0\right] \Big/ \bar{\lambda}^2 \quad (2.5.8)$$

By assuming separability of the spatial coherence function and incoherence of the source, we are led to separability and spatial homogeneity at the receiver. This separability will now let us consider the spatial and temporal properties of the system in an independent manner. And, in fact, for the purposes of this book, we are primarily interested in how the spatial properties of the system are evolved as we transit from source to receiver. This then leads us to the other important observation about Equation (2.5.3), which is that when we separate the spatial component out as

$$R'_{r_a}(\boldsymbol{\rho}_2) = \frac{1}{\bar{\lambda}^2 R^2} \int_{\Sigma_s} I_{r_s}(\mathbf{R}_1)\, e^{j(k_0/R_1)\mathbf{R}_1 \cdot \boldsymbol{\rho}_2}\, d^2\mathbf{R}_1 \quad (2.5.9)$$

it has the familiar form of a Fourier transform, which immediately implies a wealth of experience in the characterization of linear transmission systems. It is this analogy which we quantify in the following section.

2.6. System Coherence

Although Equation (2.5.9) can clearly be recognized as a Fourier transform, there are a few steps which should be taken to make the association rigorous. First, we see that the domain of integration is the surface of the source, which is finite in extent. Therefore, we must incorporate the finite size of the source into the limits of the function $I_{r_s}(\mathbf{R}_1)$ and then extend the integration over the whole domain. Returning to the notation of $\mathbf{R}_1$, Equation (2.3.5), we define the new variable $\mathbf{f} = \mathbf{R}_1/\bar{\lambda}R$, and Equation (2.5.9) can then be written as

$$R_{r_a}(\boldsymbol{\rho}_2) = \int_{-\infty}^{\infty} I_{r_s}(\mathbf{f})\, e^{j2\pi\mathbf{f}\cdot\boldsymbol{\rho}_2}\, d^2\mathbf{f} \tag{2.6.1}$$

with an inverse given by

$$I_{r_s}(\mathbf{f}) = \int_{-\infty}^{\infty} R_{r_a}(\boldsymbol{\rho}_2)\, e^{-j2\pi\mathbf{f}\cdot\boldsymbol{\rho}_2}\, d^2\boldsymbol{\rho}_2 \tag{2.6.2}$$

and where we have dropped the dependence on $\mathbf{R}_2$. The projection of $\mathbf{f}$ onto $\boldsymbol{\rho}_2$ is called the spatial frequency and takes the form

$$\frac{\mathbf{f}\cdot\boldsymbol{\rho}_2}{|\boldsymbol{\rho}_2|} = |\mathbf{f}|\cos\theta \tag{2.6.3}$$

The function $R_{r_a}(\boldsymbol{\rho}_2)$ is the spatial portion of the coherence function at the receiver aperture. As such, it is a measure of the spatial coherence of the field. While there are several ways open to us with regard to normalization, we will adopt the convention used in Ref. 4. We will define the function $\tilde{R}_{r_a}(\boldsymbol{\rho}_2)$ as

$$R_{r_a}(\boldsymbol{\rho}_2) = \frac{R_{r_a}(\boldsymbol{\rho}_2)}{R_{r_a}(0)}\cdot R_{r_a}(0) = \tilde{R}_{r_a}(\boldsymbol{\rho}_2)R_{r_a}(0) \tag{2.6.4}$$

with

$$C_s\left[\left(t-\frac{R}{c}\right), 0\right] R_{r_a}(0) = I_a(t) \tag{2.6.5}$$

where we have incorporated the remaining factor $R_{r_a}(0)$ into the temporal function, which will then have the dimension of intensity or irradiance.† This definition will then yield a range of spatial coherence for $\tilde{R}_{r_a}(\boldsymbol{\rho})$ between zero and one. Thus we see that unity corresponds to perfect coherence while zero corresponds to perfect incoherence. In fact, more physical systems lie somewhere in between and are partially coherent. Since $\tilde{R}_{r_a}(\boldsymbol{\rho}_2)$ is maximum at $\boldsymbol{\rho}_2 = 0$, we see the degree of coherence will vary according to the distance between two points in the aperture plane and will strongly depend on the functional form of $\tilde{R}_{r_a}(\boldsymbol{\rho}_2)$.

2.7. Linear Channels

We will define a linear channel as one in which we can describe the effect upon the source element $f_{r_s}(\mathbf{r}_s, \omega)$ as measured at the points $\mathbf{r}_a$ in the receiver aperture, at a frequency ω, by a multiplication with a frequency transfer function. This is the basic assumption in linear system theory. We will further assume that this transfer function may be time dependent and include a dependence on both the point of origin $\mathbf{r}_s$ and the point of reception $\mathbf{r}_a$. We will designate such a transfer function by $g(\mathbf{r}_s, \mathbf{r}_a; \omega, t)$. Formally computing the mutual coherence function at the aperture plane, assuming the narrowband conditions of Equations (2.3.1) and (2.3.2) and using Equations (2.2.4)–(2.2.7), yields

$$\Gamma_{r_a}(\mathbf{R}_2, \boldsymbol{\rho}_2; t, \tau) = \frac{\bar{\omega}^2}{(2\pi Rc)^2} \int_{-\infty}^{\infty} \int_{-\infty}^{\infty} \int_{\Sigma_s} \int_{\Sigma_s} R_{r_s}(\mathbf{R}_1, \boldsymbol{\rho}_1; \omega_+, \omega_-) M(\ ) \, e^{j\{\omega_+\tau + \omega_-[t-(R/c)]\}} \times e^{-j\bar{\omega}(\mathbf{R}_1-\mathbf{R}_2)\cdot(\boldsymbol{\rho}_1-\boldsymbol{\rho}_2)/Rc} \, d\omega_+ \, d\omega_- \, d^2\mathbf{R}_1 \, d^2\boldsymbol{\rho}_1 \tag{2.7.1}$$

where the function $M(\ )$ is the channel correlation function defined by the ensemble average

$$M(\ ) = E[g(\mathbf{r}_a, \mathbf{r}_a; \omega, t) g^*(\mathbf{r}'_s, \mathbf{r}'_a; \omega', t')] = M(\mathbf{R}_1, \mathbf{R}_2, \boldsymbol{\rho}_1, \boldsymbol{\rho}_2; \omega_+, \omega_-, t, \tau) \tag{2.7.2}$$

It is usually safe to assume that the frequency and time dependences in this ensemble average are not a function of the source surface variables $\mathbf{R}_1$ and $\boldsymbol{\rho}_1$ or the aperture surface variables $\mathbf{R}_2$ and $\boldsymbol{\rho}_2$ and, consequently, $M(\ )$ separates into‡

$$M(\ ) = M_c[\mathbf{R}_1, \mathbf{R}_2, \boldsymbol{\rho}_1, \boldsymbol{\rho}_2] \cdot R_T(\omega_+, \omega_-; t, \tau) \tag{2.7.3}$$

†With this normalization it can be shown that the sum of the spatial eigenvalues associated with $\tilde{R}_{r_a}(\boldsymbol{\rho}_2)$ sums to the receiver area A_r (see Section 5.4).[14]

‡Notice that this is a weaker condition than requiring $g(\ )$ to be time-space separable.

We recognize $R_T(\omega_+, \omega_-; t, \tau)$ as the correlation function of the time variant transfer function of a channel, as given by Bello.[5] If we assume the separation of the mutual power spectrum, Equation (2.4.1), then Equation (2.7.1) can be written as the product of two sets of integrals

$$\Gamma_{r_a}(\mathbf{R}_2, \boldsymbol{\rho}_2; t, \tau) = R'_{r_a}[R_2, \boldsymbol{\rho}_2] C_a\left[\left(t - \frac{R}{c}\right), \tau\right] \qquad (2.7.4)$$

where

$$R'_{r_a}[\mathbf{R}_2, \boldsymbol{\rho}_2]$$
$$= \frac{\bar{\omega}^2}{(Rc)^2} \int_{\Sigma_s} \int_{\Sigma_s} J_s(\mathbf{R}_1, \boldsymbol{\rho}_1) M_c[\mathbf{R}_1, \mathbf{R}_2; \boldsymbol{\rho}_1, \boldsymbol{\rho}_2]\, e^{-j\bar{\omega}(\mathbf{R}_1 - \mathbf{R}_2)\cdot(\boldsymbol{\rho}_1 - \boldsymbol{\rho}_2)}\, d^2\mathbf{R}_1\, d^2\boldsymbol{\rho}_1 \qquad (2.7.5)$$

and

$$C_a\left[\left(t - \frac{R}{c}\right), \tau\right]$$
$$= \frac{1}{(2\pi)^2} \int_{-\infty}^{\infty} \int_{-\infty}^{\infty} R_s(\omega_+, \omega_-) R_T(\omega_+, \omega_-; t, \tau)\, e^{j\{\omega_+ \tau + \omega_-[t - (R/c)]\}}\, d\omega_+\, d\omega_- \qquad (2.7.6)$$

Notice that once again the received mutual coherence function separates into temporal and spatial components. Since we are primarily interested in the spatial properties of channels in this book, we will not dwell long on Equation (2.7.6). However, it is important to point out the maturity of our knowledge in treating these channels.[5] We will make two observations. For the important condition of a quasi-wide-sense stationary uncorrelated scattering (QWSSUS) channel where the variation of $R_T(\omega_+, \omega_-; t, \tau)$ about ω_+, t are small compared to the input signal bandwidth and the output signal duration, respectively, Equation (2.7.6) can be written as

$$C_a\left[\left(t - \frac{R}{c}\right), \tau\right] \cong \frac{1}{2\pi} \int_{-\infty}^{\infty} X(\omega_-, \tau) R_T(\omega_+, \omega_-; t, \tau)\, e^{j\omega_-[t - (R/c)]}\, d\omega_- \qquad (2.7.7)$$

where

$$X(\omega_-, \tau) = \frac{1}{2\pi} \int_{-\infty}^{\infty} R_s(\omega_+, \omega_-)\, e^{j\omega_+ \tau}\, d\omega_+ \qquad (2.7.8)$$

Notice that $C_a[(t - R/c), \tau]$ is implicitly dependent on the time separately and the carrier frequency ω_+. The variations are assumed small enough to be neglected in the analysis.

When the temporal portion of the field is deterministic, with frequency response $f_s(\omega)$, as is usually the case in communications systems, $X(\omega_-, \tau)$ takes the form

$$X(\omega_-, \tau) = \frac{1}{2\pi}\int_{-\infty}^{\infty} f_s^*\left(\omega_+ - \frac{\omega_-}{2}\right) f_s\left(\omega_+ + \frac{\omega_-}{2}\right) e^{j\omega_+\tau}\, d\omega_+ \qquad (2.7.9)$$

and it called the signal ambiguity function. Notice that in Equation (2.7.7) the complex envelope of the signal is linearly filtered by the complex envelope of the channel correlation function. This becomes more apparent if we recall that the intensity of the temporal process can be evaluated by setting $\tau = 0$. Then Equation (2.7.7) becomes

$$\begin{aligned} I_a(t) &= R'_{r_a}(\mathbf{R}_1, 0) C_a\left(t - \frac{R}{c}\right) \\ &= \frac{R'_{r_a}(\mathbf{R}, 0)}{2\pi}\int_{-\infty}^{\infty} X(\omega_-, 0) R_T(\omega_-, 0)\, e^{j\omega_-[t-(r/c)]}\, d\omega_- \end{aligned} \qquad (2.7.10)$$

a simple linear time-invariant filter. Notice, however, that a product in frequency yields a convolution in time, so that

$$C_a(t) = C_s\left(t - \frac{R}{c}\right) \otimes Q\left(t - \frac{R}{c}\right) \qquad (2.7.11)$$

where $\otimes$ denotes convolution and

$$Q(t) = \frac{1}{2\pi}\int_{-\infty}^{\infty} R_T(\omega_-, 0)\, e^{j\omega_- t}\, d\omega_- \qquad (2.7.12)$$

is the delay power spectrum of the channel. It is the intensity of the channel scattering as a function of path delay. For the given conditions of low dispersion in both time and frequency, the channel is referred to as a "flat-flat" fading channel.

From a physical point of view, we see that the transfer function $R_T(\omega_-, \tau)$ filters the ambiguity function $X(\omega_-, \tau)$ in the variables ω_- and τ. If the transfer is perfect, $R_T(\omega_-, \tau) = 1$, and the signal emerges unscathed. Filtering in the variable ω_- is called frequency dispersion, which limits the bandwidth over which the signal remains coherent. Similarly, filtering in the variable τ is called time dispersion and limits the time over which the signal can remain coherent. For example, echoes çreate a multipath time spread which stretches the signal of interest. When this stretching is small, we are in the condition considered above and little distortion of the modulating signal is encountered. A dual effect is observed in the frequency domain in the variable ω_-.

Again assuming that the source is spatially incoherent, $J_s(\mathbf{R}_1, \boldsymbol{\rho}_1) = R_{r_s}(\mathbf{R}_1)\delta(\boldsymbol{\rho}_1)$, we find that the spatial coherence function at the aperture becomes

$$R_{r_a}(\boldsymbol{\rho}_2) = e^{j(\bar{\omega}/cR)\mathbf{R}_2\cdot\boldsymbol{\rho}_2} R'_{r_a}(\mathbf{R}_2, \boldsymbol{\rho}_2)$$
$$= \frac{\bar{\omega}^2}{(Rc)^2} \int_{\Sigma_s} R_{r_s}(\mathbf{R}_1) M(\mathbf{R}_1, \mathbf{R}_2, \boldsymbol{\rho}_2)\, e^{j(\bar{\omega}/cR)\mathbf{R}_1\cdot\boldsymbol{\rho}_2}\, d^2\mathbf{R}_1 \tag{2.7.13}$$

$M(\mathbf{R}_1, \mathbf{R}_2, \boldsymbol{\rho}_2)$ is often referred to as the point source coherence function. The significance of this can be seen if we allow the source to be modeled as an incoherent point source. Then

$$R_{r_s}(\mathbf{R}_1) = R_s\delta(R_1 - R_0) \tag{2.7.14}$$

$$R_{r_a}(\boldsymbol{\rho}_2) = \frac{R_s}{\bar{\lambda}^2 R^2} M(R_2, \boldsymbol{\rho}_2, \mathbf{R}_0) \tag{2.7.15}$$

with $R_{r_a}(\boldsymbol{\rho}_2)$ taking the form of $M(\mathbf{R}_2, \boldsymbol{\rho}_2, \mathbf{R}_0)$. The transform of $R_{r_a}(\boldsymbol{\rho}_2)$ is the angular distribution, Equation (2.6.2), or an effective source brightness of

$$N(\mathbf{R}_1, \theta) = C_s\left[\left(t - \frac{R}{c}\right), 0\right] \int_{-\infty}^{\infty} R_{r_a}(\boldsymbol{\rho}_2)\, e^{-j2\pi\mathbf{f}\cdot\boldsymbol{\rho}_2}\, d^2\boldsymbol{\rho}_2/\bar{\lambda}^2; \qquad \mathbf{f} = \frac{\mathbf{R}_1}{\bar{\lambda} R_1} \tag{2.7.16}$$

Since we assume passive channels, the function $M(\mathbf{R}_2, \boldsymbol{\rho}_2, \mathbf{R}_0)$ is always bounded by unity and, in the absence of the channel disturbances, takes on the value unity for all values of $\mathbf{R}_2, \boldsymbol{\rho}_2$. Thus, by our definition of coherence, we see that an incoherent point source illumination will produce a spatially coherent coherence function at the receiver aperture in the absence of channel distortion. Furthermore, we would not expect the point source coherence function, which is a measure of the randomness associated with the channel, to have a strong dependence on the location of the source, $\mathbf{R}_0$, in the object plane. Thus suppressing the $\mathbf{R}_0$ dependence and using Equation (2.6.1), we have [by taking $M(\mathbf{R}_2, \boldsymbol{\rho}_2)$ outside the integral]

$$\tilde{R}^M_{r_a}(\mathbf{R}_2, \boldsymbol{\rho}_2) = M(\mathbf{R}_2, \boldsymbol{\rho}_2)\tilde{R}_{r_a}(\boldsymbol{\rho}_2) \tag{2.7.17}$$

And further assuming that the channel is spatially homogeneous (independent of $\mathbf{R}_2$), we are led to our main result

$$\tilde{R}^M_{r_a}(\boldsymbol{\rho}_2) = M(\boldsymbol{\rho}_2)\tilde{R}_{r_a}(\boldsymbol{\rho}_2) \tag{2.7.18}$$

where $\tilde{R}^M_{r_a}(\boldsymbol{\rho})$ is the normalized spatial coherence function in the presence of channel effects. The significance of Equation (2.7.18) lies in the fact that since the point source coherence function is bounded by unity, its effect is to decrease the spatial coherence at a receiver aperture. The variable $\boldsymbol{\rho}_2$ is a measure of the lateral extent in the aperture plane over which we have coherence. Any effect by $M(\boldsymbol{\rho}_2)$ that filters the spatial coherence function $\tilde{R}_{r_a}(\boldsymbol{\rho}_2)$ decreases the area of spatial coherence. The effective source brightness then becomes

$$N(\mathbf{r}_s, \theta) = C_s\left[\left(t - \frac{R}{c}\right), 0\right] \int_{-\infty}^{\infty} R_{r_a}(\boldsymbol{\rho}_2) M(\boldsymbol{\rho}_2)\, e^{-j2\pi \mathbf{f}\cdot\boldsymbol{\rho}_2}\, d^2\boldsymbol{\rho}_2/\bar{\lambda}^2 \tag{2.7.19}$$

It is common to define a coherence area as that area in $\boldsymbol{\rho}_2$ over which $\tilde{R}_{r_a}(\boldsymbol{\rho}_2)$ maintains a value greater than some prescribed number, and we denote this area by A_c. It can also be shown that the spatial dimension of a system, which we have earlier denoted as D (Section 1.4), can also be obtained as A_r/A_c.[†] It is also convenient to define a coherence length ρ_c, such that

$$A_c = \pi\rho_c^2 \tag{2.7.20}$$

and to model the coherence function by

$$\tilde{R}_{r_a}(\boldsymbol{\rho}_2) = \begin{cases} 1 & \text{for } |\boldsymbol{\rho}| < \rho_c \\ 0 & \text{elsewhere} \end{cases} \tag{2.7.21}$$

In some channels we find that the spatial coherence function is not only spatially homogeneous, but rotationally invariant. This means that the dependence in Equation (2.5.9) is only a function of the magnitude $\boldsymbol{\rho}_2$. When this is the case, the spatial coherence function (and the medium) is referred to as isotropic.

It is instructive to point out that in the absence of channel effects the simple Fourier relationship in Equation (2.6.1) implies that as the extent of the source brightness $I_s(\mathbf{R}_1)$ increases, the spatial coherence of the received field decreases. On the other hand, as we limit the extent of the source, it

[†] Since A_c is the area of coherence, we can associate a field of view for the system as λ^2/A_c. Thus $D = (\lambda^2/A_c)/(\lambda^2/A_r) = A_r/A_c$.

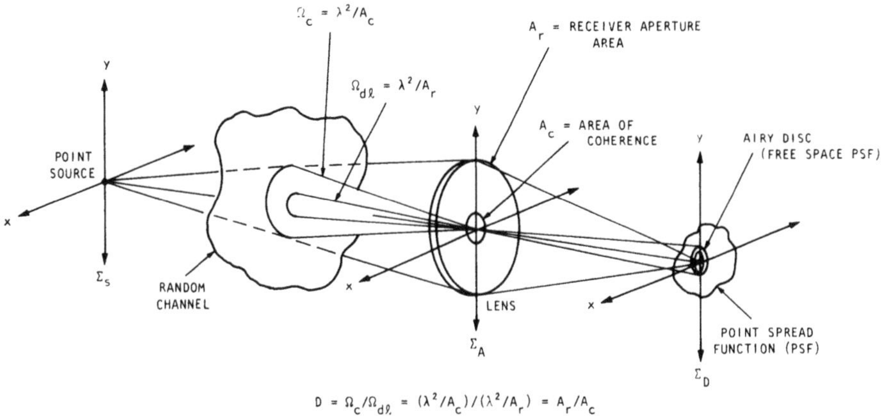

Figure 2.3. Relationship between coherence area, field of view, and spatial dimension.

appears to have a greater spatial coherence at the receiver; the limit, of course, being the point source which yielded perfect coherence. We notice, however, that a point source transmitted over a random channel yields an effect [Equation (2.7.15)] that could just as easily have come from an extended source. Consequently, we would expect a random channel to blur or smear a point source and make its effect equal to that of an extended source. The extent of this smeared source will be a solid angle $\Omega_s = \lambda^2/A_c$ and will contain $A_r/A_c = D$ degrees of freedom. Or, since we can approximate A_c by $\pi\rho_c^2$, we have that the effective or coherent field of view Ω_c is

$$\Omega_c = \frac{\lambda^2}{A_c} \approx \frac{\lambda^2}{\pi\rho_c^2} = \frac{1}{\pi}\left(\frac{\lambda}{\rho_c}\right)^2 \tag{2.7.22}$$

and the dimension D becomes

$$D = \frac{A_r}{A_c} \approx \frac{A_r}{\pi\rho_c^2} = \left(\frac{r}{\rho_c}\right)^2 \tag{2.7.23}$$

where r is the radius of the aperture; see Figure 2.3.

2.8. Receiver Optics

In almost all optical systems the field that is received at the receiver aperture is propagated through appropriate optics to a detector plane (Figure 2.3). The equations governing this propagation are precisely those that we have considered when transmitting from the source to the receiver.

The differences lie in the fact that the distance from source to aperture, R, is much larger than the distance from aperture to detector plane, f_c, the focal length of the optical system. For the purposes of this book, we are interested in optical systems which image the object (or the smeared image) over the detector plane. It can be shown using thin-lens theory[1] that one can construct a coherent optical system which projects a linearly filtered version of the object onto the image plane whenever the "lens law" is satisfied. Thus if the object is a distance R from the aperture, and the detector is a distance d behind the aperture, this law takes the form

$$\frac{1}{R} + \frac{1}{d} - \frac{1}{f_c} = 0 \tag{2.8.1}$$

Notice that whenever the object distance is very far from the aperture, the source is effectively at infinity and the image plane coincides with the focal length. The magnification, $1/M$, of the system is defined as R/d. The relationship between these variables can be seen in Figure 2.4.

We determine the linear filter response of the optical system in the following way. If we designate the field at the detector as $f_{r_d}(\mathbf{r}_d, \omega)$, then Equation (2.1.6) can be rewritten as

$$f_{r_d}(\mathbf{r}_d, \omega) = \frac{e^{jkf_c}}{jf_c\lambda} e^{j(k/2f_c)|\mathbf{r}_d|^2} \int_{\Sigma_a} f_{r_a}(\mathbf{r}_a, k)\, e^{j(k/f_c)(\mathbf{r}_d \cdot \mathbf{r}_a)}\, d^2\mathbf{r}_a; \qquad k = \frac{2\pi}{\lambda} = \frac{\omega}{c} \tag{2.8.2}$$

where we have used thin-lens theory to eliminate the quadratic term inside the integral. We next define the pupil function $P(\mathbf{r}_a)$ as the domain of

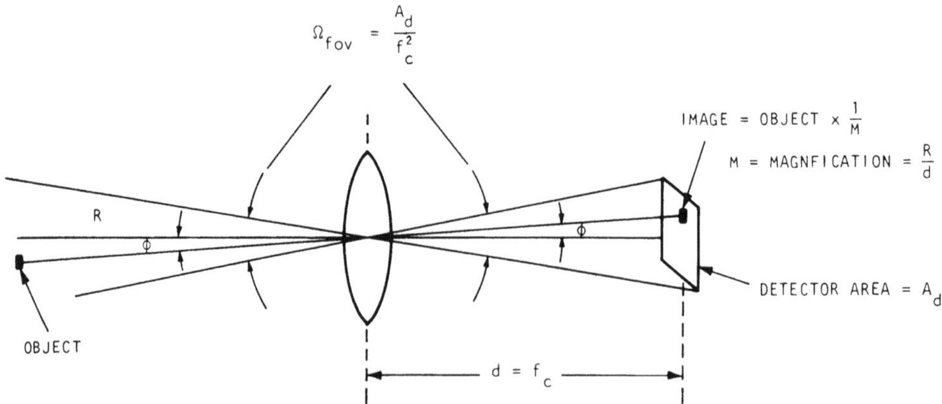

Figure 2.4. Magnification by receiver optics.

integration representing the aperture and extend the integration to infinity yielding

$$f_{r_d}(\mathbf{r}_d, k) = \frac{e^{jkf_c}}{j\lambda f_c} e^{j(k/2f_c)|\mathbf{r}_d|^2} \int_{-\infty}^{\infty} f_{r_a}(\mathbf{r}_a, k) P(\mathbf{r}_a)\, e^{j(k/f_c)(\mathbf{r}_d \cdot \mathbf{r}_a)}\, d^2\mathbf{r}_a \quad (2.8.3)$$

We can ignore the phase terms outside the integral since we will only be interested in the intensity at the detector and write Equation (2.8.3) as

$$f_{r_d}(\mathbf{r}_d, k) = \int_{-\infty}^{\infty} f_{r_a}(\lambda f_c \mathbf{f}_o, k) P(\lambda f_c \mathbf{f}_o)\, e^{j2\pi \mathbf{f}_o \cdot \mathbf{r}_d}\, d^2\mathbf{f}_o \quad (2.8.4)$$

Notice that we have inherently defined the system transfer function as the pupil function evaluated at $\mathbf{r}_a = \lambda f_c \mathbf{f}_o$. We then obtain the intensity at the detector plane as the expected value of $|f_{r_d}(\mathbf{r}_d, t)|^2$ or

$$I_a(t) I_d(\mathbf{r}_d) = I_a(t) \int_{-\infty}^{\infty} \int_{-\infty}^{\infty} \tilde{R}_{r_a}(\lambda \mathbf{f}_o f_c, \lambda f_c \mathbf{f}'_o) P(\lambda f_c \mathbf{f}_o) P(\lambda f_c \mathbf{f}'_o)\, e^{j2\pi \mathbf{r}_d (\mathbf{f}_o - \mathbf{f}'_o)}\, d^2\mathbf{f}_o\, d^2\mathbf{f}'_o \quad (2.8.5)$$

again using separability to consider only the spatial coherence. Finally, assuming spatial homogeneity and inserting Equation (2.7.18) yields

$$I_d(\mathbf{r}_d) = \int_{-\infty}^{\infty} \hat{R}_{r_a}(\lambda f_c \boldsymbol{\xi}) M(\lambda f_c \boldsymbol{\xi}) \theta(\lambda f_c \boldsymbol{\xi})\, e^{j2\pi \boldsymbol{\xi} \cdot \mathbf{r}_d}\, d^2\boldsymbol{\xi} \quad (2.8.6)$$

where $\mathbf{f}_o - \mathbf{f}'_o = \boldsymbol{\xi}$, $\lambda f_c \boldsymbol{\xi} = \boldsymbol{\rho}_2$, and

$$\theta(\lambda f_c \boldsymbol{\xi}) = \int_{-\infty}^{\infty} P(\lambda f_c \mathbf{f}_o) P[\lambda f_c (\mathbf{f}_o + \boldsymbol{\xi})]\, d^2\mathbf{f}_o \quad (2.8.7)$$

is the correlation of the pupil function with itself (autocorrelation). $\theta(\lambda f_c \boldsymbol{\xi})$ is called the optical transfer function, or the OTF, of the system. The modulus of the optical transfer function is called the modulation transfer function, or the MTF. Notice that the OTF reduces the coherence of the signal in preeisely the same manner as the point source coherence function, both of which act as cascaded filters. Thus we see that even the best optical system will limit the spatial coherence of the received signal because of its finite aperture. This will act to blur or smear the image in the image plane. The resulting blur circle for a coherent optical system is equal to $\lambda^2 f_c^2 / A_r$. We can see this by inserting Equation (2.6.1) into Equation (2.8.6) to obtain

$$I_d(\mathbf{r}_d) = \int_{-\infty}^{\infty} \tilde{R}_{r_s}(\lambda R \mathbf{f})\, d^2\mathbf{f} \int_{-\infty}^{\infty} M(\lambda f_c \boldsymbol{\xi}) \theta(\lambda f_c \boldsymbol{\xi})\, e^{j2\pi \boldsymbol{\xi} \cdot [\lambda f_c \mathbf{f} + \mathbf{r}_d]}\, d^2\boldsymbol{\xi} \quad (2.8.8)$$

where $\tilde{R}_{r_s}(\lambda R\mathbf{f}) = R_{r_s}(\lambda R\mathbf{f})/R_{r_s}(0)$. Notice that when we have no coherence distortion, $M(\lambda f_c \boldsymbol{\xi})\theta(\lambda f_c \boldsymbol{\xi}) = 1$ and the inner integral becomes $\delta[\lambda f_c \mathbf{f} + \mathbf{r}_d]$ and the resulting intensity on the detector becomes

$$I_d(\mathbf{r}_d) = \int_{-\infty}^{\infty} \tilde{R}_{r_s}(\lambda R\mathbf{f})\delta(\lambda f_c \mathbf{f} + \mathbf{r}_d)\, d^2\mathbf{f}$$
$$= \left(\frac{1}{\lambda f_c}\right)^2 \tilde{R}_{r_s}\left(-\frac{r_d}{M}\right) \tag{2.8.9}$$

This, as expected, is a reversed replica of the image reduced in scale by the magnification M. Notice from Figure 2.4 that the image can fall anywhere on the detector, or anywhere within the solid angle A_d/f_c^2 subtended by the detector. This solid angle is the receiver field of view which we denote by Ω_{fov}. The minimum field of view is determined by the blur circle. For the coherent OTF this is $\lambda^2 f_c^2/A_s$ and we have that

$$\Omega_{\text{fov}} = \frac{\lambda^2 f_c^2/A_r}{f_c^2} = \frac{\lambda^2}{A_r} = \Omega_{\text{dl}} \tag{2.8.10}$$

is the minimum field of view.

More generally, of course, the inner integral of Equation (2.8.8) is not a delta function. Rather, it takes on a general form of $\tilde{H}(\lambda f_c \mathbf{f} + \mathbf{r}_d)$ and Equation (2.8.10) becomes

$$I_d(\mathbf{r}_d) = \int_{-\infty}^{\infty} \tilde{R}_{r_s}(\lambda R\mathbf{f})\tilde{H}(\lambda f_c \mathbf{f} + \mathbf{r}_d)\, d^2\mathbf{f} \tag{2.8.11}$$

Since this is in the form of a convolution of the image intensity distribution with $\tilde{H}(\lambda f_c \mathbf{f} + \mathbf{r}_d)$, the net result is to cause a blurring or smearing of the image on the image plane. The extent of this distortion will be a direct consequence of the channel traversed and the optical system employed. $\tilde{H}(\mathbf{r}_d)$ is called the point spread function of the system.

For the case of no channel distortions, the OTF of a circular aperture has the well-known form

$$\theta(\boldsymbol{\rho}_2) = \begin{cases} \dfrac{2}{\pi}\left[\cos^{-1}\left(\dfrac{\rho_2}{2\rho_0}\right) - \dfrac{\rho_2}{2\rho_0}\sqrt{1-\left(\dfrac{\rho_2}{2\rho_0}\right)^2}\right]; & \rho_2 \geqslant 2\rho_0 = \dfrac{2r}{\lambda f_c} \\ 0 & \text{elsewhere} \end{cases} \tag{2.8.12}$$

and the corresponding point spread function is

$$\tilde{H}(\mathbf{r}_d) = \frac{kr^2}{2f_c}\left[\frac{2J_1\left(\dfrac{krr_d}{f_c}\right)}{\left(\dfrac{krr_d}{f_c}\right)}\right]^2 \tag{2.8.13}$$

where $J_1(x)$ is a Bessel function, and r is the radius of the aperture. This yields the blur circle in Equation (2.8.10).

Finally, notice that integrating $I_d(\mathbf{r}_d)$ over the detection area yields

$$P_{r_d} = \int_{-\infty}^{\infty} R_{r_a}(\lambda f_c \boldsymbol{\xi}) M(\lambda f_c \boldsymbol{\xi}) \theta(\lambda f_c \boldsymbol{\xi}) \tau(\boldsymbol{\xi})\, d^2\boldsymbol{\xi} \tag{2.8.14}$$

where the detector "footprint" satisfies the Fourier relation

$$\tau(\boldsymbol{\xi}) = \int_{A_d} e^{j2\pi\boldsymbol{\xi}\cdot\mathbf{r}_d}\, d^2\mathbf{r}_d \tag{2.8.15}$$

and filters its source.

2.9. Clutter

Up to this point we have assumed that the total effect of noise could be treated either as an emission from a source (generally a blackbody) or even a reflection off some other surface which would preserve the radiant nature of the emitting source and merely scale it by an attenuation factor, called an albedo. This noise field then was shown to produce an intensity at the detector equal to $I_a(t)I_d(\mathbf{r}_d)$[(Equation (2.8.5)]in the same manner as a signal would. In Chapter 3, we will show how the detector responds to the intensity incident on it to produce a current, and how one can sort out the signal portion from the noise portion to compute a signal-to-noise ratio.

Clutter, on the other hand, is a form of noise that will have all the appearances of a signal. Sometimes it is caused by the direct movement of the background. Such would be the case if a cloud drifted through the field of view of the receiver. The detector, in addition to recording the time-varying intensity of the source, would also record the time-varying intensity of the cloud radiation. Another common cause of clutter comes from the line-of-sight movement of the optical system, caused by servo noise. If the background is highly space dependent in its radiant properties, line-of-sight drift would cause the field of view to dither through this structured background causing a time-varying noise much like the drifting cloud. A similar effect can happen with active mirrors, where the effects of vibrations cause modal shifts and surface deformation which in turn create an apparent line-of-sight drift. The general treatment of the problem is beyond the scope of this book. We will, however, treat a simple case to give the reader some insight into how this particular problem can be addressed.

We begin this treatment by assuming that the intensity in object space for the clutter source has its coordinate transformed to

$$\mathbf{r}_s + \boldsymbol{\zeta}(t) \tag{2.9.1}$$

where $\boldsymbol{\zeta}(t)$ represents some arbitrary source motion. From Equation (2.9.1) it is easy to show that the net effect is to multiply $R_{r_a}(\boldsymbol{\rho}_2)$ by the factor

$$\exp^{j2\pi\boldsymbol{\zeta}(t)\cdot\boldsymbol{\rho}_2/\lambda R}$$

which in turn alters Equation (2.8.6) to

$$I_d(\mathbf{r}_d) = \int_{-\infty}^{\infty} \tilde{R}_{r_a}(\lambda f_c\boldsymbol{\xi})M(\lambda f_c\boldsymbol{\xi})\theta(\lambda f_c\boldsymbol{\xi})\, e^{j2\pi\boldsymbol{\xi}\cdot\mathbf{r}_d}\, e^{j2\pi(\boldsymbol{\zeta}(t)\cdot\boldsymbol{\xi})(f_c/R)}\, d^2\boldsymbol{\xi} \tag{2.9.2}$$

We will be interested in the total current at the detector so we integrate $I(\mathbf{r}_d)$ over the detector area A_d. Since the current produced by the clutter at the detector output is now time dependent, we will also Fourier transform the detector current output yielding

$$\alpha I_d(\omega) = \alpha \int_{-\infty}^{\infty} \tilde{R}_{r_a}(\lambda f_c\boldsymbol{\xi})M(\lambda f_c\boldsymbol{\xi})\theta(\lambda f_c\boldsymbol{\xi})\tau(\boldsymbol{\xi}) \int_{-\infty}^{\infty} e^{-j[\omega t-(2\pi f_c/R)\boldsymbol{\zeta}(t)\cdot\boldsymbol{\xi}]}\, dt\, d^2\boldsymbol{\xi} \tag{2.9.3}$$

where

$$\tau(\boldsymbol{\xi}) = \int_{A_d} e^{j2\pi\boldsymbol{\xi}\cdot\mathbf{r}_d}\, d^2\mathbf{r}_d \tag{2.9.4}$$

is the Fourier transform of the detector shape and is analogous to the pupil function in its behavior. This is again the detector footprint (2.8.15), which multiplies the scene in object space.

At this point one has to know more about $\boldsymbol{\zeta}(t)$ to proceed further. We will assume a linearly drifting scene so that we can simplify $\boldsymbol{\zeta}(t)$ to

$$\boldsymbol{\zeta}(t) = \mathbf{V}t; \qquad \mathbf{V} = (V_x, V_y) \tag{2.9.5}$$

where $\mathbf{V}$ is the velocity vector of the scene. With this assumption the time integral becomes a delta function

$$\int_{-\infty}^{\infty} e^{-j[\omega-(2\pi f_c/R)\mathbf{V}\cdot\boldsymbol{\xi}]t}\, dt = \delta\left(\omega - \frac{2\pi f_c}{R}\mathbf{V}\cdot\boldsymbol{\xi}\right) \tag{2.9.6}$$

which allows us to integrate one of the variables in $\boldsymbol{\xi}$

$$\xi_x = \frac{R}{2\pi f_c V_x}\omega - \xi_y \frac{V_y}{V_x} \tag{2.9.7}$$

yielding

$$\alpha I_d(\omega) = \frac{\alpha R}{2\pi V_x f_c}\int_{-\infty}^{\infty} d\xi_y [R_{r_a}(\lambda f_c \boldsymbol{\xi}) \times M(\lambda f_c \boldsymbol{\xi})\tau(\boldsymbol{\xi})\theta(\lambda f_c \boldsymbol{\xi})]|_{\boldsymbol{\xi}=\{[(R\omega/2\pi V_x f_c)-\xi_y(V_y/V_x],\xi_y\}} \tag{2.9.8}$$

Notice that all the spatial filtering functions have been transformed into the frequency domain as bandpass filters, together with the clutter frequency spectrum $R_{r_a}(\lambda f_c \boldsymbol{\xi})$. Finally, assuming that the statistics for the clutter generated by the moving scene are stationary allows us to calculate the power density spectrum for the current produced by the clutter as

$$\alpha^2\overline{|I_d(\omega)|^2} = \left(\frac{\alpha R}{2\pi V_x f_c}\right)^2 \int_{-\infty}^{\infty} d\xi_y [\phi(\lambda f_c \boldsymbol{\xi}) \times M^2(\lambda f_c \boldsymbol{\xi})\theta^2(\lambda f_c \boldsymbol{\xi})\tau^2(\boldsymbol{\xi})|_{\boldsymbol{\xi}=\{[(R\omega/2\pi V_x f_0)-\xi_y(V_y/V_x)],\xi_y\}}] \tag{2.9.9}$$

where

$$\phi(\lambda f_c \boldsymbol{\xi}) = \overline{R_{r_a}(\lambda f_c \boldsymbol{\xi}) R_{r_a}(\lambda f_c \boldsymbol{\xi})} \tag{2.9.10}$$

is the power spectral density of the clutter. From Equation (2.9.8) we see that the clutter frequency spectrum is the intensity distribution of the background scene scaled by the motion according to Equation (2.9.6). Higher velocities will spread the spectrum by $2\pi \mathbf{V} \cdot \boldsymbol{\xi}/M$ (M = magnification) whereas in the limit of zero velocity no clutter is present. Thus we see that in the presence of clutter, one would opt for a subcarrier high enough in frequency to avoid the clutter spectrum. This is a contribution to what is commonly referred to as "$1/f$" noise.

When $\boldsymbol{\zeta}(t)$ is random, the net effect will be a broadening of the clutter spectrum in proportion to the velocity spectrum of the motion. This in turn will tend to further broaden the current noise spectrum produced by this noise source. In some imaging systems, attempts are made to track the background scenes, estimate the motion, and cancel the deleterious effects.

2.10. Summary

In this chapter we have described a means for characterizing sources and linear channels. We have developed a methodology for treating the transmission of a class of random fields characterized by a mutual coherence function over this class of linear channels. By motivating a reasonable set of assumptions, we have been able to reduce a general formalism to a working set of linear equations, culminating in Equations (2.7.17) and (2.7.18). The spatial effects of the channel are contained entirely in the point source coherence function. In Appendix B we show how electromagnetic theory can be brought in consonance with radiative transport theory. This will allow us to analyze a broad class of commonly occurring channels in the context of the results developed here. With the exception of Chapter 3, the remainder of the book will be devoted to just this end. In Chapter 3 we will digress to establish a minimum set of criteria which are used in assessing system performance. By doing so, we will aid the reader in performing a first-level system evaluation when transmitting over the channels considered, with a minimum of additional reference material.

References

1. J. W. Goodman, *Introduction to Fourier Optics*, McGraw-Hill, New York (1968).
2. M. Born and E. Wolf, *Principles of Optics*, Second Edition, Pergamon Press, New York (1964).
3. J. Dugundji, "Envelopes and pre-envelopes" of real waveforms, *IEE Trans. Inf. Theory* **IT-4**, 53 (1958).
4. R. M. Gagliardi and S. Karp, *Optical Communications*, Wiley-Interscience, New York (1976).
5. P. A. Bello, Characterization of randomly tune-variant linear channels, Section V, *IEEE Trans. Commun. Systems* **CS-11**, 360 (1963).

3

Optical Receivers

In Chapter 1 we formulated the basic optical communication system model, showing the interface of the optical transmitter, the channel, and the receiver. In this chapter we examine in detail the receiver structure and the way in which the intercepted field at the receiver area is converted to detected signals. Our objective is to define the key parameters characterizing the optical receiver and to establish the relation between these parameters and desired system performance.

3.1. The Direct Detection Receiver

The block diagram of a direct detection optical receiver is shown in Figure 3.1. The received optical field at the aperture $f_r(\mathbf{r}_a, t)$ is focused onto

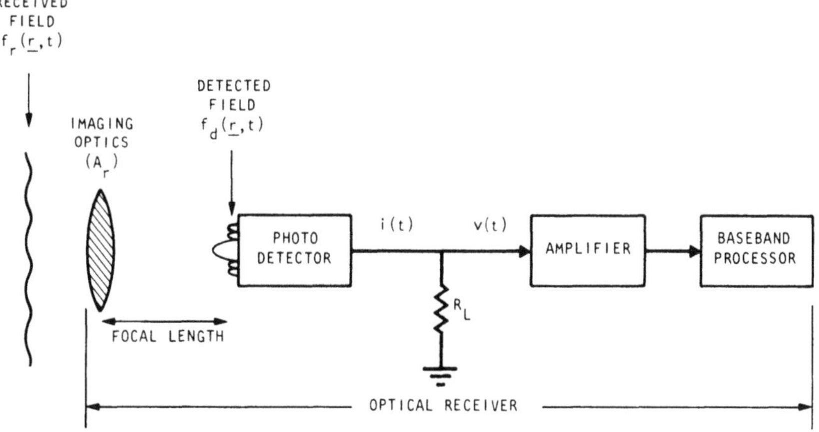

Figure 3.1. The direct detection optical receiver.

the photodetection surface by the receiver lensing system, forming the imaged field $f_d(\mathbf{r}_d, t)$. The relation between f_r and f_d is discussed in Chapter 2 and depends on the imaging properties of the receiver lensing system. The photosensitive surface of the photodetector releases photoelectrons in response to the impinging imaged field $f_d(\mathbf{r}_d, t)$ over its area. The released electrons appear as detector output current pulses that pass through the detector load impedance and combine to develop the detector output field voltage. The detector load impedance circuitry, composed of the load resistance R_L and subsequent amplifier circuitry, generates inherent thermal noise that adds directly to the detector voltage. The combined waveform $v(t)$ represents the input voltage to the receiver baseband circuitry, either an amplifier or baseband processing filter, depending on the system operation. Note that optical field detection, including all optical background noise fields, is accomplished through quantum detection at the photodetector, while all the subsequent circuitry involves electronic processing and is entirely in the hands of the communication engineer. Thus all the inherent properties of the optical channel are immersed within the detected field, and the engineer must design the appropriate electronics for properly combating or compensating for any deleterious channel effects.

An optical detector is simply a photosensitive surface that responds to incident light by releasing photoelectrons. These electrons are collected at the anode to produce a current flow at the detector output. Figure 3.2a shows a typical photodetector circuit biased so that the released electrons from the incident light are caused to flow through the load resistor R_L, producing the detector output voltage $v(t)$. Variations in light intensity cause similar variations in the number of released electrons, thereby producing corresponding voltage variations across R_L. From the circuit, we can see that the photodetector can be modeled as a current source, as shown in Figure 3.2b. The current $i_d(t)$ is the actual current flow produced from the detector. Since this current is built up from random occurrences of photoelectrons, it evolves as a shot noise process.[1-3] The capacitance C_p accounts for any detector shunt capacitance affecting the current flow through R_L.

The important communication parameters characterizing photodetectors are the surface *quantum efficiency* (or *responsivity*), *gain*, *bandwidth*, and *dark current*. The quantum efficiency of a photosensitive surface is formally defined as

$$\eta = \frac{\text{Detected field power}}{\text{Incident field power}} \tag{3.1.1}$$

The efficiency η measures the ability of the photosensitive surface to detect incident field power by converting it into a current. The efficiency η is

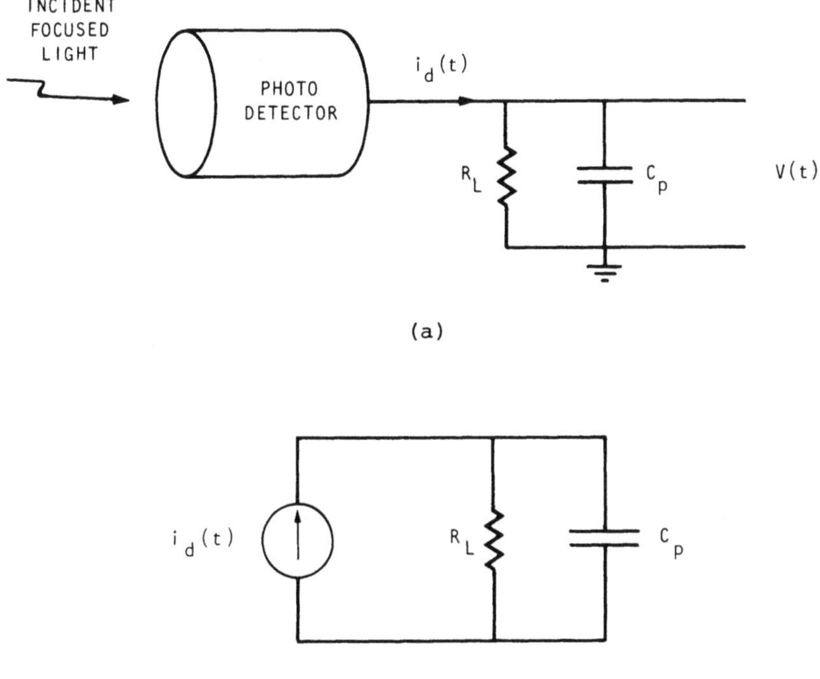

Figure 3.2. Photodetector current model.

determined by the material's work function and is frequency dependent. Figure 3.3 plots typical efficiency curves for two different photosensitive materials as a function of optical wavelength. Most detectors are efficient in the visible range ($0.5 - 1\ \mu$m).

Detected field power can be interpreted as the product of the number of released photoelectrons per second and the amount of energy hf required to release each one, where h is Planck's constant and f is the impinging field frequency. This allows us to define

$$\alpha = \frac{\eta}{hf} = \frac{\text{Number of released photoelectrons/second}}{\text{Incident field power (watts)}} \tag{3.1.2}$$

It is often convenient to deal instead with the detector *responsivity*

$$u = \frac{\text{Detector output current (amps)}}{\text{Incident field power (watts)}} \tag{3.1.3}$$

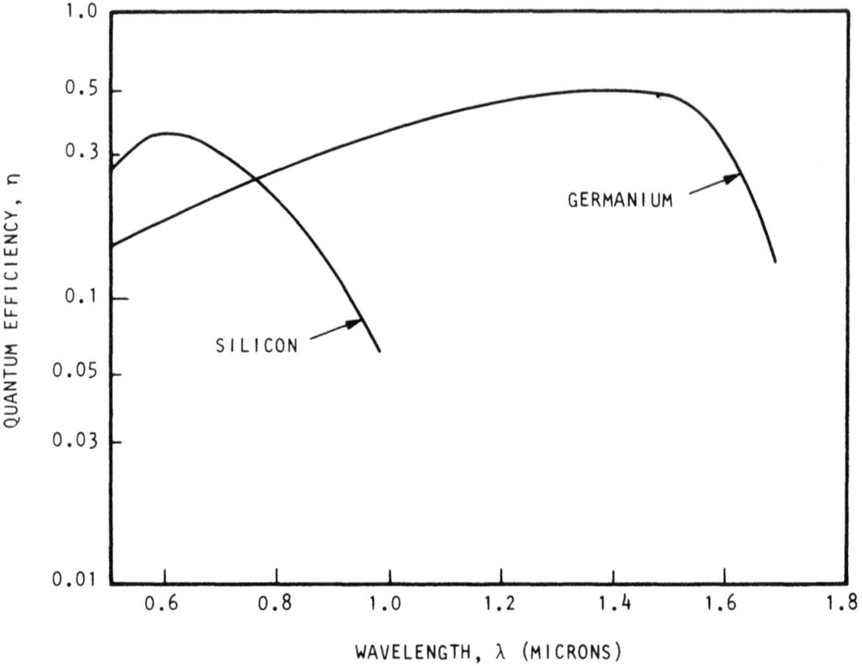

Figure 3.3. Photoemissive quantum efficiency.

Since the product of electron rate and electron charge e is current, we see that

$$u = \alpha e = \frac{\eta e}{hf} \text{ amps/watt} \tag{3.1.4}$$

Thus the responsivity, which is easier to measure, is directly related to the detector's quantum efficiency.

If more than one electron is collected at the output (due to internal regeneration) when an initial (primary) photoelectron is released from the photosensitive surface, the detector sensitivity is increased. Thus if G electrons are collected for each primary photoelectron, the sensitivity becomes $eG\alpha$, and it appears that the output current is amplified by the factor G. The latter is called the *gain* of the photodetector, and we see that it is related to the ability of the detector to provide internal photomultiplication. This internal gain can be obtained from secondary electronic emissions,[1] or by electronic recombinations within the photodetector itself.[4]

The baseband *bandwidth* of a photodetector is the highest frequency of the power variations of the incident field to which the surface can respond.

Therefore, it is the field power variations that can be detected as output current variations. As such, detector bandwidths determine the upper frequency limit to which an intensity-modulated field can be faithfully detected. Optical detectors generally have bandwidths of several gigahertz. *Dark current* is an output current flow that appears even in the absence of incident light and is caused by random emission from the detector surface due only to inherent thermal energy. Dark current appears as a randomly varying current flow about an average value of I_{dc} amps. The latter is often called the detector mean dark current. The random variations about the mean current value are due to the shot noise nature of the emissions and, therefore, have a variance dependence also proportional to I_{dc}. Since dark current is a thermally induced phenomenon, it usually can be reduced by detector cooling. This effect becomes more of a problem in the infrared where the carrier frequencies are nearer to the natural frequencies associated with blackbody emission at 300 °K.

Photodetectors are of four basic types, depending on their construction and mechanisms for electron flow. The vacuum tube detector (phototube) contains a vacuum through which electrons, released from a photoemissive surface in response to impinging light, travel to be collected at the anode. Its size is, therefore, that typical of most vacuum tubes. The photodiode, or PIN diode, is a semiconductor device using the incident light as a PN junction to release electrons. Such devices can be millimeters in size, and have relatively high efficiencies, with the bandwidth limited by recombination times of electrons in transiting the PN gap. Photomultiplier tubes (PMT) are vacuum tubes that produce photomultiplication via secondary emissions from internal dynodes within the vacuum tube and, therefore, have inherent gain. Since several such dynodes can be inserted in standard phototubes, the PMT gain is generally quite large (10^4 to 10^8). The avalanche photodetector (APD) is a semiconductor photodetector device that utilizes gap doping to trigger further free-electron regeneration from an initial primary electron (avalanching) to produce gain. APDs are small in size, and their useful gain values are limited to the range of 50 to 200. Higher gain values cause gain instability and gain sensitivity to bias circuit power variations.

A basic problem with any photomultiplying device is that the number of regenerated electrons from a primary released photoelectron is actually random. This means that the gain G of such a device is in fact a random variable, having a prescribed gain probability density, a mean gain $\bar{G}$, mean square gain $\overline{G^2}$, etc. Probability densities on high-gain PMT have generally been measured to be Gaussian-shaped about the mean gain, with a variance proportional to the mean gain,

$$\text{Var } G = (\zeta\bar{G})^2 \tag{3.1.5}$$

where ζ is called the PMT *spreading* factor. Gain statistics on avalanche devices also have been studied in detail[4-6] and will be considered later. Mean square gains of APD devices are also related to the mean gain; the relationship is usually characterized as

$$\overline{G^2} = (\bar{G})^\sigma \tag{3.1.6a}$$

or in terms of the variance,

$$\text{Var } G = \overline{G^2}[\bar{G}^{(\sigma-2)} - 1] \tag{3.1.6b}$$

where σ is a factor between 2 and 3, depending on the material used. The primary effect of Equations (3.1.5) and (3.1.6) is that the higher the mean gain of the detector, the greater the randomness in the true gain. This additional randomness must be properly accounted for in subsequent detector output analyses.

The optical receiver can be analyzed by the circuit model in Figure 3.2b. The current $i_d(t)$ is the shot noise current process produced at the photodetector output, composed of the combination of the detector response to the focused received fields and the inherent dark current. These current waveforms are filtered by the postdetection circuitry that interconnects the photodetector and the amplifier processor. To carry out the required circuit analysis, it is first necessary to understand the properties of the detector shot noise process itself. Although a complete statistical study of such processes[1-3] is outside the scope of the present discussion, its power spectrum can be directly computed. The spectrum of the photodetected current is given by

$$S_{i_d}(\omega) = [S_n(\omega) + N_{\text{sn}}(\omega)]|H_d(\omega)|^2 \tag{3.1.7}$$

where $H_d(\omega)$ is the normalized baseband transfer function of the photodetector, normalized with $|H_d(0)| = 1$, and $S_n(\omega)$ is the spectrum of the normalized collected field intensity $n(t)$, as it is exhibited by the detector current,

$$n(t) = \bar{G}u \int_{A_d} |f_d(\mathbf{r}, t)|^2 \, d^2\mathbf{r} \tag{3.1.8}$$

Here $|f_d(\mathbf{r}, t)|^2 = I_d(\mathbf{r}, t)$, the field intensity over the detector surface A_d, $\bar{G}$ is the photodetector average gain, and u the responsivity. The field $f_d(\mathbf{r}, t)$ is the field imaged on the detector by the receiver lensing system and is, therefore, random in nature due to the background, and possibly the source signal. The time process produced in Equation (3.1.8) by integrating this random field intensity over the detector area generates the power spectral density that is the first term in Equation (3.1.7). The spectrum $N_{sn}(\omega)$ is a flat shot noise spectrum due to the discreteness of the released electrons and is a basic characteristic of all photodetecting surfaces. The shot noise spectral level is proportional to the long-term time-average field power collected over the detector area. Hence, we can write

$$N_{sn}(\omega) = \overline{G^2}eu \int_{A_d} \overline{|f_d(\mathbf{r}, t)|^2} \, d^2\mathbf{r} \tag{3.1.9}$$

where the overbar denotes time averaging, and $\overline{G^2}$ is the mean square detector gain described earlier. Note that the integral in Equation (3.1.9) corresponds to the spatial integral of the mean field intensity over the detector collecting area. This corresponds to the mean detector field power over the area A_d. By Parceval's theorem, this is equal to the received field power, integrated over the receiver aperture. Hence we can denote the received power P_r as

$$\begin{aligned} P_r &= \int_{A_d} \overline{|f_d(\mathbf{r}, t)|^2} \, d^2\mathbf{r} \\ &= \int_{A_r} \overline{|f_r(\mathbf{r}, t)|^2} \, d^2\mathbf{r} \end{aligned} \tag{3.1.10}$$

where f_r and A_r refer to the receiver apertures. This means in Equation (3.1.9) we can write simply

$$N_{sn}(\omega) = \overline{G^2}euP_r \tag{3.1.11}$$

as the shot noise spectral level.

To compute the spectral contribution of the first term in Equation (3.1.7), it is necessary to first compute the correlation function of $n(t)$, and then transform to obtain $S_n(\omega)$. Denoting this correlation as $R_n(\tau) = E[n(t)n(t+\tau)]$, with the $E[\cdot]$ operator indicating statistical averaging, we obtain

$$R_n(\tau) = (\bar{G}u)^2 \int_{A_d} \int_{A_d} \Gamma_I(\mathbf{r}_1, \mathbf{r}_2, \tau) \, d^2\mathbf{r}_1 \, d^2\mathbf{r}_2 \tag{3.1.12}$$

where

$$\Gamma_I(\mathbf{r}_1, \mathbf{r}_2, \tau) = E[I_d(\mathbf{r}_1, t) I_d(\mathbf{r}_2, t+\tau)] \tag{3.1.13}$$

Thus, the correlation function of the detected current process is dependent on the integrated mutual coherence function of the detected field *intensity.* This requires knowledge of the statistical properties of the field intensity, rather than the field itself. When the impinging field is a Gaussian field, however, the intensity coherence can be related to the field coherence discussed in Chapter 2. In this Gaussian case, the complex field can be written in terms of its real and imaginary field components, which are statistically independent due to the Gaussian assumption. The intensity in Equation (3.1.13) can then be expanded into sums of products of these components which, when averaged, produce the result

$$\Gamma_I(\mathbf{r}_1, \mathbf{r}_2, \tau) = \Gamma_f^2(\mathbf{r}_1, \mathbf{r}_2, 0) + 2\Gamma_f^2(\mathbf{r}_1, \mathbf{r}_2, \tau) \tag{3.1.14}$$

where $\Gamma_f(\mathbf{r}_1, \mathbf{r}_2, \tau)$ is the field coherence function defined in Chapter 2. Thus, the narrowband optical Gaussian field assumption allows us to write $R_n(\tau)$ in Equation (3.1.12) directly in terms of the integrals of the *field* coherence function. This is why it is important to characterize the behavior of this field coherence as it propagates to the receiver. This also means that many of the properties of field coherence carry over to the intensity coherence.

In the typical communication situation, the detector field is composed of the sum of a signal source field $f_s(\mathbf{r}, t)$ and an independent, zero mean background noise field $f_b(\mathbf{r}, t)$. Thus, we can define the combined detected field as

$$f_d(\mathbf{r}, t) = f_s(\mathbf{r}, t) + f_b(\mathbf{r}, t) \tag{3.1.15}$$

over the detector surface. The detected field current intensity in Equation (3.1.8) then expands as

$$\begin{aligned} n(t) = \bar{G}u\bigg\{ &\int_{A_d} |f_s(\mathbf{r}, t)|^2\, d^2\mathbf{r} + \int_{A_d} |f_b(\mathbf{r}, t)|^2\, d^2\mathbf{r} \\ &+ 2\,\mathrm{Re}\bigg[\int_{A_d} f_s(\mathbf{r}, t) f_b^*(\mathbf{r}, t)\, d^2\mathbf{r}\bigg]\bigg\} \end{aligned} \tag{3.1.16}$$

The combined field in Equation (3.1.15) has a field coherence function given by the sum of the field coherence functions due to f_s and f_b. Hence

$$\Gamma_f(\mathbf{r}_1, \mathbf{r}_2, \tau) = \Gamma_{f_s}(\mathbf{r}_1, \mathbf{r}_2, \tau) + \Gamma_{f_b}(\mathbf{r}_1, \mathbf{r}_2, \tau) \tag{3.1.17}$$

When Equation (3.1.17) is used in Equations (3.1.14) and (3.1.12), we obtain a correlation function $R_n(\tau)$ that transforms to the spectral density

$$\begin{aligned} S_n(\omega) = (\bar{G}u)^2\{&S_{|f_s|^2}(\omega)A_r + S_{|f_b|^2}(\omega)A_r + [S_{f_s}(\omega) \otimes S_{f_b}(\omega)]2A_r^2 \\ &+ 2P_sP_b[2\pi\delta(\omega)\} \end{aligned} \tag{3.1.18}$$

where $S_{|f_s|^2}(\omega)$ is the spectrum of $|f_s(t)|^2$, $S_{|f_b|^2}(\omega)$ is the spectrum of $|f_b(t)|^2$ $S_{f_s}(\omega) \otimes S_{f_b}(\omega)$ is the convolution of spectra of $f_s(t)$ and $f_b(t)$, P_s is the received signal power, and P_b is the receiver background noise power. Thus the first term in Equation (3.1.7) contains spectral components due to the signal intensity, the background intensity, convolutions of each, and a spectral line at the origin (delta function). In the usual operating condition, the background noise arises from diffused light that appears to arrive from all directions and illuminates the detector area uniformly. Its field temporal spectrum is that of blackbody radiation and generally is a flat, wide spectrum spread over an optical bandwidth B_0 Hz, with a spectral level N_{ob}, given in Equation (1.4.1) as

$$N_{ob} = N(f)\Omega_{fov}A_r \quad \text{watts/Hz} \tag{3.1.19}$$

where $N(f)$ is the background radiance, Ω_{fov} the receiver field of view, and A_r the receiver lens area. This produces a background intensity spectrum that contains a continuous portion spread over a bandwidth $2B_0$ with peak value $N_{ob}^2B_0$, and a delta function at $\omega = 0$ with strength

$$P_b = N_{ob}B_0 \quad \text{watts} \tag{3.1.20}$$

Since the optical bandwidth B_0 is generally much wider than the signal field bandwidth, the continuous part appears as a white (flat) background intensity spectrum with level equal to this peak value. Likewise, the convolution in Equation (3.1.18) corresponds to that of a narrow spectrum convolved with a wideband spectrum. This convolution also appears as a flat spectrum, with level $P_s^2N_{ob}$. The resulting signal-plus-background intensity spectrum in Equation (3.1.18) therefore simplifies to

$$S_n(\omega) = (\bar{G}u)^2\{S_{|f_s|^2}(\omega)A_r + (N_{ob}^2B_0 + P_s^2N_{ob}) + (2P_sP_b + P_b)[2\pi\delta(\omega)]\} \tag{3.1.21}$$

The shot noise spectral level in Equation (3.1.7) becomes, using Equation (3.1.11),

$$N_{sn}(\omega) = \overline{G^2}eu(P_s + P_b) \tag{3.1.22}$$

Hence the shot noise spectrum level is directly proportional to the received signal and background average powers.

In addition to the shot noise at the detector output, detector dark current must be included. Since dark current is itself a shot noise process with mean average current of I_{dc} amps, its power spectrum will appear as

$$N_{dc}(\omega) = eI_{dc} + 2\pi I_{dc}^2\delta(\omega) \tag{3.1.23}$$

This corresponds to a flat (shot) level of eI_{dc} amp^2/Hz and a "dc" value of I_{dc} amps. Although the dc contribution theoretically can be removed the shot noise spectrum level, due to the randomness of the dark current, must be added to the detector output shot noise spectrum in Equation (3.1.22).

The previous equations describe the frequency characteristics of the detector output shot noise and dark current processes in Figure 3.2b. To this must be added the spectrum of the thermal noise current induced by the output circuitry. The latter is a white Gaussian noise current process with spectral level

$$N_{oc} = \frac{2kT^\circ}{R_L} \quad \text{amp}^2/\text{H}_z \tag{3.1.24}$$

where k is Boltzmann's constant and T° is the effective temperature of the receiver circuitry in degrees Kelvin. The combination of the intensity spectrum in Equation (3.1.21), the shot noise spectrum in Equation (3.1.22), the dark current spectrum in Equation (3.1.23), and the thermal noise spectrum in Equation (3.1.24) produces the overall photodetection frequency spectrum shown in Figure 3.4. The result shows the frequency spectrum of the source intensity imbedded in the wideband noise spectrum and filtered by the inherent frequency characteristics of the detector itself, $H_d(\omega)$.

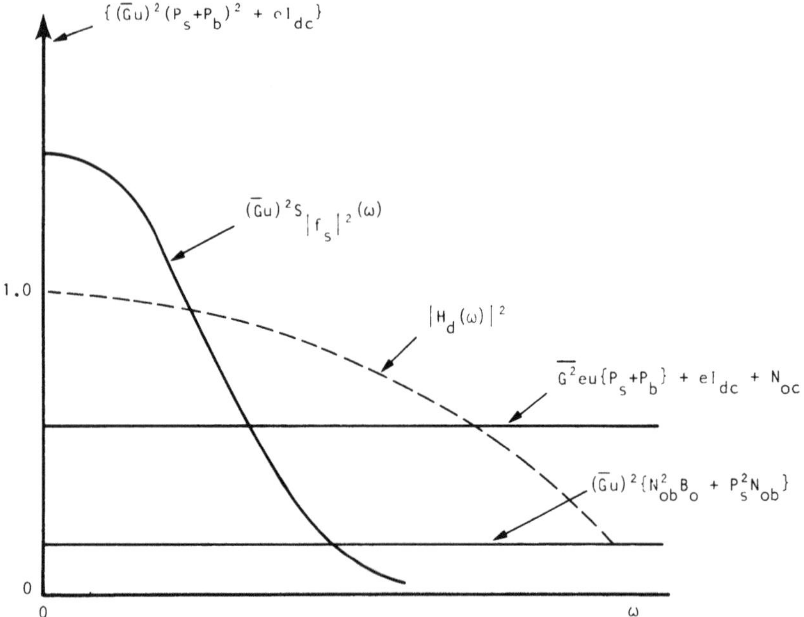

Figure 3.4. Power spectral densities at the output of a direct detection photodetector.

The above detector current process $i_d(t)$ then generates the voltage at the amplifier input in Figure 3.2 as

$$v(t) = i_d(t) \otimes h_c(t) \tag{3.1.25}$$

where $\otimes$ denotes time convolution, and $h_c(t)$ is the impulse response of the circuitry loading the detector output. In the frequency domain, the Fourier transform of $h_c(t)$ is

$$\begin{aligned} H_c(\omega) &= \frac{1}{\left(\frac{1}{R_L}\right) + \left(\frac{1}{R_{\text{in}}}\right) + j\omega C_p} \\ &= \frac{R_L}{\left(1 + \frac{R_L}{R_{\text{in}}}\right) + j\omega c_p R_L} \end{aligned} \tag{3.1.26}$$

where R_{in} and C_p are the input resistance and shunt capacitance, respectively, in Figure 3.1. We see that the total current output (field current plus dark current plus thermal circuit noise) is filtered by $H_c(\omega)$ in converting the impinging field to the amplifier electronic input voltage. Note that the signal portion of the voltage (i.e., the contribution to $v(t)$ due to the transmitted optical source field alone) is obtained from Equations (3.1.7), (3.1.8), and (3.1.25) as

$$v_s(t) = \bar{G}u[P_r(t) \otimes h_d(t) \otimes h_c(t)] \tag{3.1.27}$$

where again

$$P_r(t) = \int_{A_d} \overline{|f_d(\mathbf{r}, t)|^2}\, d^2\mathbf{r} \tag{3.1.28}$$

Thus, the detected signal intensity is filtered by the photodetector response function and the receiver circuitry at the amplifier input. Any further analysis of $v_s(t)$ requires specific identification of the focused signal field in both its time and spatial variation at the detector. Because of the receiver focusing effects, the spatial distribution of $f_d(\mathbf{r}, t)$ tends to image the source distribution, as discussed in Chapter 2. Any channel effects on the transmitted signal field must be incorporated into the description of Equation (3.1.28). It is precisely this point that connects the receiver and the characteristics of the optical channel. In the subsequent chapters dealing with specific optical channels, the properties of $P_r(t)$ will be further developed in detail.

We emphasize that the equations developed here aid primarily in understanding the frequency characteristics of the direct detection optical receiver. In situations where a frequency description is important, such as receiver filter designs, bandwidth selections, and signal-to-noise ratio studies, this model is adequate. In later work where we become intimately involved with the actual statistics of the recovered voltage for digital communications, a more detailed analysis of the detector output is needed. This we shall examine in Section 3.5.

3.2. The Heterodyne (Photomixing) Receiver

In an optical heterodyne, or photomixing, receiver, an optical field from a local light source is added to the received field prior to photodetection. This is generally accomplished by a receiver using the mirror arrangement shown in Figure 3.5. The local field is reflected by the half-mirror combined with the received field and the sum superimposed onto the detector. If we denote the local field produced at the detector by $f_L(\mathbf{r}, t)$, then the combined field on the detector is

$$f_d(\mathbf{r}, t) + f_L(\mathbf{r}, t) \tag{3.2.1}$$

where $f_d(\mathbf{r}, t)$ is the detector received field. Following Equations (3.1.1)-(3.1.4), we generate the detector current as

$$\begin{aligned} i_d(t) = \bar{G}u\bigg\{ \int_{A_d} |f_d(\mathbf{r}, t)|^2\, d^2\mathbf{r} + \int_{A_d} |f_L(\mathbf{r}, t)|^2\, d^2\mathbf{r} \\ + 2\int_{A_d} \mathrm{Re}[f_d(\mathbf{r}, t) f_L^*(\mathbf{r}, t)]\, d^2\mathbf{r}\bigg\} \end{aligned} \tag{3.2.2}$$

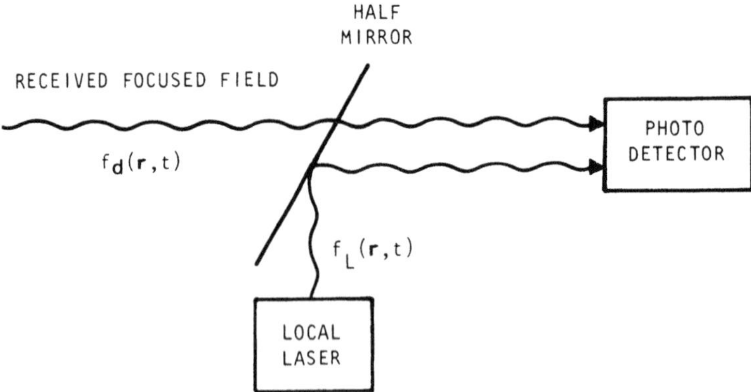

Figure 3.5. The heterodyne optical receiver.

The first two terms are the individual intensities of the received and local fields. The third term is the effective beat, or mixing, term of the two fields. If we assume the local field is due to a stable laser source, and the received field is a modulated plane wave plus a background noise field, we can write the individual complex fields at the detector as

$$f_L(r, t) = a_L e^{j\omega_L t} f_L(\mathbf{r}) \tag{3.2.3a}$$

$$f_r(r, t) = a_s e^{j(\omega_s t + \phi)} f_s(\mathbf{r}) + b(t) e^{j\omega_s t} f_b(\mathbf{r}) \tag{3.2.3b}$$

where a_s and ϕ represent modulation functions, ω_L and ω_s are optical frequencies of the local and received field, and $b(t)$ is the complex noise envelope of the local and received field. The functions $f(\mathbf{r})$ describe the field distributions of the focused fields on the detector surface and are normalized to have an integrated magnitude squared value equal to A_r. Equation (3.2.2) then becomes

$$\begin{aligned} i_d(t) = \bar{G}u\bigg[a_L^2 \int_{A_d} |f_L(\mathbf{r})|^2 \, d^2\mathbf{r} \\ + a_s^2 \int_{A_d} |f_s(\mathbf{r})|^2 \, d^2\mathbf{r} + b^2(t) \int_{A_d} |f_b(\mathbf{r})|^2 \, d^2\mathbf{r} \bigg] + i_{\text{IF}}^1(t) \end{aligned} \tag{3.2.4}$$

where

$$\begin{aligned} i_{\text{IF}}(t) = \bar{G}u 2 a_s a_L \cos(\omega_{\text{IF}} t + \phi) \bigg[\int_{A_d} f_s(\mathbf{r}) f_L^*(\mathbf{r}) \, d^2\mathbf{r} \bigg] \\ + \bar{G}u 2 b(t) a_L \cos(\omega_{\text{IF}} t) \bigg[\int_{A_d} f_b(\mathbf{r}) f_L^*(\mathbf{r}) \, d^2\mathbf{r} \bigg] \end{aligned} \tag{3.2.5}$$

and $\omega_{\text{IF}} = \omega_s - \omega_L$ is the difference frequency of the two light fields. We immediately see that the beat term $i_{\text{IF}}(t)$ contains the mixing effect of the signal source and local field, and produces a detector output component with the amplitude and phase of the received optical field at the beat frequency ω_{IF}. This term can be bandpass filtered and demodulated to recover the amplitude or phase modulation. Hence, the heterodyne receiver is much like a conventional heterodyne system and can, theoretically, demodulate amplitude or phase modulation on the received optical carrier. (Recall the direct detection receiver only recovers intensity modulation.) We also see that this signal term is proportional to the integrated product of the field distributions of the two fields on the detector surface.

First, let us consider the case where $f_L(\mathbf{r})$ is exactly equal to $f_s(\mathbf{r})$. That is, the local oscillator is spatially matched with the received signal field. Then, the spatial portion of the signal beat frequency term becomes

$$\int_{A_d} f_L^*(\mathbf{r}) f_s(\mathbf{r}) \, d^2\mathbf{r} = \int_{A_d} |f_s(\mathbf{r})|^2 \, d^2\mathbf{r} = A_r \tag{3.2.6}$$

as in Equation (3.1.10). Furthermore, we know by the Schwarz inequality that

$$\operatorname{Re}\left[\int_{A_d} f_L^*(\mathbf{r}) f_s(\mathbf{r})\, d^2\mathbf{r}\right] \leqslant \left[\int_{A_d} |f_L(\mathbf{r})|^2\, d^2\mathbf{r}\right]^{1/2}\left[\int_{A_d} |f_s(\mathbf{r})|^2\, d^2\mathbf{r}\right]^{1/2} = A_r \tag{3.2.7}$$

so that any deviation from perfectly matched patterns will result in an effective area less than A_r. For a more rigorous derivation of this result the reader is referred to Ref. 1.

In order for $f_L(\mathbf{r})$ to be exactly equal to $f_s(\mathbf{r})$, two conditions must pertain. First, it is necessary for the two patterns, $f_s(\mathbf{r})$ and $f_L(\mathbf{r})$, to have precisely the same functional form, thus requiring identical optical focusing systems. Second, it is necessary that the two fields be precisely aligned, since any misalignment will cause a relative offset of the two patterns. Let us suppose, as an example, that each pattern at the detector was the result of focusing with a square aperture of dimension d, but misaligned by the angles ψ_x and ψ_y, as shown in Figure 3.6. It is then straightforward to show (Ref. 1, Chapter 6) that

$$\left|\int_{A_d} f_L^*(\mathbf{r}) f_s(\mathbf{r})\, d^2\mathbf{r}\right|^2 = A_r\left(\frac{\sin\phi_x}{\phi_x}\right)^2\left(\frac{\sin\phi_y}{\phi_y}\right)^2 \tag{3.2.8}$$

where

$$\phi_x = \frac{\pi d}{\lambda}\sin\psi_x$$
$$\phi_y = \frac{\pi d}{\lambda}\sin\psi_y \tag{3.2.9}$$

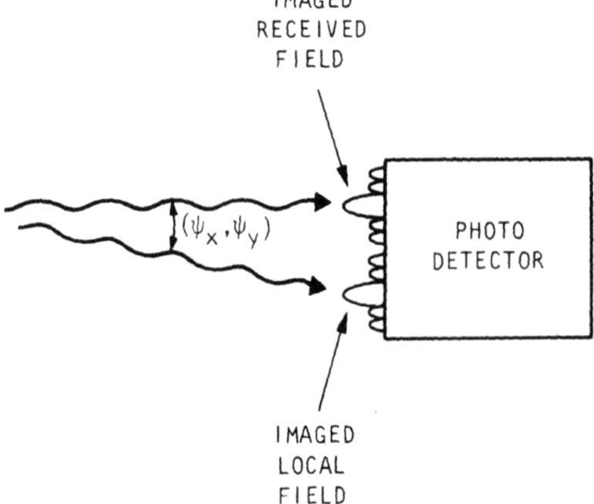

Figure 3.6. Heterodyne alignment errors.

It is evident from Equation (3.2.9) that for minimal reduction in the effective receiving area due to misaligned fields it is necessary that both angles ψ be sufficiently small so that

$$\sin \psi \ll \frac{\lambda}{\pi d} \tag{3.2.10}$$

Since λ is on the order of about 10^{-4} cm and d is generally fractions of a meter, this requires that the maximum offset angles ψ be on the order of 10^{-4} radians. Hence, rather strict requirements occur on the aligning of the two fields in a heterodyne operation. Note that the requirement becomes even more severe as the aperture is increased in size. Purposely defocusing the local field aids the alignment problem, but produces the field mismatch effect described in Equation (3.2.7).

The background noise field [second term in Equation (3.2.5)] is heterodyned along with the signal field and contributes to the beat term. Only the portion of the noise spatial distribution that overlaps the local field distribution contributes to this term. Although a complete derivation of this result is beyond the scope of this book, we can think of $f_s(\mathbf{r})$ as one orthogonal function from a complete set of orthogonal functions whose domain of integration is the detector plane. If we set $f_L(\mathbf{r}) = f_s(\mathbf{r})$ in this orthogonal set, the only contribution to the integral in Equation (3.2.5) would come from the term $f_s(\mathbf{r})$, with all the other terms integrating to zero. Thus we are only affected by the noise which appears spatially similar to the source, or which comes from, or appears to come from, a local region in the vicinity of the source. Hence, the noise field is effectively collected over a smaller detection area, and the noise power is reduced from that which would be collected by the entire detector aperture.

In typical heterodyne receivers, the local field power will be many times stronger than that of the received field. In this case $a_L \gg a_s$, and Equation (3.2.4) is adequately approximated by

$$i_d(t) \simeq \bar{G}ua_L^2A_r + \bar{G}2ua_La_sA_r\cos(\omega_{\mathrm{IF}}t + \phi) + \bar{G}2ua_LA_r[b(t)\cos(\omega_{\mathrm{IF}}t)]. \tag{3.2.11}$$

assuming perfectly aligned fields. This current has an average value due entirely to the local field. The shot noise spectrum at the detector output is therefore

$$N_{\mathrm{sn}}(\omega) = \overline{G^2}euP_L \tag{3.2.12}$$

where

$$P_L = a_L^2A_r \tag{3.2.13}$$

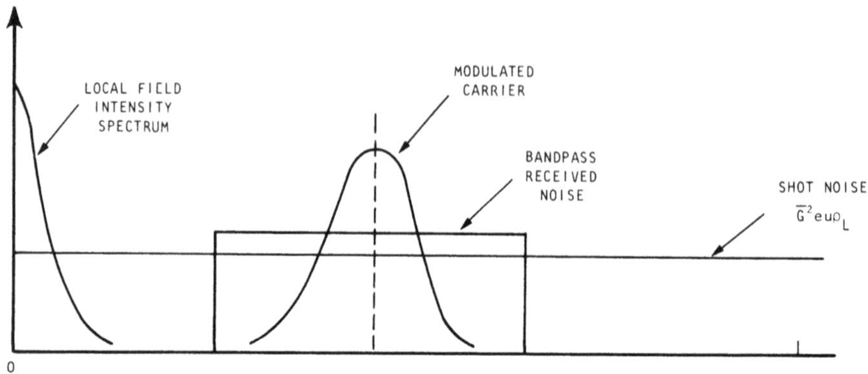

Figure 3.7. Power spectral densities at the output of a heterodyne photodetector.

Note that Equation (3.2.11) treats the local field, in defining its power value, as if it were projected onto the receiving lens. The corresponding IF terms in Equation (3.2.4) then can be written as

$$i_{\mathrm{IF}}(t) = \bar{G}2u(P_L P_s)^{1/2} \cos(\omega_{\mathrm{IF}} t + \phi) + \bar{G}2u(P_L A_r)^{1/2} b(t) \cos(\omega_{\mathrm{IF}} t) \tag{3.2.14}$$

where $P_s = a_s^2 A_r$ is the received optical signal power. Note that the spectrum of $i_{\mathrm{IF}}(t)$ is that of a modulated optical signal field plus noise shifted to ω_{IF} and multiplied by $\bar{G}2u(P_L A_r)^{1/2}$. The overall photodetector output spectrum is sketched in Figure 3.7. The actual shape of the signal spectrum at ω_{IF} will depend on the modulation used on the optical carrier. The background noise is assumed to be a wideband white noise process around the optical carrier frequency and, therefore, shifts to ω_{IF} with the same spectral shape. Any intensity modulation (amplitude variation) on the local field will appear in Figure 3.7 as spectral components at the corresponding frequency. As long as these frequency components are well below ω_{IF}, they will not affect the spectrum of the modulated carrier. The detector output spectrum, therefore, corresponds to a noise IF carrier waveform in which the desired modulation appears on the signal carrier at ω_{IF}. This can now be filtered off and demodulated for information recovery.

3.3. Signal-to-Noise Ratio in Optical Receivers

One of the classical measures of the performance of a communication receiver is the detected signal-to-noise ratio (SNR). This parameter is simply the ratio of the average signal power in the photodetected signal waveform to the power of the total interference added to it. As such, SNR is a measure of the relative strengths of the detected signal and interference, and serves as an indication of the ability of the receiver to collect the desired optical field. The SNR for the direct detection optical receiver can be obtained from the spectral discussion in Section 3.1. The recovered signal current at the processor input in Figure 3.1 is proportional to the received signal field power. Hence

$$P_I = (u\bar{G})^2 P_s^2 \tag{3.3.1}$$

where P_s is the average power of the received signal in Equation (3.1.18). The detector noise occurring in the noise bandwidth of the output circuitry is obtained from the combined spectra of the shot noise, dark current, background field, and circuit noise. From Figure 3.4 the total noise power is

$$P_n = [\overline{G^2} eu(P_s + P_b) + eI_{dc} + N_{0c}]2B_b \tag{3.3.2}$$

where B_b is the one-sided baseband noise bandwidth of the detector output circuitry,

$$B_b = \frac{1}{2\pi}\int_0^\infty |H_d(\omega)H_b(\omega)|^2\,d\omega \tag{3.3.3}$$

The resulting SNR is then

$$\begin{aligned}\text{SNR} &= \frac{P_I}{P_n}\\ &= \frac{(\mu\bar{G}P_s)^2}{[\overline{G^2}eu(P_s + P_b) + N_{0c} + eI_{dc}]2B_b}\\ &= \frac{\alpha P_s}{\left[F\left(1 + \dfrac{P_b}{P_s}\right) + \dfrac{N_{0c}}{e^2\overline{G^2}\alpha P_s} + \dfrac{I_{dc}}{\alpha \overline{G^2}\, eP_s}\right]2B_b}\end{aligned} \tag{3.3.4}$$

Here $\alpha = \eta/hf$, as defined in Equation (3.1.2), and we have denoted

$$F = \frac{\overline{G^2}}{\bar{G}^2} \tag{3.3.5}$$

This is called the detector *excess noise factor*, in that it multiplies the detector noise in contributing to the output total noise. F is simply the ratio of the mean squared to square mean gain of the photodetector. Since $\overline{G^2} = (\text{Var } G) + (\bar{G})^2$, where Var G is the gain variance about the mean gain $\bar{G}$, we see that

$$F = 1 + \frac{\text{Var } G}{\overline{G}^2} \tag{3.3.6}$$

Thus $F > 1$; $F = 1$ only if Var $G = 0$, i.e., the photodetector gain is ideal.

The excess noise factor depends on the type of photomultiplier and generally is related to the spreading factor of the mean gain, as was noted in Equation (3.1.6a). Figure 3.8 shows the resulting behavior of F with mean gain for an APD with various ionization coefficients. Note that the only effect on SNR of the random detector gain is the introduction of the noise factor F to the noise term.

The SNR parameter indicates the effect of the various noise sources on the detecting ability of the receiver. An optical receiver is often given special designation depending on which terms tend to dominate in the

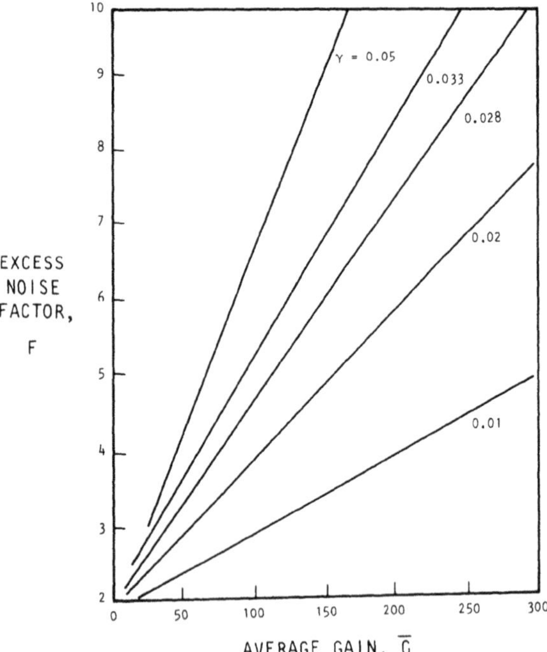

Figure 3.8. APD excess noise factors; γ = ionization coefficient.

denominator of Equation (3.3.4). Table 3.1 lists the various receiver designations and the corresponding noise conditions. Maximum SNR occurs under quantum noise conditions, and the principal interference is due to the signal shot noise alone. Note that in background, thermal noise-limited, or dark current-limited operation, the SNR varies as the square of the received signal power, but only varies linearly when the quantum-limited condition is reached.

We note from (3.3.4) the following factors. The detector gain $\bar{G}$ directly reduces the effect of thermal noise in contributing to the output SNR. Hence, high-gain photodetection is desired with noisy receivers. However, once shot noise-limited operation is achieved, there is no advantage in additional gain, and the designer must be be concerned with the excess noise factor F. Similarly, large load impedances reduce the effects of circuit noise. Note also that SNR is always directly related to $\alpha = \eta/hf$, where η is the detector efficiency, and we therefore always desire highly efficient detectors. The detector efficiency, however, depends on frequency, as does the background noise irradiance, so that SNR is implicitly related to the particular optical wavelength. The quantum-limited SNR can be written out explicitly as

$$\text{SNR} = \frac{\eta P_s}{F(hf)2B_b} \tag{3.3.7}$$

When expressed in this way, it appears that the noise portion is due to a white noise source of two-sided spectral level $F(hf)$. The latter is often called the *quantum noise* spectral level.

Often, we choose to deal with the voltage, or "rms" SNR, which we write simply as $\text{SNR}^{1/2}$. When the detector is not quantum limited, Equation

Table 3.1. Receiver Operating Conditions and Corresponding Designations and SNR

Receiver condition	Designation	SNR
$N_{oc} \gg \alpha(e\bar{G})^2(P_s + P_b), eI_{dc}$	Thermal noise limited	$\dfrac{(u\bar{G}P_s)^2}{(N_{oc})2B_b}$
$\alpha(e\bar{G})^2(P_s + P_b) \gg N_{oc}, eI_{dc}$	Shot noise limited	$\dfrac{(\alpha P_s)^2}{F\alpha(P_s + P_b)2B_b}$
$\alpha(e\bar{G})^2(P_s + P_b) \gg N_{oc}, eI_{dc};\ P_b \gg P_s$	Background limited	$\dfrac{(\alpha P_s)^2}{(F\alpha P_b)2B_b}$
$\alpha(e\bar{G})^2(P_s + P_b) \gg N_{oc}, eI_{dc};\ P_s \gg P_b$	Quantum limited	$\dfrac{\alpha P_s}{2FB_b}$

(3.3.4) can be written as

$$\mathrm{SNR}^{1/2} = \frac{(P_s)^{\frac{1}{2}}}{(\mathrm{NEP})(B_b)^{1/2}} \tag{3.3.8}$$

where

$$\mathrm{NEP} = \left[\frac{2F}{\alpha}\left(1 + \frac{P_b}{P_s}\right) + \frac{2N_{0c}}{u^2\overline{G^2}} + \frac{2I_d}{\overline{G^2}\alpha^2 eP_s}\right]^{1/2} \quad \text{volts/Hz} \tag{3.3.9}$$

The parameter NEP is called the *noise-equivalent power* of the receiver and is often listed as a figure of merit for photodetecting receivers. Note that NEP is also the value of received optical signal power needed to produce a $\mathrm{SNR}^{1/2}$ of 1 in a 1 Hz bandwidth at the detector output.† In essence, NEP is a measure of the "noisiness" of the receiver, but caution must be used in accepting NEP numbers associated with particular detectors because of their dependence on operating conditions (background noise,† receiver temperature, receiver loading, dark current, frequency, etc.). In addition, when the receiver is shot noise limited (large background), the NEP is determined primarily by the amount of collected background field power and, therefore, depends on the receiver area. For this reason, it is convenient to normalize the NEP to a 1-cm area in shot noise-limited operation. The reciprocal of this normalized NEP, NEP*, is called the D-star (D^*) parameter of the detector. That is,

$$D^* = \frac{1}{\mathrm{NEP}^*} \quad \mathrm{cm}/(\mathrm{watts}/\mathrm{Hz})^{1/2} \tag{3.3.10}$$

Like SNR, the parameter D^* depends on frequency through its dependence on α. The primary significance of knowing D^* detector values is that they can easily be converted to NEP values by unnormalizing. That is,

$$\mathrm{NEP} = \frac{A_r^{1/2}}{D^*} \tag{3.3.11}$$

The value of NEP leads directly to the operating SNR of the detector output. From Equation (3.3.8) we can write

$$\mathrm{SNR} = \frac{P_s}{(\mathrm{NEP})^2 B_b} \tag{3.3.12}$$

We see that $(\mathrm{NEP})^2$ plays the role of an effective one-sided, baseband white noise spectral level in evaluating detector output SNR.

†In measuring NEP, P_b is usually defined relative to a 300 K blackbody source.

The SNR of a heterodyned system can be similarly computed. Recall that an optical receiver using optical mixing produces a modulated carrier at the beat frequency at the detector output. The carrier can be extracted by using a bandpass filter centered at the beat frequency ω_{IF}. The output SNR can be determined at the output of this filter using the spectral characteristics sketched in Figures 3.7. The power in the desired signal current is obtained from Equation (3.2.14), with $\bar{G} = 1$, as

$$P_c = 4e^2\alpha^2 P_L A_r\left(\frac{a_s^2}{2}\right) \tag{3.3.13}$$

The noise power passing through the bandpass filter is

$$P_n = e^2(\alpha P_L + 2\alpha^2 P_L N_{ob} + N_{0c})2B_n \tag{3.3.14}$$

where B_n is the IF filter noise bandwidth. The terms in parentheses represent the two-sided noise levels due to shot noise, background, and circuit noise, respectively. The heterodyned SNR is then

$$\text{SNR} = \frac{2\alpha\alpha^2 A_r P_L a_s^2}{(\alpha P_L + 2\alpha^2 P_L N_{ob} + N_{0c})2B_n} \tag{3.3.15}$$

$$= \frac{2\alpha\alpha P_s}{\left[1 + 2\alpha N_{ob} + \dfrac{N_{0c}}{\alpha P_L e^2}\right]2B_n} \tag{3.3.16}$$

This represents the achievable IF SNR available in a bandpass bandwidth B_n around the IF frequency. We see that the SNR depends directly on the total available signal field power P_s collected over the receiver area A_r. We note from Equation (3.3.16) that the effect of circuit noise can be eliminated, and shot noise-limited operation achieved, by using a strong local source such that $e^2\alpha P_L \gg N_{0c}$. This means the local source selectively increases the signal power and a portion of the background noise and provides the amplification needed to overcome thermal noise. However, we see that the local source cannot eliminate the effect of background, since it will always amplify the noise in the same spatial "mode" as the signal. Even with a strong local source, Equation (3.3.16) becomes

$$\text{SNR} = \frac{2\alpha a_s^2 A_r/2B_n}{1 + 2\alpha N_{ob}} = \frac{2\alpha P_s}{(1 + 2\alpha N_{ob})2B_n} \tag{3.3.17}$$

and we never actually reach the quantum-limited condition of the direct detection system in Equation (3.3.7). In typical operation, however, $\alpha N_{ob} \ll 1$ (see Section 1.4.1) and Equation (3.3.17) is usually accepted as an identical quantum-limited result.

When the received signal field has been randomly distorted by the channel, its spatial distribution must be considered a random distribution in Equation (3.2.5). The resulting signal power in $i_{IF}(t)$ now becomes

$$P_c = 2\alpha a_s^2 A_c \tag{3.3.18}$$

where A_c is the mean squared value of the spatial integral,

$$\begin{aligned} A_c &= E\left[\int_{A_d} f_s(\mathbf{r}) f_L^*(\mathbf{r})\, d^2\mathbf{r}\right]^2 \\ &= \int_{A_d}\int_{A_d} \Gamma_{f_s}(\mathbf{r}_1, \mathbf{r}_2) f_L(\mathbf{r}_1) f_L^*(\mathbf{r}_2)\, d^2\mathbf{r}_1, d^2\mathbf{r}_2 \end{aligned} \tag{3.3.19}$$

Hence the heterodyned signal power depends on the spatial coherence function Γ_{f_s} of the distorted signal field. The parameter A_c plays the role of a "coherence area" of the received field, and in essence determines the portion of the detector area available for collecting signal power. It is this concept of a coherence area in Equation (3.3.19) that links together the properties of the channel, through its effect on the mutual coherence function Γ_{f_s} and the detection capability of the receiver, as exhibited by Equation (3.3.18). The resulting heterodyned SNR in Equation (3.3.17) now takes the form

$$\begin{aligned} \text{SNR} &= \frac{2\alpha P_s}{(1 + 2\alpha N_{ob}) 2B_n} \\ &= \frac{(2\alpha a_r^2) A_c}{(1 + 2\alpha N_{ob}) 2B_n} \end{aligned} \tag{3.3.20}$$

with A_c replacing the full aperture area A_r. Since $A_c \leqslant A_r$, the random channel always reduces the heterodyned SNR, with the equality occurring only if $\Gamma_{f_s}(\mathbf{r}_1, \mathbf{r}_2)$ is a delta function in the detector plane; i.e., the field is spatially coherent at the receiver. By reconstructing the local oscillator it is theoretically possible to increase the coherence area and effectively utilize the full aperture area A_r.[1] This procedure is often called heterodyne diversity combining, and requires phase alignment of the heterodyned field in each coherence area. This will be elaborated on in Section 5.4.

3.4. The Free-Space Optical Channel

The basic optical communication channel is the free-space channel, where the modulated optical field generated at the transmitter propagates to the receiver with no additional transmission losses other than the dispersive (beam spreading) loss. For this channel, performance can be determined directly from the power flow equations of Chapter 1, and from the background and receiver noise analysis of this chapter. The receiver SNR values of the previous section can then be computed and serve as the primary indication of the performance capability of the overall link.

The free-space system model is shown in Figure 3.9. The transmitter uses a laser with average power P_T watts and an aperture of diameter d_t, and transmits a modulated light field at wavelength λ. The receiver, with receiving aperture d_r located a distance z away, collects the source power

$$P_s = \frac{P_T d_t^2}{\pi\lambda^2\left(\frac{z^2}{4}\right)}\left(\frac{\pi d_r^2}{4}\right)$$

$$= \frac{P_T(d_t d_r)^2}{\lambda^2 z^2} \tag{3.4.1}$$

which is Equation (1.2.10b) derived earlier. The background is designated by its radiance function $N(f)$ and produces background noise power at the receiver according to Equations (3.1.19) and (3.2.20) as

$$P_b = N(f)\Omega_{fv} B_0 A_r \tag{3.4.2}$$

The receiver photodetector has a known efficiency η, responsivity u, dark current I_{dc}, noise temperature $T°$, gain $\bar{G}$, and noise factor F. The direct

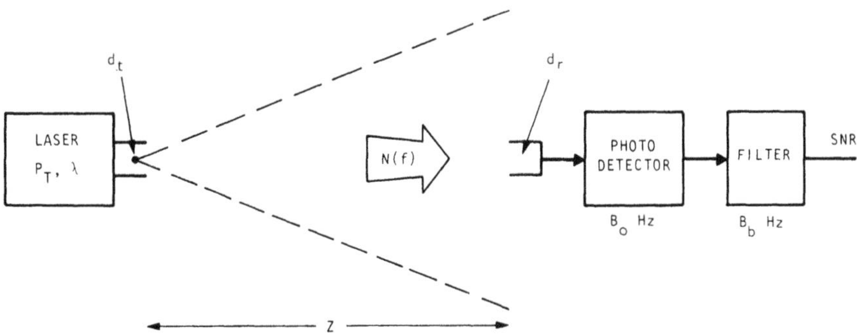

Figure 3.9. The free-space optical channel.

Table 3.2. Example Calculation of SNR of Direct Detection and Heterodyne Receivers

Parameters	Equation	Calculation
Link		
$P_T = 100\ \text{mW}$ $z = 10^7\ \text{m}$ $d_t = 10\ \text{cm}$ $d_r = 10\ \text{cm}$ $\lambda = 1\ \mu\text{m}$	$P_s = \frac{P_T d_t^2 d_r^2}{z^2 \lambda^2}$	$\frac{(10^{-1})(10^{-1})^4}{(10^{14})(10^{-12})} = 10^{-7}\ \text{W}$
Background		
$N(f)$ at $\lambda = 10^{-6} = 2\ \text{W/cm}^2\ \text{s}\ \mu\text{m}$	$N_{ob} = N(f)\Omega_{fov}A_r\ \text{W}/\mu\text{m}^2$ $= N(f)\Omega_{fov}A_r(\lambda^2/c)\ \text{W/Hz}$	$2(10^{-5})^2(100) = 2 \times 10^{-8}\ \text{W}/\mu\text{m}$ $= 0.6 \times 10^{-27}\ \text{W/Hz}$
Receiver		
$A_r = (0.1)^2\ \text{m}^2$ $\Omega_{fov} = 10\ \mu\text{rad}$ $B_0 = 10^{-3}\ \mu\text{m}$	$P_b = N_{ob}B_0$	$(2 \times 10^{-8})(10^{-3}) = 2 \times 10^{-11}\ \text{W}$
$I_{dc} = 10^{-12}\ \text{A}$	$eI_{dc}\ \text{A/Hz}$	$(1.97 \times 10^{-19})(10^{-12}) = 1.97 \times 10^{-31}\ \text{A/Hz}$
$T^\circ = 300\ \text{K}$ $R_L = 100\ \Omega$	$N_{oc} = \frac{4kT^\circ}{R_L}$	$\frac{4(1.4 \times 10^{-23})300}{100} = 1.7 \times 10^{-22}$
$\eta = 0.1$	$\alpha = \frac{\eta}{hf}$	$\frac{0.1}{(6.6 \times 10^{-34})(3 \times 10^{14})} = 5 \times 10^{17}$
	$u = e\alpha$	$(1.97 \times 10^{-19})(5 \times 10^{17}) = 9.8 \times 10^{-2}$
$\bar{G} = 100$	$\bar{G}u$	9.8
$F = 3$	$\bar{G}^2 = F(\bar{G})^2$	3×10^4
$B_n = 10^9\ \text{Hz}$		
Direct detection SNR		
$\frac{(\bar{G}uP_s)^2}{[\overline{G^2}eu[P_s + P_b] + N_{oc} + eI_{dc}]2B_b}$		$(9.8 \times 10^{-7})^2/\{(3 \times 10^4)(1.93 \times 10^{-26})[10^{-7} + 2 \times 10^{-11}] + (1.7 \times 10^{-22}) + (1.97 \times 10^{-31})\}10^9 = 14.04\ (11.4\ \text{dB})$
Heterodyne detection SNR		
$\frac{2\alpha P_s}{[1 + 2\alpha N_{ob}]2B_n}$		$\frac{(5 \times 10^{17})(10^{-7})}{[1+2(5 \times 10^{17})(0.6 \times 10^{-27})]10^9} = 50\ (17\ \text{dB})$

detection and heterodyne SNR can then be determined by evaluating Equations (3.3.4) and (3.3.16) for a specific postdetection circuit bandwidth B_b. Table 3.2 summarizes an example calculation of these SNR.

3.5. Counting Statistics

In optical receivers we often integrate, or filter, the photodetector output prior to receiver processing in the filter-amplifier circuitry, as shown in Figure 3.10. We assume the integration time is T seconds, or, conversely, a filter is used whose time constant is approximately T seconds. The integrator generates a voltage sample

$$v = \int_0^T v(t)\, dt \tag{3.5.1}$$

where $v(t)$ is the photodetector output voltage, and $(0, T)$ is the specified integration interval. If we separate $v(t)$ into the part due to the photodetector output voltage, $v_d(t)$, and the part due to the receiver thermal noise voltage, $v_T(t)$, we can write

$$\begin{aligned} v &= \int_0^T v_d(t)\, dt + \int_0^T v_T(t)\, dt \\ &= eR_L m + n_T \end{aligned} \tag{3.5.2}$$

Since the parameter eR_L corresponds to the amount of voltage generated across R_L for each photoelectron released during $(0, T)$, the parameter m is, therefore, the number of photoelectrons emitted from the photodetector surface during this interval. This number is called the "count" of the photodetector for the $(0, T)$ interval. As such, m in Equation (3.5.2) is a random variable whose statistical properties depend on those of the received optical field. The integrated thermal noise variable n_T in Equation (3.5.2)

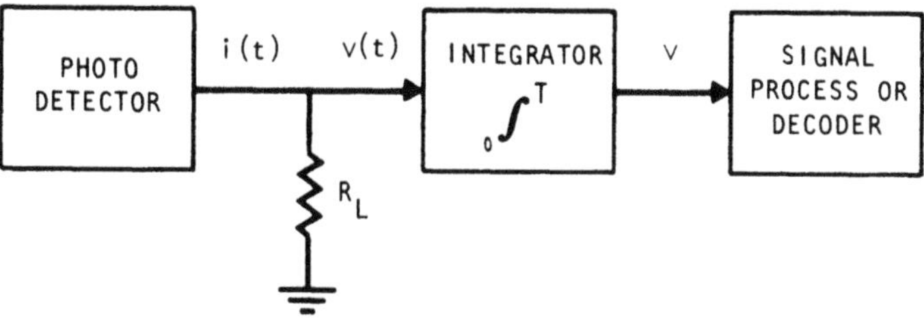

Figure 3.10. Photodetector integration.

appears as the integral of a Gaussian noise process and, therefore, has zero mean and variance

$$\sigma_n^2 = (4kT^\circ R_L)T \tag{3.5.3}$$

Detailed studies of the count variable m have been extensively pursued elsewhere,[5-7] and a repeat of this analysis is beyond our scope here. If the photodetector is assumed ideal with unit gain, the probabilities governing m can be obtained from the field statistics at the photodetector input. Table 3.3 summarizes the common count probabilities used to model the statistics of m for various optical field models. Also included is the mean of the count m in each case. Note that the well-known Poisson statistics for m, which are exact for the deterministic field case, serve as an accurate approximation for the cases involving wideband random fields. The count probabilities in Table 3.3 depend specifically on the mean (average) count n produced by the received field over the integration interval. This is given by

$$n = \alpha \int_{A_r} \int_0^T \overline{|f_r(\mathbf{r}, t)|^2}\, d^2\mathbf{r}\, dt \tag{3.5.4}$$

where α is the scale parameter in Equation (3.1.2). Using Equation (3.1.16) this becomes

$$n = K_s + K_b \tag{3.5.5}$$

where

$$K_s = \alpha \int_0^T P_s(t)\, dt \tag{3.5.6a}$$

$$K_b = \alpha P_b T \tag{3.5.6b}$$

and $P_s(t)$ and P_b are the average power variations over the receiver area due to the signal and background field, respectively. For a blackbody background source of temperature T° (Section 1.4.1) we can write explicitly

$$\begin{aligned} K_b &= \alpha\left[\frac{hf}{e^{hf/kT^\circ} - 1}\right] B_0 TD \\ &= \eta\left[\frac{1}{e^{hf/kT^\circ} - 1}\right] B_0 TD \end{aligned} \tag{3.5.7}$$

with

$$D = \frac{\Omega_{\text{fov}}}{\lambda^2 / A_r}$$

Table 3.3. Summary of Count Probabilities for Optical Receiver of Area A_r and Time Interval $(0, T)$

Received optical field	$P(m)$	Mean
Deterministic field $I(\mathbf{r}, t)$	$\dfrac{n^m e^{-n}}{m!}$ (Poisson)	$n = \alpha \int_{A_r} \int_0^T I(\mathbf{r}, t)\, dt\, dr$
Narrowband optical Gaussian field $(B_0 T < 1)$ spectral level N_{ob} W/Hz	$\left(\dfrac{1}{1+n}\right)\left(\dfrac{n}{1+n}\right)^m$ (Bose–Einstein)	$n = \alpha N_{ob}$
Wideband optical Gaussian field $(M = B_0 T \gg 1)$, spectral level N_{ob}	$\dbinom{M+m}{m}\left(\dfrac{1}{1+\alpha N_{ob}}\right)^{M+1}\left(\dfrac{\alpha N_{ob}}{1+\alpha N_{ob}}\right)^m$ (negative binomial) Approximation: $\dfrac{(\alpha N_{ob} M)^m e^{-\alpha N_{ob} M}}{m!}$	$\alpha N_{ob} M$
Deterministic field $I(\mathbf{r}, t)$ + narrowband Gaussian field	$\left(\dfrac{1}{1+n_2}\right)\left(\dfrac{n_2}{1+n_2}\right)^m e^{-n_1/1+n_2} L_m\left(\dfrac{-n_1/n_2}{1+n_2}\right)$ (Laguerre)	$n_1 + n_2$; $n_1 = \alpha \int_{A_r} \int_1^T I(\mathbf{r}, t)\, dt\, dr$; $n_2 = \alpha N_{ob}$
Deterministic field $I(\mathbf{r}, t)$ + wideband Gaussian field $(M = B_0 T)$	$\dfrac{(n_2)^m}{(1+n_2)^{M+m+1}} e^{-n_1/1+n_2} L_m^{m-1}\left(\dfrac{-n_1/n_2}{1+n_2}\right)$ (Laguerre) Approximation: $\dfrac{(mn_2+n_1)^m e^{-(mn_2+n_1)}}{m!}$	$Mn_2 + n_1$

Here D is the number of spatial degrees of freedom in the system. Recall that if one considers $B_0 T$ as the number of temporal modes, then $(e^{hf/kT} - 1)^{-1}$ plays the role of the number of background counts per space-time mode produced at the receiver. Thus Equation (3.5.7) implies that background noise is being collected by summing mode noise counts over all received modes.

If photomultiplication is involved, the count statistics are altered by the additional randomness introduced by the photoelectron multiplication mechanism. For the case of an avalanche photodetector (APD), the probabilities governing the distribution of the secondary (output) photoelectrons occurring after avalanche multiplication were derived by McIntyre,[(4)] and experimentally verified by Conradi,[(8)] to have the form

$$P(m|n) = \sum_{q=1}^{\infty} \frac{q\Gamma\left(\dfrac{m}{1-\gamma}+1\right)}{m(m-q)!\Gamma\left(\dfrac{\gamma m}{1-\gamma}+1+q\right)} \left[\frac{1+\gamma(\bar{G}-1)}{\bar{G}}\right]^{q+\gamma m/(1-\gamma)}$$

$$\cdot\left[\frac{(1-\gamma)(\bar{G}-1)}{\bar{G}}\right]^{m-q}\left(\frac{n^q}{q!}\right)\exp -n \tag{3.5.8}$$

where $\Gamma(x)$ is the gamma function, and γ and $\bar{G}$ are the avalanche ionization coefficient and mean gain, respectively. Typically, γ has a value of 0.01 to 0.1 for silicon detectors and 0.1 to 0.5 for germanium detectors, while $\bar{G}$ is usually in the range of 10 to 200. Statistical analysis associated with the probability function in (3.5.8) is hampered by its complexity. A useful approximation is given by Webb *et al.*[(9)] as

$$P(m|n) = \frac{1}{(2\pi c_1^2)^{1/2}}\left(\frac{1}{1+\dfrac{(m-n\bar{G})^{3/2}}{c_1 c_2}}\right)\exp\left\{-\frac{(m-n\bar{G})^2}{2c_1^2\left(1+\dfrac{m-n\bar{G}}{c_1 c_2}\right)}\right\} \tag{3.5.9}$$

where

$$F = \gamma\bar{G} + \left(2-\frac{1}{\bar{G}}\right)(1-\gamma)$$

$$c_1^2 = n\bar{G}^2 F$$

$$c_2 = (nF)^{1/2}/F - 1$$

If $n/F \gg 1$, then Equation (3.5.9) approaches the Gaussian-shaped probability

$$P(m|n) \cong \frac{1}{(2\pi n\bar{G}^2F)^{1/2}} \exp\left\{-\frac{(m-n\bar{G})^2}{2n\bar{G}^2F}\right\} \tag{3.5.10}$$

We emphasize that $P(m|n)$ is a discrete probability on the output counts m and has values only at the integers $m = 0, 1, 2, \ldots$. Figure 3.11 compares the probability envelopes in Equations (3.5.8), (3.5.9), and (3.5.10) at two specific values of n, showing their range of comparison. In particular, we note the departure at the lower values of m, and the relatively accurate approximation of Equation (3.5.9) for m values greater than about 50. Since signal counts usually exceed this value, Equation (3.5.9) indeed serves as a valid approximation to Equation (3.5.8). The Gaussian approximation in Equation (3.5.10) tends to depart from the exact over the tails.

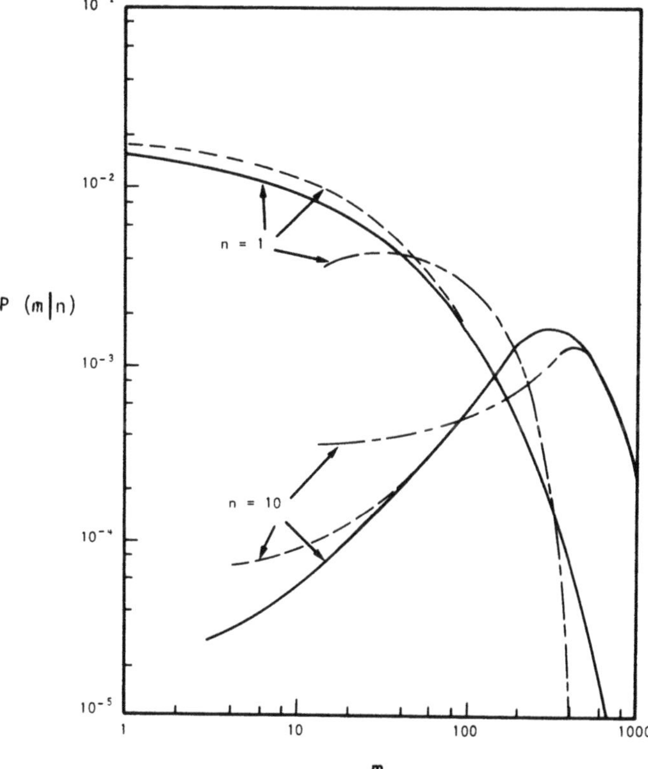

Figure 3.11. Comparison of APD output counting statistics; n = average number of primary electrons, $\gamma = 0.028$, $\bar{G} = 50$. —— Equation (3.5.8); – – –, Equation (3.5.9); — — — Equation (3.5.10).

Since m is discrete, the probability density of the voltage variable v in (3.5.2) evolves as a mixture of a discrete and a continuous Gaussian variable. The probability density of the voltage v can be determined from Equation (3.5.2) by first conditioning on the number of photodetected counts m, then averaging over the avalanche statistics. Since n_T is Gaussian, the conditioned variable v is also Gaussian with mean $eR_L m$ and variance σ_n^2 in Equation (3.5.3). The probability density on the detected receiver voltage v is then

$$p(v|n) = \sum_{q=0}^{\infty} \Phi(v, eR_L q, \sigma_n^2) P(q|n) \tag{3.5.11}$$

where

$$\Phi(v, a, b) \triangleq (2\pi b)^{-1/2} \exp\left\{-\frac{(v-a)^2}{2b}\right\} \tag{3.5.12}$$

Thus, the receiver thermal noise converts the *discrete* counting variable m to the *continuous* observation variable v. Note that the probability density in Equation (3.5.11) depends on the average field count n [Equation (3.5.5)] and, therefore, requires specification of the particular signal field received from the transmitter. When the receiver temperature is low enough to be negligible [i.e., $\sigma_n \ll eR_L m$], the density in Equation (3.5.11) approaches a discrete value, and the receiver is shot noise-limited. In this case, statistical performance based on observations of v can be equivalently stated directly in terms of counting statistics on m. In the following sections, the statistical characteristics of the observed count voltage are used to generate operating performance parameters of various digital modulation systems.

3.6. Binary Digital Performance

Information can be sent over an optical link as digital symbols. This is accomplished by encoding the source information into binary symbols (bits) and transmitting the bits as some type of coded optical field. The encoding can be on a bit-by-bit basis (binary encoding) or on a bit word basis (block encoding). We consider binary encoding in this section and block encoding in the next section.

In binary encoding, each bit is sent individually by transmitting one of two optical fields to represent each bit. In direct detection systems, the standard binary encoding procedure is to pulse an optical source (light-emitting diode, flashlamp, laser, etc.) on or off depending on the data bit. This is a form of intensity corresponding to a zero waveform if a zero bit is to be sent. This encoding is referred to as on–off keying (OOK) in the literature. Decoding is achieved by determining either the presence or absence of a pulse during the time interval it should occur. This is accomplished by integrating over the pulse interval at the photodetector output and determining if the integration voltage sample is above or below some threshold value v_T. Since an integration of a photodetector output is a measure of the received field energy, OOK decoding is based on a decision as to whether the pulse slot time has high enough field energy or not. The threshold v_T is selected to yield the best performance in decoding the correct signal, i.e., yield the lowest probability of making a bit decision error.

The bit error probability depends on the receiver characteristics. An error occurs if a zero bit (no optical pulse) is sent and the detected voltage sample is above the threshold, or if a one bit (optical pulse) is sent but the voltage sample is below the threshold. The probability of these events depends on the noise characteristics of the receiver model. In the following, we examine the specific receiver models.

3.6.1. Shot Noise Limited, Ideal Gain

In this case, the receiver thermal noise is negligible, and the integrated sample voltage is directly proportional to the integrated photodetector count. Thus, in Equation (3.5.2), for a specific pulse slot time

$$v = eR_L m \tag{3.6.1}$$

where m is the random count variable collected from the pulse slot integration over the T-sec slot time. For an ideal gain photodetector, the probability that $m = k$ is taken from Table 3.3 as a Poisson probability

$$\begin{aligned} P(m = k) &= \frac{(K_s + K_b)^k}{k!} e^{-(K_s+K_b)} \\ &\triangleq \text{Pos}\,(k, K_s + K_b) \end{aligned} \tag{3.6.2}$$

where K_s and K_b are the contributions to the average count from the signal field and background noise field in (3.5.6). When a zero is sent, $K_s = 0$ and only background noise contributes to the count probability. A threshold comparison of v in (3.6.1) to a voltage v_T is equivalent to a comparison of the count m to a threshold count $m_T = v_T$. The average probability of making a decoding error, assuming equal likely bits are sent, is then

$$\text{PE} = \tfrac{1}{2} \sum_{k=0}^{m_T-1} \text{Pos}\,(k, k_b) + \tfrac{1}{2} \sum_{k=m_T}^{\infty} \text{Pos}\,(k, k_s + k_b) \tag{3.6.3}$$

The threshold m_T is to be selected so as to minimize PE. This occurs at the value of m_T where $d\text{PE}/dm_T = 0$, or where

$$\text{Pos}\,(m_T, K_b) = \text{Pos}\,(m_T, K_s + K_b) \tag{3.6.4}$$

This occurs at

$$m_T = \frac{K_s}{\log\,(K_s + K_b)} \tag{3.6.5}$$

The evaluation of Equation (3.6.3), using Equation (3.6.5), is shown in Figure 3.12 as a function of K_s values for different values of background counts K_b. These results can be generated by simply reading from known tabulations[10] of Poisson summations.

3.6.2. Shot Noise Limited, APD Gain

When the photodetector is an avalanche detector, the count statistics must be modified to account for the random gain, as discussed in Section 3.5. A comparison test for bit decoding decides a pulse or no pulse is received, based on an observation count m, depending on whether

$$P(m\,|\,K_s + K_b) \gtrless P(m\,|\,K_b) \tag{3.6.6}$$

where $P(m\,|\,n)$ is the APD output count probability in (3.5.8). Equation (3.6.6) is simply a comparison as to whether the observed count m was more probable from a field with average count $K_S + K_b$ or from one with count K_b. Using Equation (3.5.10) to represent $P(m\,|\,n)$, this reduces to a comparison of the integrated m count to the threshold

$$m_T = \bar{G}\left[K_s - F\log\left(\frac{K_b}{K_s + K_b}\right)\right]\left(\frac{K_s + K_b}{K_b}\right)^{1/2} \tag{3.6.7}$$

Decoding is therefore again achieved by a threshold test using the observed count m collected over each expected pulse arrival interval, and comparing to the threshold in Equation (3.6.7). The bit error probability then follows as

$$\text{PE} = \tfrac{1}{2}\sum_{m=0}^{m_T-1} P(m\,|\,K_s + K_b) + \tfrac{1}{2}\sum_{m=m_T}^{\infty} P(m\,|\,K_b) \tag{3.6.8}$$

Equation (3.6.8) was evaluated [11,12] using the exact APD statistics in Equation (3.5.8) and with the threshold in Equation (3.6.7) inserted. Figure 3.13 shows the resulting performance as a function of K_s for several APD

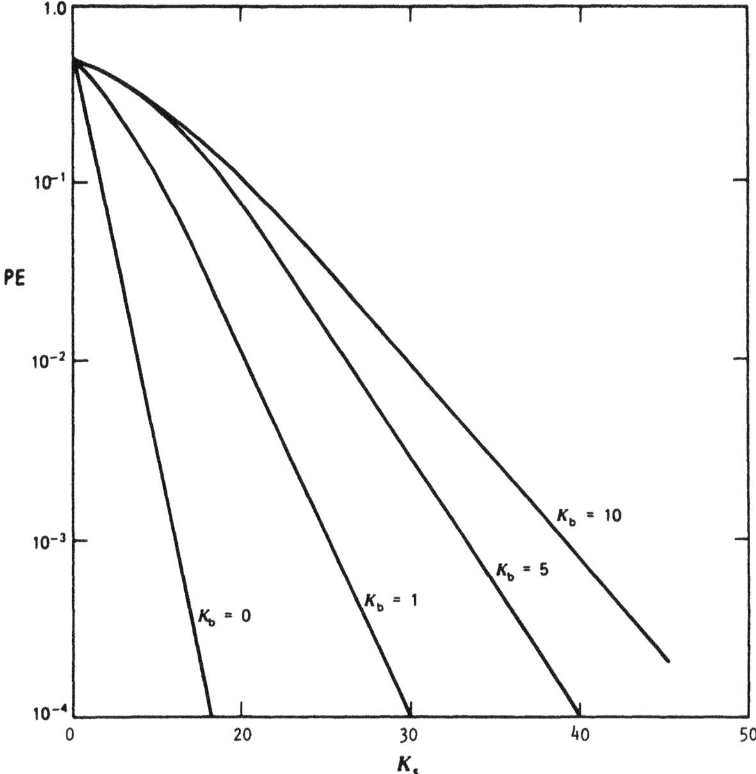

Figure 3.12. Bit error probability, OOK binary system; K_s = signal pulse count, K_b = noise pulse count.

mean gain values and a background count of 1. The pure Poisson case from Figure 3.12 is also included. The results show the degradation due to the random APD gain when receiver thermal noise is not a factor. That is, the performance is degraded by the added randomness of the counts more than it is improved by the increased average signal count. Superimposed is the corresponding PE using the Webb approximation in Equation (3.5.9) for the discrete APD density. Note that the former gives almost exact values of PE for $K_s > 50$ at $\bar{G} = 50$.

3.6.3. Thermal Noise-Limited Receivers, APD Gain

When receiver noise is included, the voltage v in Equation (3.5.2) has the continuous probability density in Equation (3.5.11). A binary detection threshold test for on-off-keyed optical pulses can be again defined by comparing the observed slot voltage v to an appropriate threshold v_T.

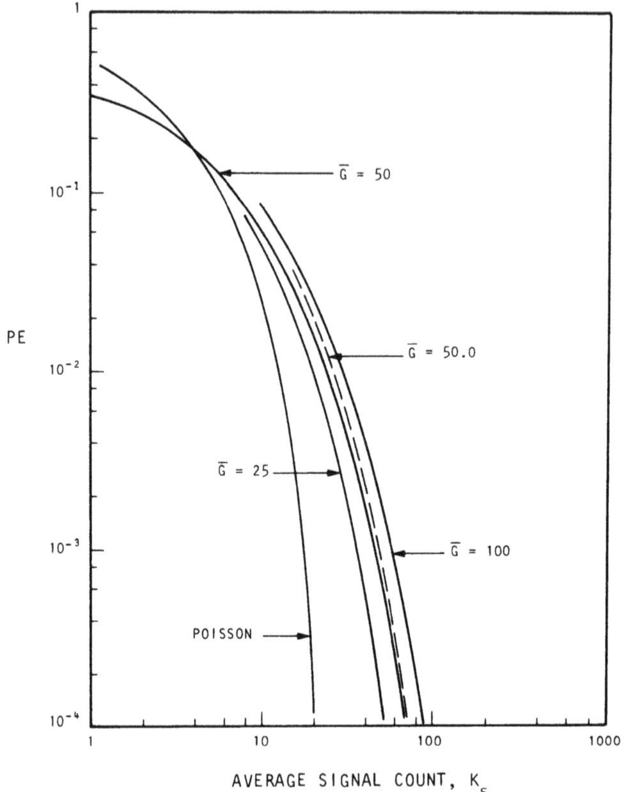

Figure 3.13. Bit error probability, OOK; shot noise limited, $K_b = 1$, $\gamma = 0.28$. Poisson curve with no gain. ———, Equation (3.5.9).

The probability of a bit error at the receiver is obtained from

$$\mathrm{PE} = \tfrac{1}{2}\int_0^{v_T} (p(v\,|\,n_1)\,dv + \tfrac{1}{2}\int_{v_T}^{\infty} P(v\,|\,n_0)\,dv \tag{3.6.9}$$

where the densities $p(v|n)$ are given in Equation (3.5.11), and v_T and n_i are

$$v_T = \frac{\sigma_n^2}{eR_L}\left(\left\{\left(\frac{x_1^2 x_0^2}{x_0^2 - x_1^2}\right)\left[4\ln\left(\frac{x_1\sqrt{n_1}}{x_0\sqrt{n_0}}\right) + \frac{2}{F}(n_1 - n_0)\right]\right\}^{1/2} - \frac{1}{\bar{G}F}\right) \tag{3.6.10a}$$

$$x_i^2 = \frac{1}{2}\left(\frac{\alpha^2}{\sigma_n^2} + \frac{1}{n_i\bar{G}^2}F\right) \tag{3.6.10b}$$

$$n_1 = K_s + K_b \tag{3.6.10c}$$

$$n_0 = K_b \tag{3.6.10d}$$

The threshold v_T is the value of v where $p(v_T|n_1) = p(v_T|n_0)$, and n_1 and n_0 are mean received intensities when a one and a zero bit, respectively, are transmitted.

When Equation (3.5.11) is used in Equation (3.6.9), we obtain

$$\begin{aligned} \mathrm{PE} &= \tfrac{1}{2}\int_{-\infty}^{v_T} \sum_{m=1}^{\infty} \Phi(v, eR_L m, \sigma_n^2) P(m|n_1)\, dv \\ &+ \tfrac{1}{2}\int_{v_T}^{\infty} \sum_{m=1}^{\infty} \Phi(v, cR_L m, \sigma_n^2) P(m|n_0)\, dv \end{aligned} \tag{3.6.11}$$

By interchanging summation and integration, Equation (3.6.11) simplifies to

$$\mathrm{PE} = \tfrac{1}{2} \sum_{m=1}^{\infty} P(m|n_1)\, \mathrm{Erf}\left(\frac{v_T - eR_L m}{\sigma_n}\right) + \tfrac{1}{2} \sum_{m=1}^{\infty} P(m|n_0)\, \mathrm{Erf}\left(\frac{v_T + eR_L m}{\sigma_n}\right) \tag{3.6.12}$$

$$\mathrm{Erf}\,(x) \triangleq \int_{-\infty}^{x} \Phi(v, 0, 1)\, dv \tag{3.6.13}$$

Further, by combining the summations in Equation (3.6.12), we can write

$$\mathrm{PE} = \tfrac{1}{2} + \tfrac{1}{2} \sum_{m=1}^{\infty} [P(m|n_1) - P(m|n_0)]\, \mathrm{Erf}\left(\frac{v_T - eR_L m}{\sigma_n}\right) \tag{3.6.14}$$

This allows direct computation of PE from the APD densities in Equation (3.5.8), using an auxiliary Erf function subroutine. The system performance will depend on the choice of the APD gain relative to the receiver noise temperature. Figure 3.14 shows a plot of the computation of Equation (3.5.14) for receiver temperatures of 300° and 3000° and two different background noise levels K_b. Figure 3.14 also contains the superimposed shot noise-limited characteristic.

Since error probability depends on gain selection, it is possible to optimize APD mean gain $\bar{G}$ with respect to PE performance. This is achieved computationally by varying $\bar{G}$ in steps, each time recomputing the required threshold v_T in Equation (3.6.10a) and the corresponding PE, and noting the minimal PE and gain at which it occurs. The optimal gain is shown in Figure 3.15. Also included is the optimal gain value for maximizing SNR directly from Equation (3.3.4).

Figure 3.16 plots the resulting PE when the optimal gain is used at each noise temperature and signal count K_s. As a comparison, the fixed-gain results of Figure 3.13 are included. Note the rather severe penalty paid in performance when the optimal gain is not used (e.g., $\bar{G}$ fixed at 100 versus

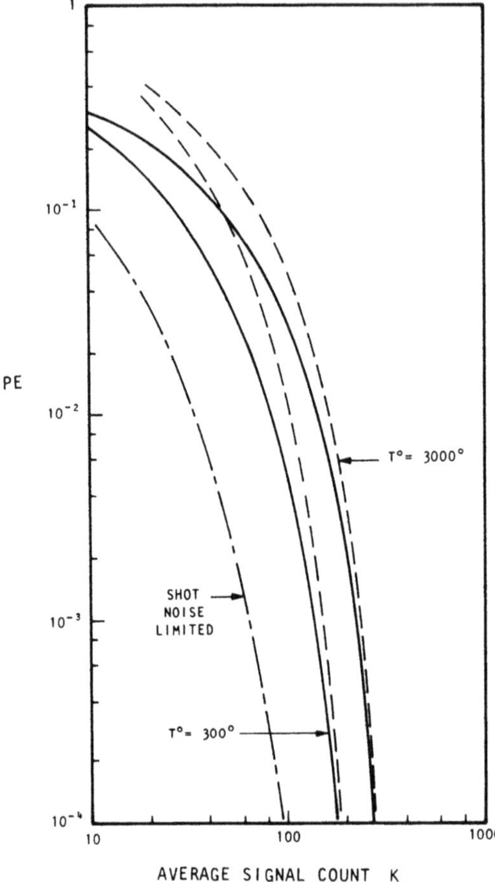

Figure 3.14. Bit error probability, OOK; additive thermal noise, $\gamma = 0.028$, $R_L = 50\ \Omega$, $T = 10^{-9}$ s, $\bar{G} = 100$. —, $K_b = 1$; — — —, $K_b = 10$; - - -, shot noise limited, $K_b = 1$.

$\bar{G}$ selected from Figure 3.15 at each K_s). Since selection of optimal gain requires knowledge of receiver signal and noise count levels, the advantages of some form of automatic power control at the receiver are immediately apparent. In addition, it should be pointed out that if the noise level is increased, the required optimal APD gain likewise increases. This could present a hardware problem since most APD devices become extremely sensitive to circuit variations when operated at relatively high gain values.

The basic disadvantage of on-off keying is that the key receiver parameters, such as signal and noise levels, must be known exactly if the decoding threshold is to be optimally set. A binary format that avoids this difficulty is the use of pulse-to-pulse comparisons for decoding. For example, an optical pulse is sent in one of two adjacent pulse intervals, and a comparison

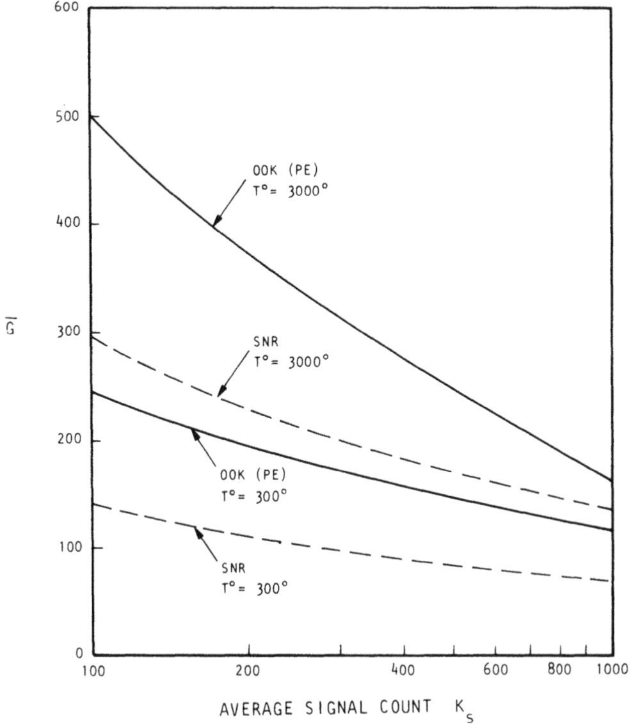

Figure 3.15. Optimal APD gain, from SNR maximization and PE minimization; $K_b = 1$, $\gamma = 0.028$, $R_L = 50\,\Omega$, $T = 10^{-9}$ s.

is made of the integrated voltage v over each interval. The bit is decoded according to which slot produces the higher voltage observable, and no threshold level need be selected. This format is often referred to as binary pulse-position modulation (PPM), or optical Manchester encoding.

The probability of a bit error based upon an observation of v in Equation (3.5.2) is given now by

$$\text{PE} = \int_{-\infty}^{\infty} p(v \mid n_1)\left[\int_{v}^{\infty} p(\alpha \mid n_0)\, d\alpha\right] dv \tag{3.6.15}$$

Substituting from Equation (3.5.11) and using the identity

$$\int_{-\infty}^{\infty} \text{Erf}\,(x)\, e^{-(ax+b)^2}\, dx = \frac{\sqrt{\pi}}{a}\,\text{Erf}\left[\frac{\sqrt{2}b}{(2a^2+1)^{1/2}}\right] \tag{3.6.16}$$

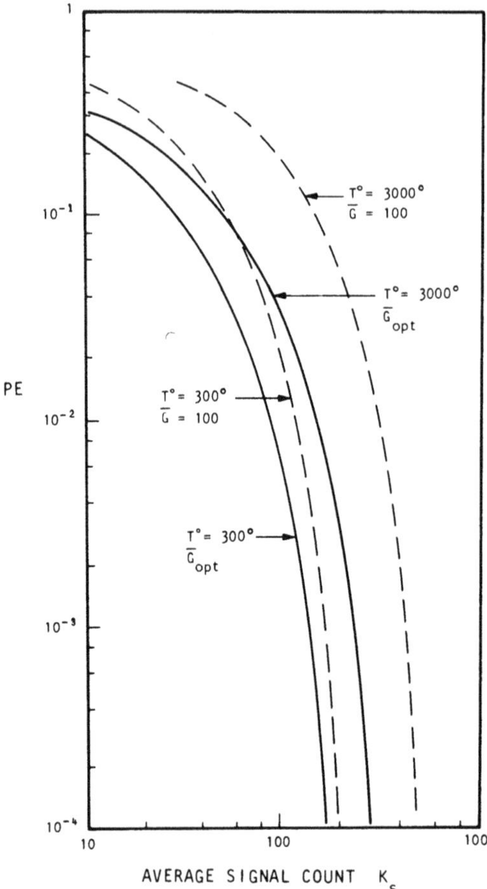

Figure 3.16. Bit error probability, OOK; additive thermal noise, with optimal gain $\bar{G}$, $K_b = 1$, $\gamma = 0.028$, $R_L = 50\,\Omega$, $\mathrm{T} = 10^{-9}$ s.

allows us to rewrite Equation (3.6.15) as

$$\mathrm{PE} = 1 - \sum_{m=1}^{\infty} \sum_{m'=1}^{\infty} P(m|n_1)P(m'|n_0)\,\mathrm{Erf}\left[\left(\frac{eR_L}{\sigma_n}\right)(m - m')\right] \tag{3.6.17}$$

Equation (3.6.17) was evaluated using the Webb approximation in Equation (3.5.9), and the results are shown in Figure 3.17 for two noise temperatures. Again, APD gain $\bar{G}$ was adjusted to give minimal PE in each case.

3.7. Block-Coded Performance

In block encoding, bits are transmitted in blocks rather than one at a time. Optical encoding is achieved by converting each block of b bits into

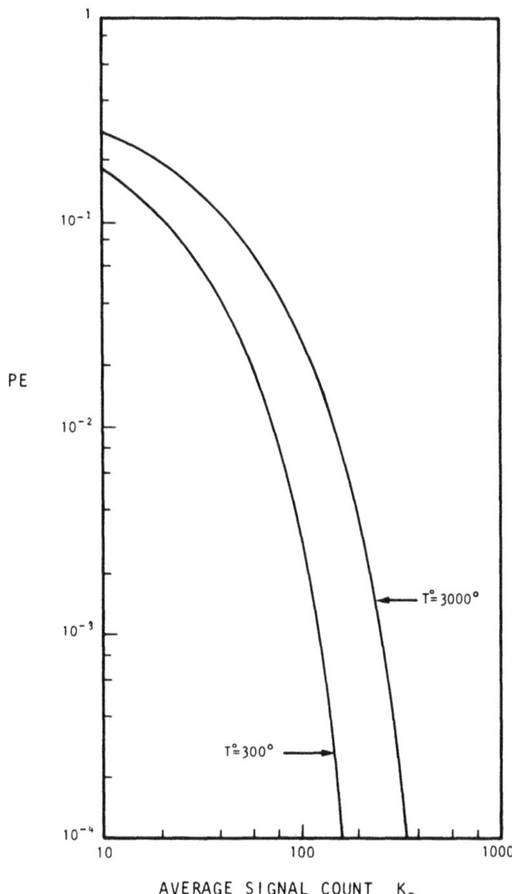

Figure 3.17. Bit error probability, binary PPM; additive thermal noise, $K_b = 1$, $\gamma = 0.028$, $R_L = 50\,\Omega$, $T = 10^{-9}$ s, optimum gain.

one of $M = 2^b$ optical intensity patterns over each block time. Decoding is theoretically achieved by computing the likelihood function of each intensity based on the observable v and selecting the largest. The practical implementation of the optimal decoder, and the associated receiver performance, depends on the type of block-encoded format. In the following, we examine several of the most common block encoding methods.

3.7.1. Multilevel Pulse Amplitude Modulation (MPAM)

In multilevel pulse amplitude intensity modulation, a binary word is sent by transmitting a laser field with one of M distinct intensity levels. Decoding is achieved after photodetection by determining which of the M intensities is being received during each word time. The decoding operation is carried out by integrating the detector output for a word time and comparing the voltage v to a set of thresholds $\{V_0, V_1, V_2, \ldots V_m\}$. If K_{si} denotes the signal mean count [Equation (3.5.6a)] when the ith field intensity is transmitted, then the mean detector count becomes

$$n_i = K_{si} + K_b \tag{3.7.1}$$

The thresholds are determined by solving

$$p(V_i | n_{i-1}) = p(V_i | n_i), \qquad i = 1, 2, \ldots (M-1) \tag{3.7.2}$$

with $V_0 = -\infty$ and $V_m = +\infty$. Intensity level i is decided if the integrated detector voltage v satisfies $V_{i-1} \leqslant v \leqslant V_i$. The solutions to Equation (3.7.2) can be determined from (3.6.10a) by replacing n_1 by n_i and n_0 by n_{i-1}. With the thresholds $\{V_i\}$ determined, receiver block (word) error probability PWE is obtained by computing

$$\text{PWE} = 1 - \sum_{i=1}^{M} \int_{V_{L-1}}^{V_i} p(v | n_i)\, dv \tag{3.7.3}$$

where $p(v | n_i)$ has the density in Equation (3.5.11).

It is possible to determine the optimal intensity level spacings (choice of K_{si}) so as to minimize PWE. Formally we require the solution to the set of equations

$$\frac{\partial \text{PWE}}{\partial K_{si}} = 0, \qquad i = 1, 2, 3, \ldots M \tag{3.7.4}$$

for each K_{si}. As a means for solution, the Gaussian assumption in Equation (3.5.10) was used to evaluate the above densities and to simplify the relation between thresholds and intensity levels. The peak intensity level K_{sM} was initially selected, and Equation (3.7.4) was solved for the remaining K_{si} by computer iteration. From the observed computational results, the optimal pulse level spacings were close to the form

$$K_{si} \cong \left(\frac{i-1}{M-1}\right)^2 K_s \tag{3.7.5}$$

for all gain levels. Hence, it appears that a nonlinear intensity level spacing for multilevel intensity word encoding is nearly optimal in terms of minimizing PWE with APD detection. The resulting PWE in Equation (3.7.3), using Equation (3.7.5), is shown in Figure 3.18 for several values of M and temperature, $T°$, as a function of the maximum intensity level K_{sM}. Points computed for each curve were determined at the gain values that produced minimal PWE.

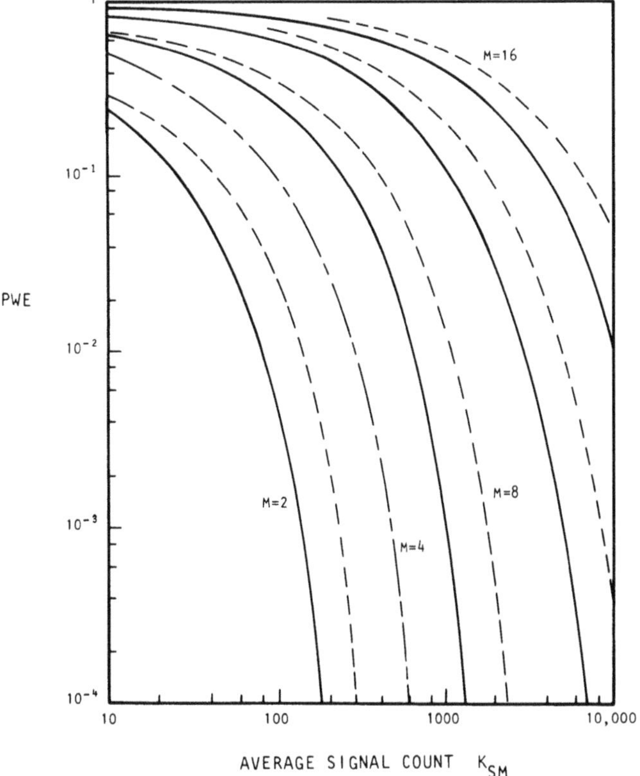

Figure 3.18. M-level pulse intensity modulation; additive thermal noise, K_{sm} = peak signal count, $K_b = 1$, $\gamma = 0.028$, $R_L = 50\ \Omega$, $T = 10^{-9}$ s. —, $T° = 300$ K; — — —, $T° = 3000$ K; — - —, shot noise limited, $M = 4$.

3.7.2. Multislot Pulse-Position Modulation (MPPM)

In multilevel pulse-position encoding, an optical pulse is placed in one of M adjacent pulse slots to represent each word. Decoding is achieved by comparing the voltage v_i generated from each pulse slot and selecting the largest. The correct pulse slot has a receiver count intensity of $n_1 = K_s + K_b$, while the incorrect slots have count intensity n_0. Hence, the probability of decoding incorrectly with APD detectors is

$$\mathrm{PWE} = 1 - \int_{-\infty}^{\infty} P(v \mid n_1)\left[\int_{-\infty}^{v} P(v' \mid n_0)\, dv'\right]^{M-1} dv \qquad (3.7.6)$$

Proceedings as in Equations (3.6.15)-(3.6.17), Equation (3.7.6) becomes

$$\text{PWE} = \sum_{m=1}^{\infty} P(m|n_1) \int_{-\infty}^{\infty} \Phi(v, eR_L m, \sigma^2) \times \left[\sum_{m=1}^{\infty} P(m|n_0) \operatorname{erf}\left(\frac{v - eR_L m}{\sigma}\right) \right]^{M-1} dv \qquad (3.7.7)$$

Using the Webb approximation for $P(m|n)_e$ in Equation (3.5.9), Equation (3.7.7) is plotted in Figure 3.19 for several values of M and two different receiver noise temperatures.

Since a $\log_2 M$ bit word can be encoded into one of M time slots, a PPM system transmits $\log_2 M$ bits with each pulse. The data rate of a PPM

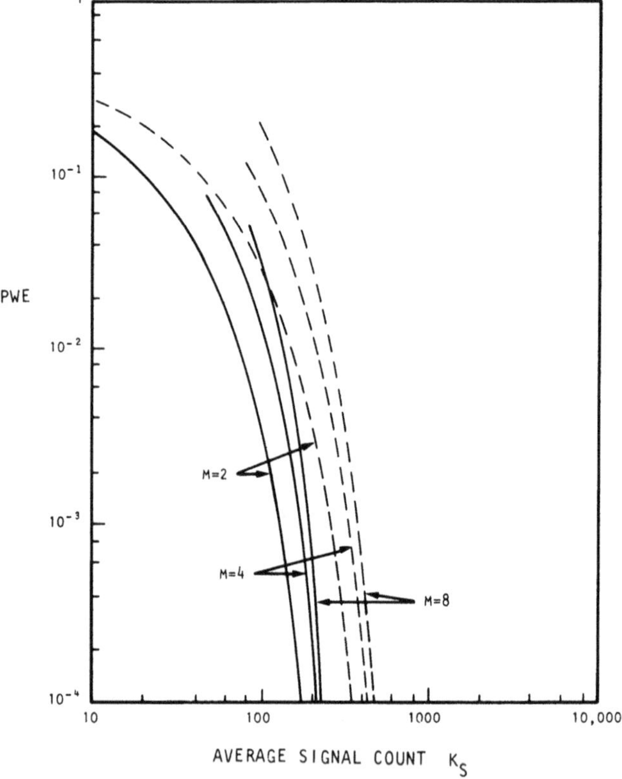

Figure 3.19. M-ary pulse position modulation; additive thermal noise, $K_b = 1$, $\gamma = 0.028$, $R_L = 50\,\Omega$, $T = 10^{-9}$ s, optimum gain. — $T° = 300$ K; ——— $T° = 3000$ K.

system operating with τ-sec pulses and a pulse repetition rate PRF is then

$$R = \frac{\log_2 M}{M\tau}$$

$$= (\log_2 \mathrm{M})\mathrm{PRF} \quad \text{bits/s} \tag{3.7.8}$$

Thus, the data rate can be increased in PPM either by increasing the number of slots per frame or by increasing the PRF of the laser source. The former however requires a reduction in pulse width τ; i.e., transmitting narrower (higher bandwidth) optical pulses. If the pulse count K_s is to remain constant in Equation (3.7.7) with narrower pulses, then the peak pulse power must likewise increase. Hence, data rate increases in PPM must be accompanied by a proportional increase in peak power levels and bandwidth if decoding error probabilities are to remain the same.

3.8. Digital Signaling with Heterodyne Detection

Digital information can be encoded directly on the phase or frequency of the laser carrier if heterodyne detection is used at the receiver. Heterodyning allows the received laser carrier to be translated to a lower RF frequency, where the digital modulation can be then decoded via standard RF decoding methods. Furthermore, if a strong local laser field is used for the heterodyning, the photodetector output current appears with a relatively high power level, so that it can be accurately modeled as a Gaussian process.[1] Hence, both the RF design procedures and the associated decoding performance can be directly obtained from the available literature on RF Gaussian systems.

The most common RF digital modulation is binary phase shift keying (BPSK) and noncoherent frequency shift keying (FSK). In the former, the RF carrier is phase shifted between two phase states 180° apart to represent a data one or zero. In FSK, the RF carrier is frequency shifted between two frequencies to represent each bit. For the laser link, this would correspond to direct phase or frequency shifting of the laser spectral line itself. Such modulation can be achieved by basic phase shifting circuitry, or by piezoelectric control of the laser cavity length. Direct heterodyning at the receiver via the local offset laser mixes to RF, producing the RF carrier with the same modulation. This can then be decoded via RF phase-detecting or frequency-detecting circuits, reproducing the transmitted bit sequence.

The decoding bit error probability performance for a white Gaussian noise channel, with BPSK or FSK modulation, has been shown to depend only on the ratio E_b/N_0, where E_b is the bit energy at the decoder input and N_0 is the corresponding Gaussian noise spectral level. The bit error probabilities are given by

$$\text{PE} = \begin{cases} Q(2E_b/N_0)^{1/2} & \text{BPSK} \quad (3.8.1\text{a}) \\ \frac{1}{2}\, e^{-E_b/2N_0} & \text{FSK} \quad (3.8.1\text{b}) \end{cases}$$

where $Q(x)$ is the Gaussian tail integral

$$Q(x) = \frac{1}{2\pi} \int_x^\infty e^{-y^2/2}\, dy \tag{3.8.2}$$

Since the RF carrier is generated from photomixing, the energy levels of the RF decoder input depend on the photodetecting parameters. From Equation (3.3.13) the photodetected carrier energy in a bit time T_b seconds is given by

$$\begin{aligned} E_b &= P_c T_b \\ &= 4e^2\alpha^2 P_L (a_s^2/2) A_r T_b \end{aligned} \tag{3.8.3}$$

The heterodyned noise spectral level, from Equation (3.3.14) is then

$$N_0 \triangleq e^2[\alpha P_L + 2\alpha^2 P_L N_{\text{ob}} + N_{\text{oc}}] \tag{3.8.4}$$

Thus, heterodyned RF decoding will therefore be based on

$$\frac{E_b}{N_0} = \frac{2\alpha P_s T_b}{[1 + 2\alpha N_{\text{ob}} + (N_{\text{oc}}/\alpha P_L e^2)]} \tag{3.8.5}$$

It is again convenient to denote $K_s = \alpha P_s T_b$ as the detected signal count per bit, and to again assume $P_L \gg N_{\text{oc}}/\alpha e^2$. In this case, Equation (3.8.5) simplifies to

$$\frac{E_b}{N_0} = \frac{2K_s}{1 + 2\alpha N_{\text{ob}}} \tag{3.8.6}$$

Figure 3.20 plots PE in Equation (3.8.1) for both BPSK and FSK heterodyning decoding, as a function of the parameter $K_s/1 + 2\alpha N_{\text{ob}}$.

Although the BPSK format exhibits the better PE performance, it inherently assumes perfect phase referencing in the decoding. Phase referencing in BPSK is typically achieved by squaring or Costas loops[11,12] with specified loop bandwidths B_L. The latter must be wide enough to track all the phase noise on the RF carrier at the decoder input. In the laser system, the phase noise (laser line broadening) of both the transmitter laser and the local laser is transferred directly to the RF carrier. Hence the phase

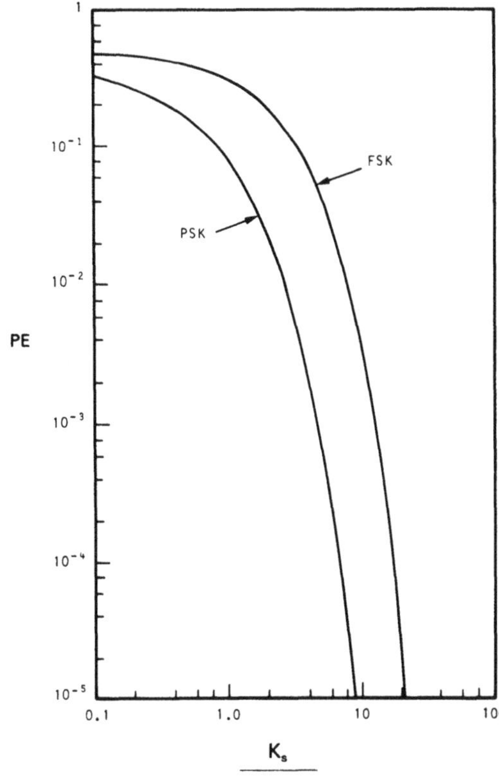

Figure 3.20. Bit error probability, heterodyne detection, strong local laser field, Gaussian statistics. N_{ob} = background noise spectral level.

referencing subsystem must have wide enough tracking bandwidths to track out this total laser phase noise on the RF carrier. The bandwidths B_L, on the other hand, determine the amount of RF thermal noise in the tracking loop, so that accurate phase referencing requires a balance of both loop thermal noise and laser noise in the loop design. FSK systems require no phase referencing, but the signaling frequencies must be known to within $1/T_b$ Hz to achieve the performance in Figure 3.20.

References

1. R. Gagliardi and S. Karp, *Optical Communications*, Wiley-Interscience, New York (1976).
2. W. Pratt, *Laser Communication Systems*, John Wiley and Sons, New York (1969).
3. M. Ross, *Laser Receivers*, John Wiley and Sons, New York (1966).
4. R. McIntyre, The distribution of gains in uniformly multiplying avalanche photodiodes: theory, *IEEE Trans. Electron Devices* **ED-19**, 703–712.
5. P. Diament and M. Teich, Photoelectron counting distributions for irradiance modulated radiation, *J. Opt. Soc. Am.* **60**, 682–689 (1970).

6. G. Bedard, Photon counting statistics of Gaussian light, *Phys. Rev.* **151**, 1038 (1966).
7. S. Karp and J. Clark, Photon counting—a problem in noise theory, *IEEE Trans. Inf. Theory* **IT-21** (1970).
8. J. Conradi, The distribution of gains in uniformly multiplying avalanche photodiodes: experimental, *IEEE Trans. Electron Devices* **6**, 713-718 (1972).
9. D. Webb, R. McIntyre, and J. Conradi, Properties of avalanche photodiodes, *RCA Review* **35**, 234-287 (1974).
10. M. Abramowitz and I. Stegun, *Handbook of Mathematical Functions*, NBS Applied Mathematics Series, Volume 55 (1964).
11. N. Sorensen, Optical system design with avalanche photodetection, Ph.D. Dissertation, Department of Electrical Engineering, University of Southern California (September, 1978).
12. N. Sorensen and R. Gagliardi, Performance of optical receivers with APD, *IEEE Trans. Commun.* **COM-27** (1979).

4

The Fiber-Optic Channel

Perhaps the most important optical communication channel is the optical fiber. The fiber is a thin "pipe" of glass through which one can shine an optical beam to transmit optical energy from one point to another. The fiber is the optical equivalent of a coaxial cable or waveguide commonly used for microwave transmission. Decades ago attempts to communicate by fiber over long distances were hampered by the severe attenuation of this channel. However, in the early 1970s the demonstration of a fiber with 20 dB/km of loss indicated the potential of this link, coupling the high data rates of the optical carriers with the small spatial occupancy of the fiber. Fiber losses have now been reduced to about 0.1 dB/km, and the technological development of solid-state sources and detectors has further advanced the fiber communication channel. In this chapter we attempt to outline the basic communication characteristics of this type of channel.

4.1. The Optical Fiber

The basic construction of an optical fiber is shown in Figure 4.1. The core of the fiber is made of high-quality glass over which the light field can

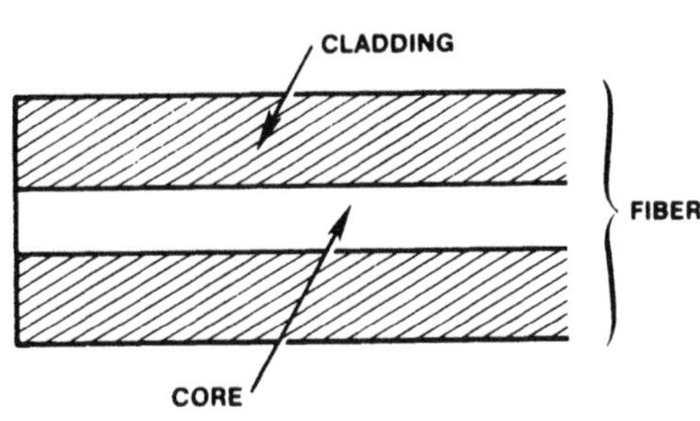

Figure 4.1. Optical fiber construction.

propagate. Glass or a similar silicon material is used since it offers minimal attenuation to the optical transmission. The glass core is supported by a shielding, or cladding, which provides mechanical strength, isolates the core from external interference and radiation, and aids in confining the light propagation to only the internal glass paths. The cladding is opaque, usually constructed of a form of silicon plastic, and attempts to reflect any escaping light back into the core. Typically, the core diameter is on the order of microns, while the fiber cross section (core plus cladding) is on the order of millimeters. Thus the fiber is literally a thread of transmission path over which the modulated light field can be propagated. The ability to pack large amounts of modulated data over an extremely small spatial area is an overwhelming advantage for optical fiber communications.

As a communication channel, the most important characteristics of the fiber are its attenuation (loss) and its field propagation distortion. The key parameter describing these propagation properties for the fiber core and cladding material is their refractive index. The cladding is designed to have its index slightly smaller than that of the core. (Glass typically has an index of about 1.5, and the cladding index is generally selected to be within a few percent of this value.) Those indices determine the propagation angles of the light rays in the core. The slightly lower cladding index is needed to reflect the off-axis core rays back into the core.

Consider the ray diagram in Figure 4.2 showing an optical ray (optical field propagating in the principal ray direction) being inserted into the fiber core at angle θ_p. If θ_p is too large, the ray will be absorbed into the cladding and will not propagate. Propagation will occur down the fiber only if the incident ray angle at the cladding boundary is shallow enough to be reflected

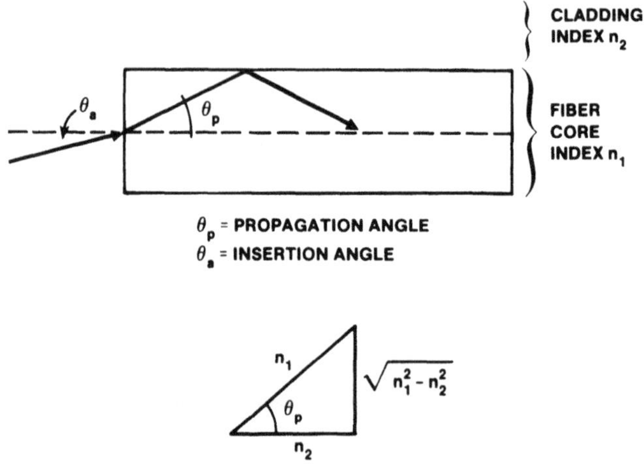

Figure 4.2. Fiber ray diagram.

back into the core. The ray will then continue to propagate down the core as it continually reflects off the core walls. The application of Snell's law at the boundary requires that

$$\theta_p \geqslant \cos^{-1}\left(\frac{n_2}{n_1}\right) \tag{4.1.1}$$

for the reflection to occur, where n_1 is the core index and n_2 the cladding index. The right-hand side of Equation (4.1.1) is called the *critical angle* of the fiber and represents the maximum ray angle at which light will propagate down the fiber. Note the angle depends only on the index ratio of the cladding and core. By adjusting these indices, one can control these light flow angles. In particular, by having $n_2 \lesssim n_1$, the propagation angles can be made quite small, so the light that propagates does so at small angles (i.e., will flow approximately down the center of the core). This means fiber boundary losses will be reduced, and less leakage will occur from the core into the cladding. Also, if more than one ray is launched into the fiber simultaneously, each angle must satisfy Equation (4.1.1), and the various propagating field rays (called fiber modes) all propagate at almost identical angles. Since n_2 will necessarily be close in value to n_1, we often deal with the fractional difference

$$\Delta = \frac{n_1 - n_2}{n_1} \tag{4.1.2}$$

of the two indices. For small propagation angles we generally desire a Δ value of a few percent.

In order to propagate a light ray in the core at angle θ_p, it is necessary that the light field be properly inserted. Let θ_a be the fiber insertion angle; i.e., the angle at which the light is fed into the fiber from an external source, as shown in Figure 4.2. Again, by Snell's law applied to the index change from the fiber external medium (assumed to be free space with index of unity) to the core material, we have

$$n_1 \sin\theta_p = (1)\sin\theta_a = \sin\theta_a \tag{4.1.3}$$

The numerical aperture of the fiber is defined as

$$\begin{aligned} \text{NA} &\triangleq \sin\theta_a \\ &= n_1 \sin\theta_p \end{aligned} \tag{4.1.4}$$

We see that NA is simply an indication of the allowable insertion angle θ_a. For a given core index, we see that a small NA corresponds to a small propagation angle θ_p, and therefore a highly collimated field. Note that Equations (4.1.1) and (4.1.3) suggest the triangle relationship in Figure 4.2, and from simple geometry, we see that

$$\begin{aligned} \mathrm{NA} &= n_1 \sin \theta_p \\ &= \frac{n_1 (n_1^2 - n_2^2)^{1/2}}{n_1} \\ &= (n_1^2 - n_2^2)^{1/2} \\ &= n_1 \left[\left(1 + \frac{n_2}{n_1}\right)\left(1 - \frac{n_2}{n_1}\right) \right]^{1/2} \end{aligned} \tag{4.1.5}$$

Thus the numerical aperture is dependent only on the indices n_1 and n_2. Since $n_1 \approx n_2$, we can substitute from Equation (4.1.2) and use the approximation

$$\begin{aligned} \mathrm{NA} &= n_1 \left[2\left(1 - \frac{n_2}{n_1}\right) \right]^{1/2} \\ &\approx n_1 \sqrt{2\Delta} \end{aligned} \tag{4.1.6}$$

Thus, as an alternative interpretation, when the core index is fixed, NA is actually a measure of the fractional difference of the two indices. Typically, NA takes on values between 0.05 and 0.2.

Our discussion of Figure 4.2 was in terms of a single light ray inserted at angle θ_p. A more typical situation is that shown in Figure 4.3. Here we have a Lambertian light source (light emitted in all directions) shining into the fiber core of diameter *d*. However, only the light rays with insertion angle θ_a will produce propagating light rays within the critical angle. This

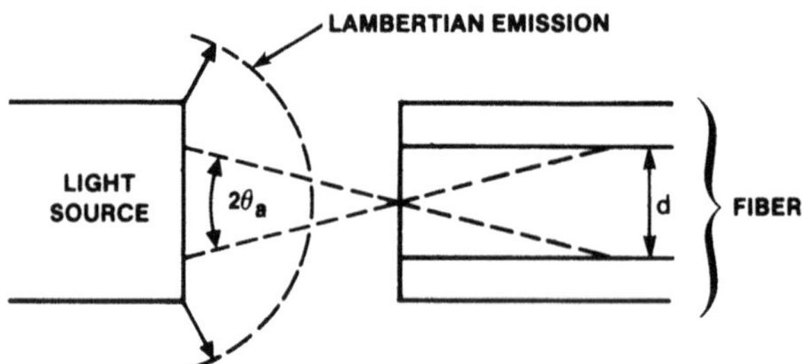

Figure 4.3. Light source feeding into the fiber.

means the fiber only collects the source emissions over the angle $2\theta_a$. It is therefore natural to define

$$\text{Fiber field of view (sr)} = \frac{\pi}{4}(2\theta_a)^2 = \pi\theta_a^2 \tag{4.1.7}$$

On the other hand, if we treat the fiber entrance as a collecting area of diameter d for the light at wavelength λ, its diffraction-limited field of view (as with any collecting antenna) is then $\lambda^2/(\pi d^2/4)$. We can now define the number of received field modes (distinct directions of arrival) per degree of polarization as

$$\begin{aligned}\text{Number of received field modes} &= \frac{\pi\theta_a^2}{\lambda^2/(\pi d^2/4)} \\ &= \frac{\pi^2 d^2 \theta_a^2}{4\lambda^2} \\ &\cong \pi^2\left(\frac{d}{2\lambda}\right)^2(\text{NA})^2\end{aligned} \tag{4.1.8}$$

That is, the light from the external source being collected by the fiber core can be considered to be divided into distinct, independent directions of arrival, each corresponding to a separate diffraction-limited field of view. The number of such modes is given by Equation (4.1.8). Each such mode, since it corresponds to a proper insertion angle, will produce a separate independent mode (propagating ray line) within the core. Hence Equation (4.1.8) also gives the number of propagating modes within the core per degree of polarization. For $\theta_a \ll 1$, and two degrees of polarization, we therefore have

$$\begin{aligned}\text{Number of propagating fiber modes} &= \frac{2\pi^2 d^2 \theta_a^2}{4\lambda^2} \\ &= \frac{1}{2}\left(\frac{\pi d}{\lambda}\right)^2(\text{NA})^2\end{aligned} \tag{4.1.9}$$

We see for a given NA, which depends only on the fiber indices, the number of propagating modes in the core depends only on the core diameter d. The larger d is, the more modes propagate simultaneously down the fiber. In particular, only one mode will propagate if

$$d \leqslant \frac{\sqrt{2}\lambda}{\pi(\text{NA})} = 0.45\left(\frac{\lambda}{\text{NA}}\right) \tag{4.1.10}$$

Single-mode fibers, therefore, require core diameters to be only several times the optical wavelength.

Mode propagation in a fiber can also be related to the so-called V-number of guided fields.[(1)] From Maxwell's equations for cylindrical guides, it is known that waves propagate with longitudinal components described in terms of Bessel functions, and the V-number is defined as

$$V = (u^2 + w^2)^{1/2} \tag{4.1.11}$$

where

$$u = \frac{d}{2}\left[\left(\frac{2\pi n_1}{\lambda}\right)^2 - k^2\right]^{1/2}$$

$$w = \frac{d}{2}\left[k^2 - \left(\frac{2\pi n_2}{\lambda}\right)^2\right]^{1/2}$$

and k is the longitudinal propagation constant ($k = 2\pi/\lambda$). By direct substitution we see that

$$V = \left(\frac{\pi d}{\lambda}\right)[n_1^2 - n_2^2]^{1/2} \tag{4.1.12}$$

Using Equations (4.1.5) and (4.1.9) shows that

$$\text{Number of fiber modes} = \frac{V^2}{2} \tag{4.1.13}$$

Hence the waveguide V-number is actually a measure of the mode number. Since each mode of the guide corresponds to a zero of a zero-order Bessel function,[2] single-mode operation requires V be less than the first zero, or

$$V \leqslant 2.4 \tag{4.1.14}$$

which roughly corresponds to Equation (4.1.10). We should also point out that the value of V determines the distribution of the propagating field power between core and cladding. Thus, with V satisfying Equation (4.1.14), a single mode will propagate down the fiber, but the actual value of V determines how the power of this mode is distributed radially within the fiber. Figure 4.4 sketches the normalized radial diameter of the single mode as a function of V for $V \leqslant 2.4$. When V is close to 1, the power is widely spread radially and most of the mode power flows in the cladding. As V

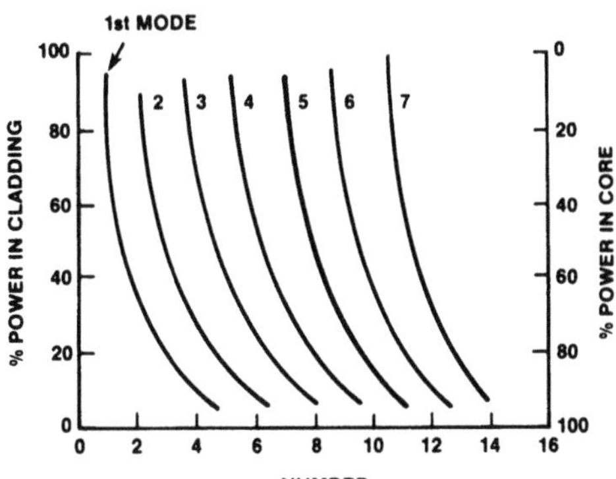

Figure 4.4. Power distribution in core and cladding versus fiber V-number.

increases, more of the power is in the core. For $V = 2$, about 80% of the field power is in the core. For this case Equation (4.1.12) requires

$$\begin{aligned} d &= \frac{\lambda V}{\pi n_1 \sqrt{2\Delta}} \\ &\approx 6\lambda \end{aligned} \tag{4.1.15}$$

for $n_1 = 1.5$ and $\Delta = 0.003$. This means a core diameter of several microns must be maintained. When V increases beyond 2.4, new modes are excited and the core power is now redistributed among all modes, as shown in Figure 4.4.

Having V small ($V \leqslant 1$) means that all the power is confined to a single mode, and the propagating field is spread significantly into the cladding. This has the advantage of making fiber splicing (connecting two fibers) easier, since exact matching of the cores is not required. However, power flow in the cladding is loosely guided and susceptible to increased attenuation and both leakage and bending losses. Preferred operation is with most of the power in the core, with serious consideration to reducing the V number only in the region of splices. Notice also from Chapter 1 that we can determine the efficiency of coupling into the fiber as

$$\gamma = \frac{\iint N(\mathbf{r}, \theta) \cos\theta \, d\Omega \, dA_s}{P_s} \tag{4.1.16}$$

For a point source, γ becomes

$$\gamma = \frac{\int_0^{\theta_a} N(\theta) \cos\theta \, d\Omega}{\int_0^{\pi/2} N(\theta) \cos\theta \, d\Omega} \tag{4.1.17}$$

For an isotropic source $N(\theta) = N$ for θ defined over a hemisphere, and

$$\gamma = \sin^2 \theta_a = (\text{NA})^2 \tag{4.1.18}$$

Hence increasing the NA of a fiber increases the fraction γ of source power that will be coupled into the fiber.

4.2. Attenuation and Dispersion in Fibers

In addition to the modal description of a fiber field, one is also interested in the amount of power loss and field dispersion occurring as the field propagates. For a single-mode fiber there is a strong dependence of attenuation on wavelength. Figure 4.5 plots the attenuation characteristics of a single mode (dB loss per km) for a glass fiber. This loss is a combination of intrinsic absorption by the atoms which constitute the core and scattering due to core impurities. Absorption loss increases with wavelength, and impurity atoms (iron, chromium, cobalt, copper, etc.) account for the rapid increase in attenuation at the higher wavelengths. Attenuation due to scattering is principally a Rayleigh scattering effect and falls off as $1/\lambda^4$, accounting for the increased attenuation at lower wavelengths. This scattering arises from compositional fluctuations which occur during fabrications. The combination of the two attenuation effects produces an optical window, which in Figure 4.5 occurs around 1.1 to 1.3 μm. In this vicinity, attenuation can be reduced well below 1 dB/km. As absorption impurities are reduced by fabricating higher-quality glass cores, the molecular scattering loss becomes the intrinsic lower loss limit. Since this effect favors higher wavelengths,

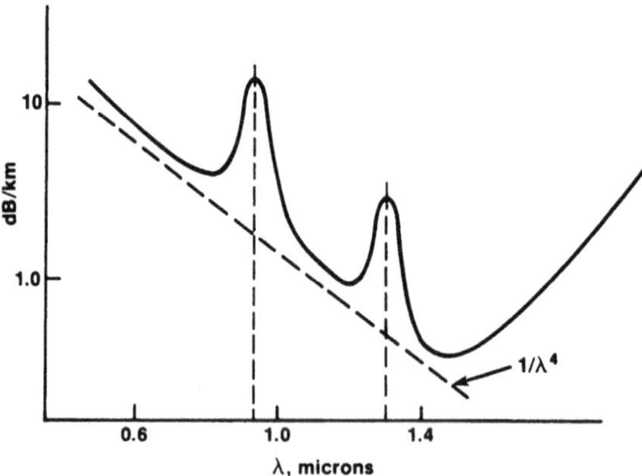

Figure 4.5. Single-mode attenuation versus wavelength for glass fibers.

the fiber window will tend to move toward the longer wavelengths as better fibers are built. Although the minimal attenuation is low in the vicinity of the window, the single-mode loss increases significantly at other wavelengths. Hence single-mode fibers require careful matching of the operating optical wavelength and the attenuation characteristics of the core material.

As more modes are produced, the power in the fiber is distributed over these modes. As a result, scattered mode energy is often simply coupled into other modes, while modes with different wavenumbers will have different absorption coefficients in the same core. As a result, when there are many modes propagating, the attenuation effects tend to be averaged over all modes. As a result, the multimode fibers have attenuation effects that are basically constant with wavelength. This means multimode attenuation is essentially independent of wavelength, and the attenuation coefficient lies somewhere between that predicted by either absorption or scattering alone. Additional losses may have to be included for fiber splicings, cable connections, and fiber bends.[(2)]

In describing propagating guided fields as a communication channel, one is interested in not only power loss but field dispersion as well. Field dispersion is basically relative displacement of the propagating field components, and leads to waveform distortion in the channel.

There are two main causes of dispersion in a fiber: material dispersion and modal dispersion. Material dispersion is dispersion that occurs within a mode and is caused by the fact that the fiber core material causes the different frequencies that constitute a mode waveform to travel at different velocities within the mode. Material dispersion is given in terms of time difference in propagating a unit length between two wavelengths $\Delta\lambda$ apart in the vicinity of a wavelength λ. This dispersion is basically proportional to $\Delta\lambda$ and is usually normalized to the percent wavelength difference $\Delta\lambda/\lambda$. Typical material dispersion is usually stated in nanoseconds per kilometer for a percent bandwidth in wavelengths, as a function of λ for a silicon fiber. The dispersion is about 1 ns/km multiplied by the percent bandwidth at 0.86 μm. An optical bandwidth of about 1000 Å would therefore produce about 0.1 ns/km while a 10-Å bandwidth produces about 10^{-3} ns/km. Since most microwave modulation formats are in this range, material dispersion of a single mode should have negligible effect on communication performance.

Most dispersion is caused by dispersive interaction between two modes. Two modes propagating at different ray angles within the core but at the same velocity will arrive at a point down the fiber at different times. Consider the ray diagram in Figure 4.6, showing an outer mode (maximum propagation angle) traveling a different path than the center ($\theta_p = 0$) mode. While the center mode travels a distance L, the outer mode travels $L/\cos\theta_p$, both

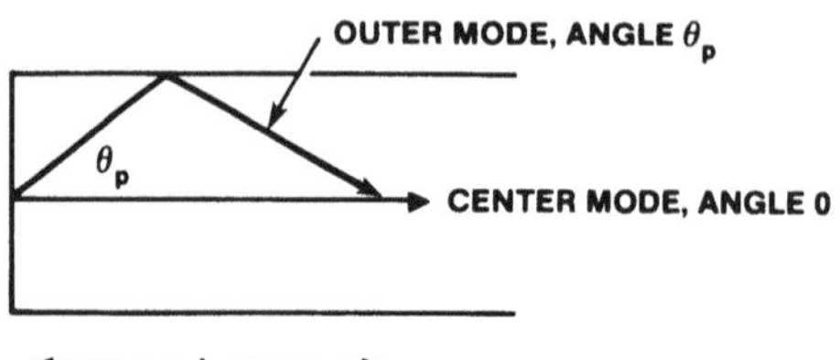

Figure 4.6. Two propagating modes.

at the wave velocity c/n_1. The difference between the arrival times of the modes at the plane at L is then

$$t_d = \left(\frac{L}{\cos\theta_p} - L\right)\frac{n_1}{c} \text{ seconds} \tag{4.2.1}$$

Using Equation (4.1.1) this corresponds to a time differential per unit length of

$$\begin{aligned}\frac{t_d}{L} &= \left(\frac{n_1}{n_2} - 1\right)\frac{n_1}{c} = \frac{n_1\Delta}{2c} \\ &= \frac{(\text{NA})^2}{2n_1c} \text{ seconds/length}\end{aligned} \tag{4.2.2}$$

Hence, worst-case mode dispersion depends only on the fiber indices and varies as the square of the numerical aperture. A fiber with a numerical aperture of 0.2 will have a dispersion of about 50 ns/km. That is, light impulses launched at the same time in these two modes will arrive 50 ns apart for each kilometer of fiber.

The derivation of Equation (4.2.2), using Figure 4.6, immediately suggests that the t_d/L value for these two modes can be decreased by reducing the time differential. This requires speeding up the travel time of the outer ray relative to the inner ray. In order to accomplish this, it is necessary to reduce the core index n_1 at the core edges. Thus the core would have to exhibit a variable index along its radial direction, from n_1 at its center to a decreasing (smaller) value of index grading; this is exhibited as an index profile. Figure 4.7 shows some graded index profiles. Figure 4.7a is the standard nongraded fiber, with the core index n_1 maintained throughout the core. (This is often called a step-index fiber due to the shape of its profile.) Figure 4.7b is an example of a graded profile, showing the gradual reduction of core index for the mode rays with the increasing propagation angles. In essence, the index profile acts like a continuous lensing action that continually refocuses the light beam along the fiber path.

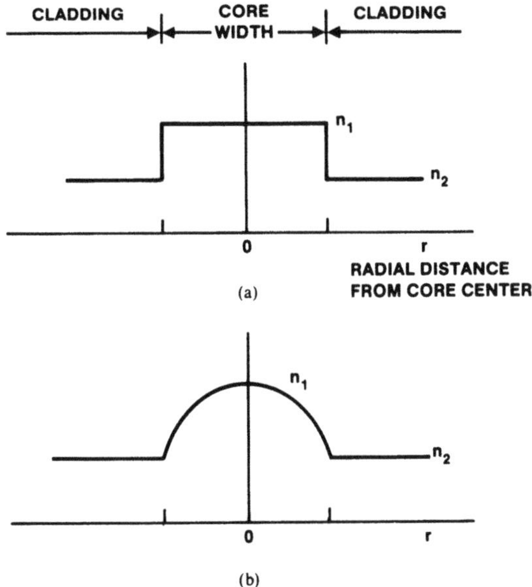

Figure 4.7. Fiber index profile: (a) Step index; (b) parabolic index.

A common form of this grading along the fiber radius r is the parabolic profile

$$n_1(r) = n_1\left[1 - q\left(\frac{r}{d/2}\right)^2\right] \tag{4.2.3}$$

where r is the radial distance from the core center, d is the core diameter, and q is the rate of fiber grading. This profile has a maximum delay spread of approximately

$$\begin{aligned}\frac{t_d}{L} &= \left(\frac{1}{\cos\theta_p}\right)\frac{n_1(1-q)}{2c} - \frac{n_1}{2c} \\ &\approx \frac{n_1\Delta^2}{2c}\end{aligned} \tag{4.2.4}$$

This is about $1/\Delta$ times smaller than that for the step-index profile in Equation (4.2.2). For Δ equal to a fraction of a percent, this corresponds to a reduction in dispersion by several orders of magnitude. Thus, while a standard step-index fiber will have dispersion of about 50 ns/km, graded fibers may be well below 1 ns/km. We emphasize that if the single-mode condition can be maintained, mode dispersion does not occur and field dispersion is limited to only that of the core media. The trade-off, of course, is a higher-quality, thinner, low-NA fiber, making splicing and fiber field control more difficult. By using multimode fibers, the latter problems are lessened but the field dispersion is increased.

The use of the previous dispersion analyses employing only central and outer ray lines may raise some questions as to the extent to which this dispersion is really harmful to the propagating field when many modes are involved. With many modes, the field propagates at all ray angles out to the maximum ray angles, and their contribution to dispersion is not included. In addition, the total power of the central (small) angles may be significantly greater than that of the maximum angle, so the effect of the outer angle may be overemphasized. To perform a more exact analysis of field dispersion, it is necessary to consider power flow across all propagating angles and to take into account power conversion that may occur between modes. This is done in the next section.

4.3. Dispersion and Pulse Shaping in Fibers Undergoing Diffusion

Our dispersion discussion in Section 4.2 was purely in terms of isolated ray lines and field modes. A more exact analysis must account for possible power diffusion and mode regeneration as the field propagates. To develop this approach, we must treat the optical field in the fiber as undergoing diffusion, in which propagation at one ray direction can couple into another ray angle. This can be handled by considering the ray angle θ as a continuous variable and allowing diffusion over this angle. This inherently implies a multimode condition—a large number of modes existing so that all angles between 0° (central ray angle) and θ_p (maximum ray angle) are occupied.

The modeling of power in a fiber begins with the basic diffusion equation of a propagating field in a medium. In the development here, we follow the illuminating approach of Gloge[3] which is reviewed in Appendix D. We consider the diagram in Figure 4.8 and let

$$P(z, \theta, t) \triangleq \begin{array}{l}\text{Power in the fiber at time } t,\\ \text{distance } z_1 \text{ and ray direction } \theta.\end{array} \tag{4.3.1}$$

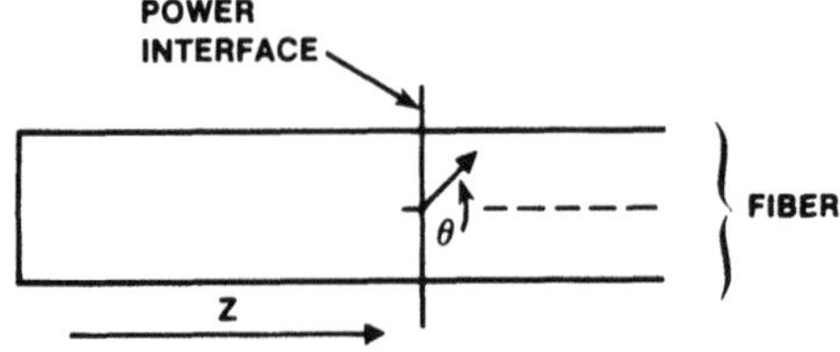

Figure 4.8. Fiber power flow.

It is known from diffusion theory that the P function, when confined to small angles θ as in a fiber, must satisfy the diffusion equation in cylindrical coordinates

$$\frac{\partial P}{\partial z}+\left(\frac{n_1 \cos \theta}{c}\right)\frac{\partial P}{\partial t} = -A\theta^2 P + \frac{1}{\theta}\frac{\partial}{\partial \theta}\left[\theta D \frac{\partial P}{\partial \theta}\right]; \qquad \theta \ll 1 \tag{4.3.2}$$

Here c is the speed of light, and A and D are called the attenuation and diffusion coefficients, respectively, of the media. These parameters are determined by the material of the fiber core and are assumed to be constant throughout the core radius.

The total power in the fiber at any time t and position z can be obtained by integrating P over all ray directions. Hence we denote

$$Q(z, t) = \begin{array}{l}\text{Integrated power over all ray angles } \theta\\ \text{at time } t \text{ and position } z \text{ due to an}\\ \text{initially launched source power distribution}\end{array}$$

$$= 2\pi \int_{\text{all } \theta} P(z, \theta, t)\, \theta \, d\theta \tag{4.3.3}$$

Thus $Q(z, t)$ defines the total integrated power that can be collected over the fiber cross-section at a distance z and time t. If the initial source power was an impulse in time [i.e., $P(0, \theta, t) = \delta(t)P_0(\theta)$, where $P_0(\theta)$ is the initial launch distribution over θ], then $Q(z, t)$ is the impulse response of the fiber at point z.

A complete solution for $Q(z, t)$, from Equations (4.3.2) and (4.3.3), was derived by Gloge[(3)] and is discussed in Appendix D. Although a completely general solution can be obtained, its form is extremely complicated and somewhat unwieldy to interpret. However, the result has some interesting limiting cases.

For the condition of a "short" fiber, defined by the condition

$$z\gamma_\infty \ll 1 \tag{4.3.4}$$

the solution reduces to

$$Q(z, t) = \frac{(2c/n_1)\pi}{z(1+\gamma_\infty z)}\, e^{-(2c/n_1\theta_\infty^2 z)t} \tag{4.3.5}$$

where

$$\gamma_\infty = (4DA)^{1/2} \tag{4.3.6}$$

$$\theta_\infty = (4D/A)^{1/4} \tag{4.3.7}$$

The parameter γ_∞ plays the role of an attenuation coefficient (due to both media loss and diffusion), while θ_∞ is like an average diffusion angle, determining the time constant of the exponential decay. Equation (4.3.5) is plotted in Figure 4.9 for several values of z. Note that as z increases (we move further down the fiber), the peak power decreases, and the impulse response spreads in time according to the time constant $\theta_\infty^2 z/(2c/n_1)$. Note that this time constant depends on both the attenuation coefficient A and the diffusion coefficient D of the fiber model.

For a "long" fiber, defined by the condition

$$z\gamma_\infty \gg 1 \tag{4.3.8}$$

the solution behaves as

$$Q(z, t) = \theta_\infty^2 \left(\frac{\pi}{T_f t}\right)^{1/2} \left(\frac{t}{\gamma_\infty z T_f} + \frac{1}{z}\right) \exp\left\{ -\frac{\gamma_\infty^2 z T_f}{4t} - \frac{t}{T_f}\right\} \tag{4.3.9}$$

where

$$T_f = \frac{n_1}{2cA} = \frac{n_1}{2c}\left(\frac{\theta_\infty^2}{\gamma_\infty}\right) \tag{4.3.10}$$

Equation (4.3.9) is included in Figure 4.9 for several values of $z\gamma_\infty$. We see that for a long fiber the power response is more pulselike, showing both an inherent delay and a pulse spreading. As the point z is further increased, the pulse delay increases and the shape widens into a wide, bell-shaped

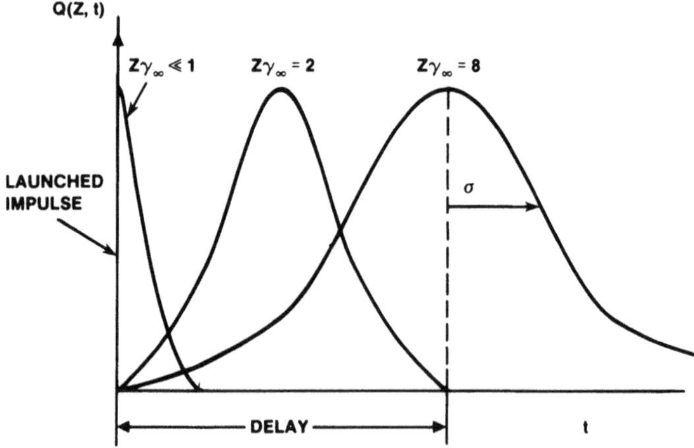

Figure 4.9. Fiber impulse response in time at various distances in fiber.

type of response function. It is convenient to describe these pulse shapes by their "location," or mean delay:

$$\delta(z) = \frac{\int_0^\infty tQ(z, t)\, dt}{\int_0^\infty Q(z, t)\, dt} \tag{4.3.11}$$

and their spread about this delay:

$$\sigma^2(z) = \frac{\int_0^\infty [t - \delta(z)]^2 Q(z, t)\, dt}{\int_0^\infty Q(z, t)\, dt} \tag{4.3.12}$$

Note that δ is like a "center of mass" of the pulse shape, and σ is like a "variance" about this center. Gloge[3] computed an exact expression for these parameters from his general solution and showed that

$$\sigma^2(z) = \frac{T_f}{2}[\gamma_\infty z(1 - 2e^{-\gamma_\infty z}) + \tfrac{3}{4} - e^{-2\gamma_\infty z} + \tfrac{1}{4}e^{-4\gamma_\infty z}] \tag{4.3.13}$$

The result is plotted in Figure 4.10 as a function of $z\gamma_\infty$, and shows how the power pulse response spreads as it moves further down the fiber. Equation (4.3.13) has the limiting forms

$$\sigma(z) = \begin{cases} (T_f\gamma_\infty)z & \text{for } z\gamma_\infty \ll 1 \\ \left(\dfrac{T_f}{2}\right)(\gamma_\infty z)^{1/2} & \text{for } z\gamma_\infty \gg 1 \end{cases} \tag{4.3.14}$$

which we see are accurate asymptotes to the actual curve. Note that for short distances, the pulse spreading is proportional to length, but for long distances the spreading eventually becomes proportional to the square root

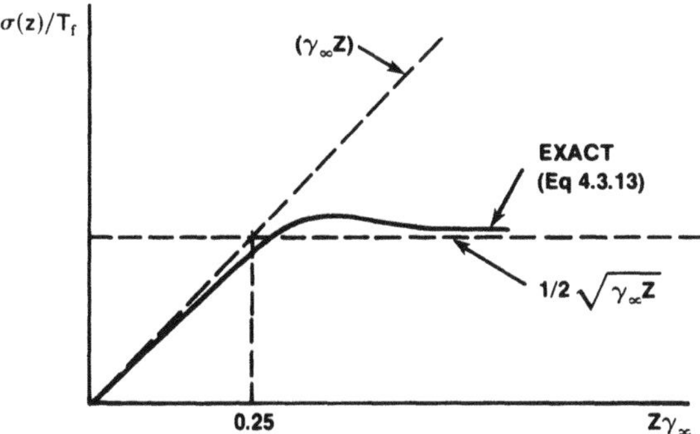

Figure 4.10. Fiber pulse spread $\sigma(z)$ as a function of distance z in the fiber.

of z. That is, the spreading increases at a slower rate at the longer distances. The transition point occurs approximately where the two asymptotes cross, which is at $z\gamma_\infty = 0.25$, or at a distance

$$z_0 \triangleq \frac{1}{4\gamma_\infty} \tag{4.3.15}$$

This is often called the "equilibrium" distance of the fiber. We see from Equation (4.3.15) that this distance z_0 is mathematically related to the attenuation and dispersion coefficients of the fiber core, and therefore depends on the fiber media. On the other hand, since pulse spreading can be readily measured, z_0 can be determined empirically. Typically, z_0 will have a value on the order of one kilometer for most fibers. A measurement of z_0 allows evaluation of the parameter γ_∞ without knowledge of the fiber coefficients.

4.4. Pulse Stretching in Multimode Fibers

The results of the previous section can now be directly applied to model the pulse stretching that occurs in a multimode fiber. Let the fiber length be L and assume a time impulse of power (short burst of light) is launched into the fiber uniformly over all angles less than θ_p. The power distribution, as a function of time, at the fiber output can be obtained by using the earlier equation with $z = L$. In particular, the output pulse width will be obtained from (4.3.14) as

$$\sigma(L) = \begin{cases} T_f\gamma_\infty L & \text{for } L \ll \dfrac{1}{4\gamma_\infty} = z_0 \\[2ex] \left(\dfrac{T_f}{2}\right)(\gamma_\infty L)^{1/2} & \text{for } L \gg z_0 \end{cases} \tag{4.4.1}$$

Furthermore, if we take the average spreading angle θ_∞ to be equal to the maximum propagation angle θ_p, we can define $T_f\gamma_\infty$ as a fiber spreading coefficient β in seconds/length,

$$\begin{aligned} \beta = T_f\gamma_\infty &= \left(\frac{n_1}{2c}\right)\left(\frac{\theta_\infty^2}{\gamma_\infty}\right)\gamma_\infty \\ &\approx \left(\frac{n_1}{2c}\right)\left(\frac{\text{NA}}{n_1}\right)^2 \\ &= \frac{(\text{NA})^2}{2cn_1} \end{aligned} \tag{4.4.2}$$

Likewise, we note

$$\frac{T_f}{2}(\gamma_\infty L)^{1/2} = \left(\frac{T_f \gamma_\infty}{2}\right)\left(\frac{L}{\gamma_\infty}\right)^{1/2}$$
$$= \beta\left(\frac{L}{4\gamma_\infty}\right)^{1/2} \tag{4.4.3}$$

We can, therefore, write Equation (4.4.1) directly in terms of β:

$$\sigma(L) = \begin{cases} \beta L & \text{for } L \ll z_0 \\ \beta z_0\left(\dfrac{L}{z_0}\right)^{1/2} & \text{for } L > z_0 \end{cases} \tag{4.4.4}$$

Thus, pulse spreading at a fiber output depends on the length of the fiber, its equilibrium distance, and the coefficient β, the latter dependent only on the fiber core index and numerical aperture in Equation (4.4.2).

Since β determines the amount of pulse spreading that will occur, let us examine it in more detail. If we rewrite the expression for the numerical aperture, we have

$$\beta = \frac{n_1^2 - n_2^2}{2cn_1}$$
$$\approx \frac{n_1 - n_2}{c}$$
$$= \frac{n_1 \Delta}{c} \tag{4.4.5}$$

Therefore, the coefficient β depends only on the core index n_1 and index difference Δ. If we relate this to our earlier result in Equation (4.2.2), we see that the spreading coefficient β, obtained from diffusion theory, is equivalent to the differential time delay per length t_d/L, obtained from ray line theory. Indeed, the maximum angular ray line contribution to the pulse dispersion is significant, and dispersion reduction via fiber grading is theoretically justified.

4.5. Communications Link Models for the Fiber Channel

The fiber analysis of the previous sections can now be used to formulate a basic fiber optic communication channel model. The procedure is to integrate together the key parameters of the communication link with the channel characteristics of the fiber itself. When viewed in this context, the only role of the fiber is to carry the modulated light field from transmitter to receiver. The characteristics of the channel will therefore depend on the manner in which the light propagates down the fiber.

A fiber optic communication channel is shown in Figure 4.11. The light source emits a modulated optical field, with time-varying power $P_t(t)$, into the fiber. The fiber, having length L, is characterized by an attenuation loss factor α (nepers or dB/unit length) and a dispersive effect that is equivalent to an effective filtering on the light modulation. The field power variation collected by the receiver at the fiber output can then be modeled by the linear baseband filtering

$$P_r(t) = P_t(t) \otimes h_f(t) \tag{4.5.1}$$

where $\otimes$ denotes time convolution and $h_f(t)$ is the effective fiber impulse response. For a single-mode fiber, $h_f(t)$ is a wideband response, limited only by the core material dispersion. For a multimode fiber undergoing diffusion, $h_f(t)$ is given by

$$h_f(t) = Q(L, t) \tag{4.5.2}$$

where $Q(z, t)$ is the response in Equation (4.3.3) to an optical impulse of power. Following the discussion in Section 3.1, a direct detection receiver collecting all fiber field modes will produce the photodetected shot noise current, whose mean time variation (signal component) is

$$i_s(t) = \bar{G}u[P_r(t) \otimes h_d(t)] \tag{4.5.3}$$

where $\bar{G}$ and u are photodetector gain and responsivity, and $h_d(t)$ is the filtering produced by the detector itself on the intensity modulation (see Figure 3.4). In the frequency domain Equations (4.5.1) and (4.5.3) yield

$$I_s(\omega) = \bar{G}u[\hat{P}_t(\omega) H_d(\omega) H_f(\omega)] \tag{4.5.4}$$

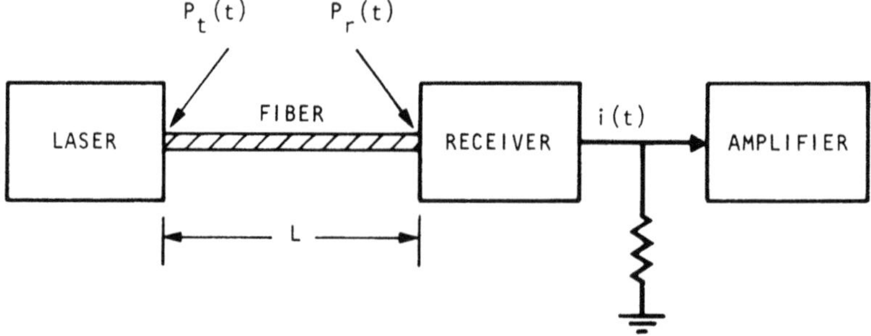

Figure 4.11. Fiber optic communication link.

where caps denote Fourier transforms. The conversion of $i_s(t)$ to a signal voltage in Figure 4.11 was shown in Equation (3.1.25) to be

$$v_s(t) = i_s(t) \otimes h_c(t) \tag{4.5.5}$$

where $h_c(t)$ is the photodetection filter response corresponding to the transfer function $H_c(\omega)$ in Equation (3.1.26). The resulting detected signal transform is then

$$V_s(\omega) = \bar{G}u[\hat{P}_t(\omega)H_f(\omega)H_d(\omega)H_c(\omega)] \tag{4.5.6}$$

This establishes the baseband equivalent link model shown in Figure 4.12, which describes the way in which light modulation is transmitted to the receiver output. The validity of linear filtering models for optical intensity modulation in fiber systems has been previously investigated.[4] For typical modulation bandwidths, $H_d(\omega)$ is relatively wideband, and the principal filtering in Equation (4.5.6) comes from the fiber and output circuitry.

To the fiber filtering model we can add the photodetector noise, as shown in Figure 4.12, where the noise has spectral level

$$N_0 = \overline{G^2}euP_r + eI_{dc} + (2kT°/R_L) \tag{4.5.7}$$

This represents the contribution of detector shot noise, dark current, and postdetection thermal noise, as derived earlier in Equations (3.1.22)-(3.1.24). The filtered version of this noise therefore appears at the receiver output.

Communication performance for the fiber link can now be determined from the channel model. For example, the detected signal-to-noise ratio (SNR) achievable at the receiver output can be computed by standard power flow analysis. We can first assume the power modulation $P_t(t)$ imposed at the transmitter is narrowband relative to the detector and circuit

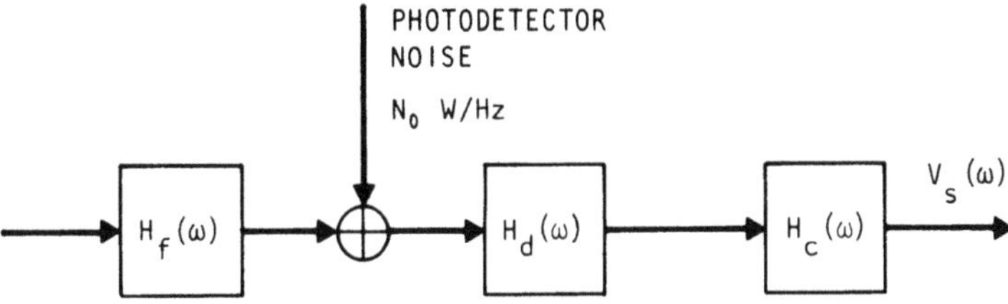

Figure 4.12. Communication model for the fiber channel.

filtering. In this case, from Equation (3.3.4),

$$\mathrm{SNR} = \frac{(\bar{G}uP_r)^2}{N_0 2B_b} = \frac{\left[\left(\frac{\eta}{hf}\right)P_r\right]^2}{\left[\frac{F\eta P_r}{hf} + \frac{I_{dc}}{\bar{G}^2 e} + \frac{2kT^0}{R_L e^2 \bar{G}^2}\right] 2B_b} \tag{4.5.8}$$

with B_b the circuit noise bandwidth defined in Equation (3.3.3), and P_r the average modulation power at the detector input. Solving for P_r gives the required power at the fiber output necessary to establish a desired SNR in a specified bandwidth B_b,

$$P_r = \frac{(\mathrm{SNR})F2B_b hf}{\eta}\left[1 + \left(\frac{2\eta I_{dc}}{hfFeB_b} + \frac{kT^0}{\bar{G}^2 e^2 R_L}\right)\frac{1}{\mathrm{SNR}}\right]^{1/2} \tag{4.5.9}$$

If we introduce the noise-equivalent power (NEP) of the receiver, as in Equation (3.3.9), this simplifies Equation (4.5.9) to

$$P_r = (\mathrm{NEP})[(\mathrm{SNR})2B_b]^{1/2} \tag{4.5.10}$$

Hence fiber power can be determined from the receiver noise characteristics. When the receiver is quantum-limited (see Table 3.1), SNR $= \eta P_r/hfF2B_b$, and the required fiber power is simply

$$P_r = (\mathrm{SNR})2B_b Fhf/\eta \tag{4.5.11}$$

Once the required fiber power is established, the source launch power can be determined from the attenuation. Thus, neglecting fiber dispension,

$$P_t = e^{\alpha L} P_r \tag{4.5.12}$$

where α is in nepers/length. Equations (4.5.9)-(4.5.12) allow an overall power analysis to be performed on the fiber channel. Note that with the fiber filtering effect neglected, the analysis is identical to that of a direct-detection free-space channel, with background noise neglected, and with fiber attenuation replacing the antenna beam spreading loss.

When fiber filtering is not negligible (i.e., the dispersive effects of the fiber propagation must be included), the degradation in performance due to the resulting distortion on the power modulation $P_t(t)$ must be taken into account. In the link model in Figure 4.12, standard Fourier transform theory can be applied to produce the actual time waveform $v_s(t)$,[4-7] from which distortion effects can be estimated. The use of baseband equalizing filters following photodetection to compensate for fiber filtering can be directly applied here.[3-6]

Of primary concern is when the modulation is transmitted in digital light pulses. The effective channel filtering in Figure 4.12 converts directly to pulse spreading, as noted in Section 4.4. Pulse distortion of this type, if not carefully designed, can produce degraded digital performance and severely limit data rate capability. This is examined in the next section.

4.6. Digital Transmission in Fiber Channels

We have seen that the fiber channel can cause transmitted light pulses to be filtered during propagation. This filtering can be equivalently integrated as a spreading of the light pulses appearing at the receiver. As shown in Figure 4.13, if a square idealized light pulse of width τ is launched into the fiber, it will appear at the output with the pulse spread in shape to an approximate width τ_s. If this transmitted light pulse corresponds to a digitally modulated pulse, and if a sequence of such pulses is used to transmit binary data, it is apparent that pulse spreading will cause (a) pulse energy loss, since a portion of the output pulse is spread outside the original pulse width interval τ that the receiver assumes is the square pulse width and (b) interpulse distortion, since the tail portion of a given spread pulse will overlap onto adjacent pulse intervals. To reduce these effects, it is necessary to use pulse spacings slightly greater than the expected spread τ_s. As a rough rule of thumb, we generally set

$$\text{pulse spacing} \geq \tau_s \tag{4.6.1}$$

to ensure small energy loss and negligible interpulse distortion. This means

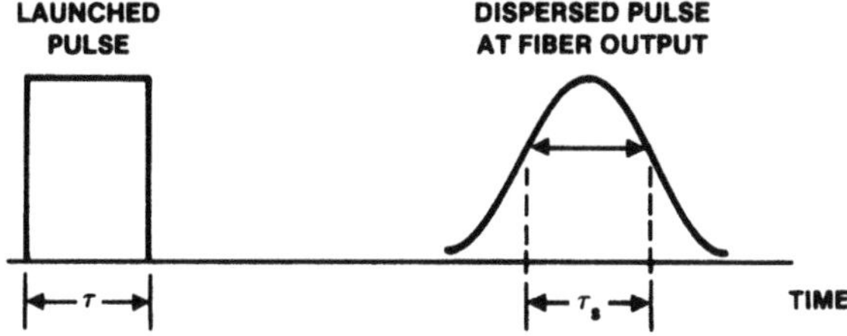

Figure 4.13. Fiber pulse spreading in digital transmission.

the pulse rate is

$$R_p = \frac{1}{\tau_s} \tag{4.6.2}$$

Since τ_s depends entirely on the fiber dispersion, and the latter depends on fiber length through Equation (4.4.4), Equation (4.6.2) implies that the maximum allowable transmitter pulse rate is related to this length. Figure 4.14 plots Equation (4.6.2) for different dispersion coefficients β, as a function of fiber length. The curve exhibits both the linear and square root behavior of the dispersion with length.

The actual transmitted digital bit rate depends on the way in which the light pulses are modulated. Table 4.1 lists the conversion from pulse rates R_p to bit rate R_b in bits/second for the modulation methods described in Chapter 3. For on–off keying and binary PPM, each pulse carries one bit of information and the bit rate equals the pulse rate. In both MPAM and MPPM, each pulse is encoded into $\log_2 M$ bits and the bit rate is $\log_2 M$ times the pulse rate. Note that such encoding raises the curves in Figure 4.14 when the ordinate displays bit rate instead of pulse rate.

As the fiber length increases, the pulse attenuation also increases. The optical light pulse is weakened in addition to being dispersed. The degraded pulse power lowers the output pulse energy in the pulse interval, and equivalently lowers decoding performance. The required pulse count K_s needed to achieve a given bit error probability PE can be determined from the equations and curves in Section 3.6, depending on the receiver characteristics (shot noise limited, quantum limited, etc.). The required pulse count K_s is related to the received pulse power P_p by $K_s = (\eta/hf)P_p\tau$. Likewise, the pulse width τ is related to the bit rate R_b, and the average power P_r is related to the peak power P_p, according to Table 4.1. We can thus determine the required average power P_r to achieve a desired bit error probability (i.e., K_s) and bit rate R_b as

$$P_r = \begin{cases} \left[\dfrac{hfK_s}{2\eta}\right] R_b & \text{for OOK} \\ \left[\dfrac{hfK_s(M+1)}{\eta(\log_2 M)2M}\right] R_b & \text{for MPAM} \\ \left[\dfrac{hfK_s}{\eta \log_2 M}\right] R_b & \text{for MPPM} \end{cases} \tag{4.6.3}$$

The required receiver power P_r can now be related to the fiber attenuation. The relation between the source launch power into the fiber P_t, the

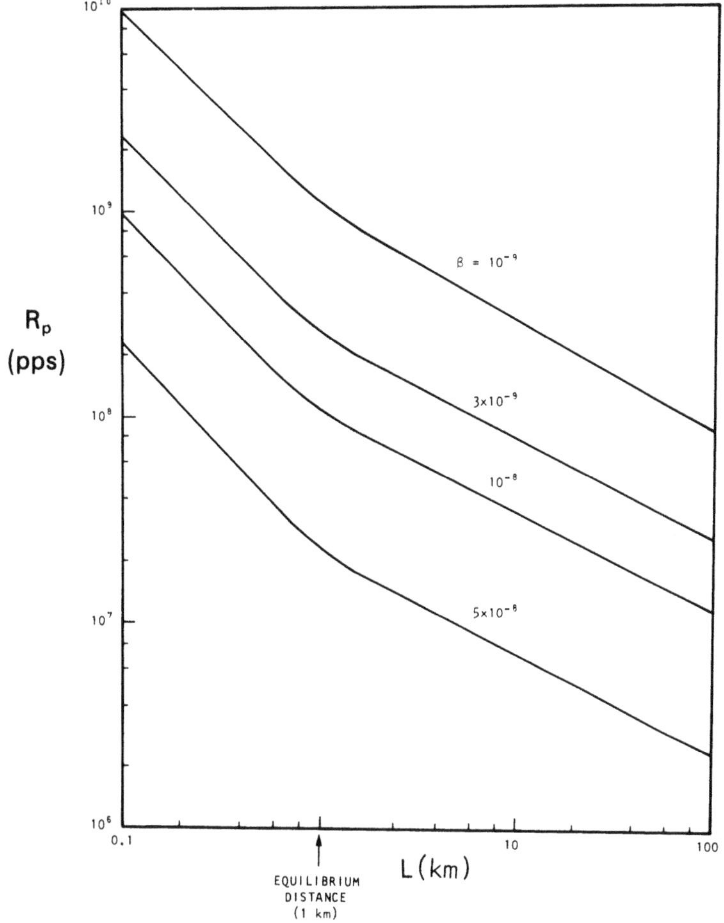

Figure 4.14. Pulse rate R_p allowed by fiber dispersion (β = dispersion coefficient, L = fiber length).

fiber length L, and the receiver power P_r is given by Equation (4.5.12):

$$P_r = P_t e^{-\alpha L} \tag{4.6.4}$$

For a desired operating performance corresponding to a specific value of K_s or SNR, the required value of P_r must first be obtained depending on the receiver type. We can then substitute into Equation (4.6.4) and solve for the allowable fiber data rate R_b corresponding to a specified launch power P_t as

$$R_b = Q e^{-\alpha L} \tag{4.6.5}$$

Table 4.1. Summary of Pulse Modulation Methods and Corresponding Bit Rates

Modulation method	Bit rate R_b,	Ratio of peak to average power
On-Off keying	R_p	2
M-level amplitude modulation (MPAM)	$(\log_2 M)R_p$	$\frac{M+1}{2M}$
Binary PPM	$\frac{1}{2}R_p$	2
M-ary PPM	$(\log_2 M)R_p$	M

where

$$Q = \begin{cases} \dfrac{2\eta P_t}{hfK_s} & \text{for OOK} \\ \dfrac{\eta(\log_2 M)2MP_t}{(M+1)hfK_s} & \text{for MPAM} \\ \dfrac{\eta(\log_2 M)P_t}{hfK_s} & \text{for MPPM} \end{cases} \tag{4.6.6}$$

Note that the receiver quality, encoding format, performance level, and launch power capability are all incorporated into the one parameter Q in Equation (4.6.6). Typically, the parameter Q will have values in the range of 10^{15} to 10^{18}. Equation (4.6.5) relates the key power flow parameters of the entire link—the fiber length L and loss α, the source launch power P_t, the encoding format M, the desired performance (K_s determines PE), the data rate R_b, and the receiver quality. Equation (4.6.5) therefore specifies the fiber data rate allowed by the link power limitations for any fiber length L and loss α. Note that the data rate R_b can be increased at any value of fiber length L by increasing Q, i.e., increasing source power (higher P_t), accepting poorer performance, or utilizing a better receiver (smaller K_s).

The data rates specified by Equation (4.6.5) correspond to the maximum power-limited data rate. This must be balanced against the data rate permitted by the dispersion of the fiber in Figure 4.14. Equation (4.6.5) is plotted in Figure 4.15 as a function of the fiber length L for a fixed set of generating parameters Q and loss coefficients α. The dispersion curves of Figure 4.14 are superimposed as shown. The results indicate the limiting data rates at each fiber length L.

As an example, suppose a digital OOK fiber link is to be operated over a 10-km path at a bit error probability of PE $= 10^{-4}$. The receiver operates

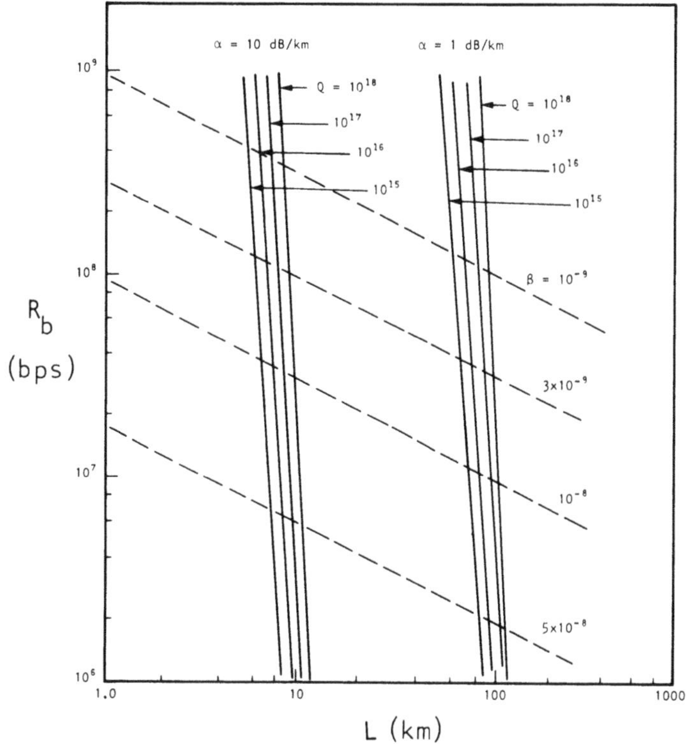

Figure 4.15. Fiber link bit rate versus fiber length L (β = dispersion coefficient, α = attenuation coefficient).

at 300 °K with an APD detector having an efficiency of 0.5, a gain of 100, a 50-ohm load resistor, and a dark current equivalent to a background of 1 count per pulse. From Figure 3.14, we see that a pulse count of $K_s = 200$ is required. At a wavelength of 0.6 μm, $\eta/hf \cong 1.5 \times 10^{18}$. A transmitting light source feeding 150 mW of pulse power into the fiber will produce a Q in Equation (4.6.6) of $(1.5 \times 10^{18}) \times (150 \times 10^{-3})/100 \cong 2.3 \times 10^{15}$. From Figure 4.15, if an $\alpha = 1$ dB/km fiber is used, with a dispersion coefficient of $\beta = 3$ ns/K_a, a bit rate of about 100 Mbps can be sustained. On the other hand, if a 10-dB/km fiber is used, the bit rate is only 1 Mbps. In the former case the fiber is dispersion limited, but it is power limited in the latter case. By resorting to pulse encoding, as for example with 16-ary PPM, with the same laser average power, the data rate in the power-limited case can be increased by $\log_2 16/2 = 2$. However, peak power P_p must be increased by 16 at the same pulse repetition rate to maintain the same K_s. This improvement in data rate is achieved, however, at the cost of additional modulation and decoding complexity.

The rate–length performance curves in Figure 4.15 can be generalized. Denote the length corresponding to the intersection of a dispersion-limited rate curve and a power-limited rate curve as L_0. That is, the length L_0 separates rate limitation due to power loss and that due to fiber dispersion. If the fiber has length $L < L_0$, the fiber has its rate limited by the inherent fiber filtering, while if $L > L_0$, the rate is instead limited by fiber attenuation. If the fiber length exceeds L_0, then a more dispersive fiber could have been used without loss of data rate. Conversely, if the fiber length was less than

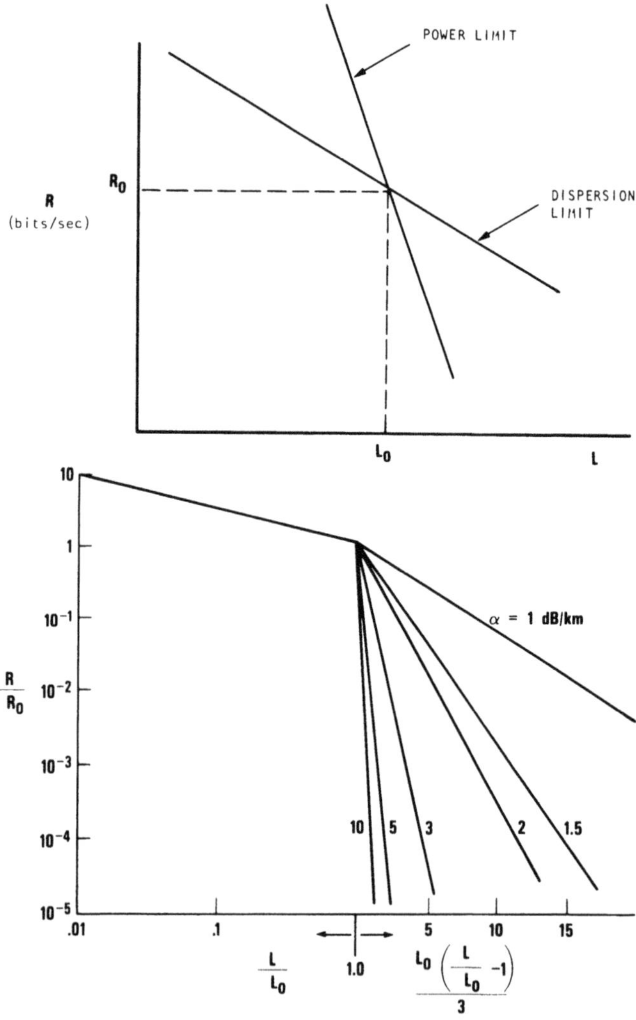

Figure 4.16. Generalized fiber optic data rate performance curves.

L_0, a lower-power performance link could have been designed without loss of rate.

The value of L_0 for any fiber system depends not only on the fiber material itself, but also on the source and receiver parameters as well. For a specific fiber, L_0 can be reduced or increased by lowering or raising the power-limited rate curves in Figure 4.16a or by increasing or decreasing the dispersion coefficient β. If we assume L_0 will always exceed the fiber equilibrium length, the parameter L_0 satisfies the relation

$$\frac{1}{\beta\sqrt{L_0}} = Q\,e^{-\alpha L_0} \tag{4.6.7}$$

Denote as R_0 the data rate corresponding to operation at fiber length $L = L_0$. That is, R_0 is the value of R_b in Equation (4.6.5) when L is replaced by L_0. We can then rewrite the general fiber rate equations as

$$R = \begin{cases} R_0\left(\dfrac{L_0}{L}\right)^{1/2} & \text{for } L < L_0 \\ R_0 2^{-(L-L_0)\alpha/3} & \text{for } L \geqslant L_0 \end{cases} \tag{4.6.8}$$

where α is the fiber loss measured in dB/km. A plot of the parameter R/R_0 versus L/L_0 is shown in Figure 4.16 and serves as a universal set of rate curves upon which fiber design can be based. That is, for a given value of R_0 and L_0, the curves relate fiber length to achievable data rates. In the dispersion-limited region, $L < L_0$, the rate only decreases as the square root of the distance. However, when attenuation is limiting, $L > L_0$, every loss of 3 dB due to increased L is offset by a rate reduction of 2. Consequently, to optimize rate as a function of fiber length it is desirable to operate in the dispersion-limited region, while to minimize the overcapacity due to delay distortion we would like to operate fairly close to $L = L_0$.

4.7. Designing Fiber Systems

The design procedure in Section 4.6 can be extended to aid in the design of fiber systems—combinations of individual fiber links interconnected to achieve an overall desired data rate. Suppose a design value data rate R and length L, (R, L), is desired for a fiber system whose components (power levels, fiber, receiver, and encoding format) produce one of the curves of Figure 4.16, redrawn in Figure 4.17. If we consider (R, L) a point in the two-dimensional space of this figure, we immediately note the following. If (R, L) falls within region I, the configuration is adequate to support the operating point. In fact, the rate can be increased to the value obtained by projecting vertically to the curve. If (R, L) lies in region II, the link must use repeaters (Figure 4.18) in order to support (R, L), since L is too long for the fiber attenuation. The minimum number q of equally spaced repeaters is then the smallest value of q such that the point $(R, L/q)$ falls in region I. The system can then be configured as q repeater sections, each involving a fiber of length L/q and each supporting a data rate R. If (R, L) falls in region III, there is no way a single-fiber system with the stated components, or even a repeater version of the system, can support the point. Instead, a parallel system of q independent fibers of length L placed in parallel between transmitter and receiver (Figure 4.18b) must be used. Here, q is the smallest integer such that the point $(R/q, L)$ falls in region I. Similarly, it can be seen that operating points in region IV can only be supported by using q_1 parallel fibers, each with q_2 repeaters, where q_1 and q_2 are the smallest integers such that $(R/q_1, L/q_2)$ fall in region I.

It is evident from Figure 4.18 that the eventual shape of the design curve is important in specifying the design regions and, therefore, influencing system fabrication. The resulting shape will depend on the selected parameters and formats. However, its shape is also indirectly influenced by the choice of the fiber numerical aperture. As the acceptance angle is increased, the attenuation curve increases (since more power is launched) while the

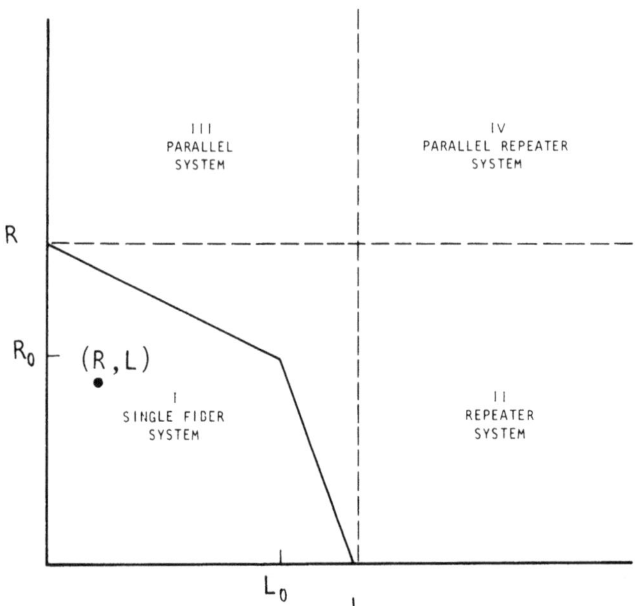

Figure 4.17. Rate-length diagrams for designing fiber systems.

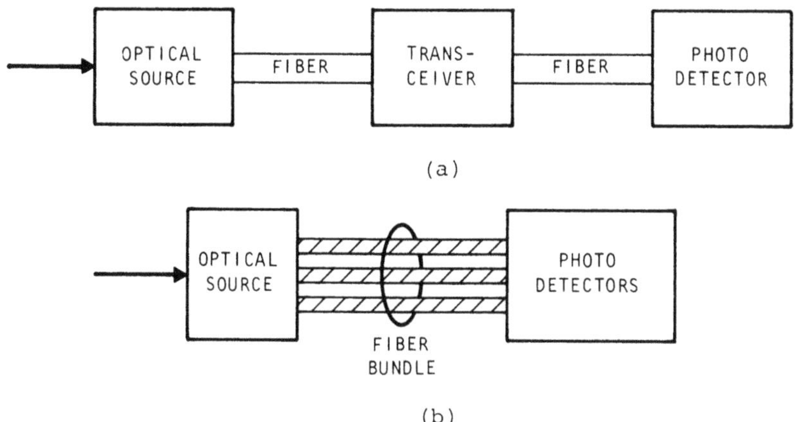

Figure 4.18. Fiber systems: (a) Serial connection with repeater; (b) parallel connection.

dispersion curve decreases, due to increased mode delay. The effect is to slide the L_0 point to a higher value of L, resulting in a squeezed-down but elongated rate characteristic in Figure 4.17. If the acceptance angle is decreased, the opposite effect occurs, and the rate characteristics is squeezed inward and upward in Figure 4.17. Thus, the numerical aperture allows a trade-off of launch power and pulse dispersion in such a manner so as to control the shape of the design curves of Figure 4.17. Whether it can be used to advantage depends on the design point (R, L) being implemented. That is, a desired operating point (R, L) may lie in region II, just to the right of region I, and would ordinarily require a repeater. By increasing the numerical aperture, it may be possible to extend region I so as to encompass the desired point, and only a single fiber is then required.

The geometrical rules above determine whether a given set of components will support the desired point (R, L) or whether modifications must be made. However, they do not allow a conclusion as to whether a point in region II, say, should be designed by repeater sections as stated, or instead by merely reselecting component parameters so as to rederive a curve in Figure 4.17 in which (R, L) lies in region I, whereby only one fiber is needed. The question involves a trade-off of multiple receiver cost versus additional fiber cost, and such queries can only be answered by taking into account component costs.

References

1. D. Marcuse, *Theory of Dielectric Optical Waveguide*, Academic Press, New York (1974).
2. C. Sandbank, *Optical Fiber Communication Systems*, John Wiley and Sons, New York (1980).

3. D. Gloge, Impulse response of clad optical multimode fibers, *Bell System Tech. J.* **52**, 801–816 (1973).
4. S. D. Personick, Baseband linearity and equalization in fiber optic digital communications, *Bell System Tech. J.* **52**, 1175–1194 (1973).
5. D. Messerschmitt, Optimum mean square equalization for digital fiber optic systems, in: International Conference on Communications, June, 1975, Vol. III, Conference Proceedings, Philadelphia.
6. S. D. Personick, Receiver design for digital fiber optic communications—Parts I and II, *Bell System Tech. J.* **52**, 843–875 (1973).
7. J. Garrett, Pulse position modulation for transmission over fibers, *IEEE Transactions Commun.* **COM-33**, 518–527 (1983).

5

The Turbulence Channel

In Chapter 4 we examined the guided optical channel or fiber link. In this chapter we consider the turbulent atmosphere as an unguided optical channel. It is well known that turbulence-induced random fluctuations in the atmosphere's temperature generate corresponding random irregularities in the index of refraction. Upon passing through these irregularities, the wavefronts associated with an optical beam become distorted, the magnitude of the distortions depending on the strength of the turbulence and the length of the atmospheric optical path. Among the effects which are attributable to wavefront distortion and which can seriously degrade the performance of an optical communication system are (1) spreading of the beam beyond that normally caused by diffraction, (2) scintillation of the received intensity, (3) a decrease in the spatial and temporal coherence, and (4) wander of the beam from position to position. Quantification of these effects requires a theoretical understanding of the relationship between the properties of the medium and the transmitted optical radiation.

There is a substantial body of literature covering this subject, and it is the intent of this chapter to distill from the myriad results presented there in an assembly of equations useful to the systems engineer. We will also provide only that theory which either provides a convenient vehicle for the introduction of terminology or is important for an understanding of underlying physical processes or certain necessary simplifying assumptions. Those readers who would benefit from a more theoretically detailed treatment can consult the comprehensive two-volume work by Ishimaru[1] or the original work of Tatarski.[2,3] The two excellent review articles by Fante[4,5] contain an exhaustive set of references. A variety of turbulence topics are covered in the collection edited by Strobehn.[6]

We will begin our description of the turbulence channel by first characterizing the physical properties of the atmosphere. The primary quantity of interest is the random index of refraction, which we will characterize statistically through the wavenumber spectrum of index of refraction fluctuations. Our efforts will then be directed toward describing the propagation of electromagnetic waves, beginning with Maxwell's equations, through turbulence in terms of this spectrum. These developments are logically and historically divided into two parts: single- and multiple-scattering theory. The single-scattering theory is analytically more tractable and, not surprisingly, was historically the first to be developed. The multiple-scattering theory focuses primarily on the moments of the field, including the mutual coherence function, and the probability density function for intensity fluctuations. While many of the useful results of single-scattering theory are relatively tractable closed-form analytical expressions, the same cannot be said of the multiple-scattering theory. Even though some asymptotic closed-form expressions are available, the complexity of the propagation integrals associated with the multiple-scattering regime usually dictates a numerical evaluation to obtain results. Although they are usually treated separately and differ in level of complexity, the single- and multiple-scattering theories are not totally disjoint. Indeed, it is found that the range of validity of the single-scatter treatment of phase fluctuations extends into the multiple-scatter regime.

5.1. Atmospheric Turbulence

In this section we will begin by describing the dependence of the atmospheric index of refraction on temperature and pressure. We will then show how the spatial fluctuations of index of refraction that are generated by turbulence-transported atmospheric temperature variations can be described in terms of structure functions and the wavenumber spectrum of index of refraction fluctuations, $\Phi_n(k)$. Models for $\Phi_n(k)$ will be presented, and it will be found that the strength of the turbulence can be characterized by a single structure parameter, $C_{n(z)}$. We will then derive the relationship between $\Phi_n(k)$ and the volume scattering function, $\beta(k)$, subsequently explaining the effects of $\Phi_n(k)$ in terms of Bragg scattering. Finally, we will present several models for the altitude dependence of the structure parameter, $C_n(z)$.

5.1.1. Atmospheric Turbulence Models

The relationship between the index of refraction of the atmosphere and temperature is given approximately by[6]

$$n - 1 = 77.6\frac{P}{T}[1 + 7.52 \times 10^{-3}\,\lambda^{-2}] \times 10^{-6} \tag{5.1.1}$$

where P = pressure in millibars, T = temperature in degrees Kelvin, and λ = light wavelength in microns. For most engineering applications we can write

$$n - 1 = 7.8 \times 10^{-5} P/T \tag{5.1.2}$$

The rate of change in index with temperature is thus given by

$$-dn/dt = 7.8 \times 10^{-5} P/T^2 \tag{5.1.3}$$

Near sea level, $-dn/dt \cong 10^{-6}$.

Random fluctuations in the atmospheric index of refraction are induced by corresponding variations in the atmospheric temperature that are transported by naturally occurring random fluctuation in wind velocity called turbulence. We can express the index of refraction as the sum of its free-space value plus a random fluctuating component due to the presence of turbulence:

$$n(\mathbf{r}, t) = 1 + n_1(\mathbf{r}, t) \tag{5.1.4}$$

where $n_1(\mathbf{r}, t)$ is the fluctuating component and is a function of position $\mathbf{r}$ within the atmosphere and time t. We will assume that the temporal dependence of the index of refraction is due primarily to atmospheric winds, so that

$$n(\mathbf{r}, t) = 1 + n_1(\mathbf{r} - vt) \tag{5.1.5}$$

where $v(\mathbf{r})$ is the component of the local wind velocity perpendicular to the line of sight. This assumption is known as "Taylor's frozen flow" hypothesis and assumes that the shape of the turbulent field of index fluctuations moves in its frozen form with the mean local wind. Clearly this is not strictly correct since we would expect the fluctuations to change as they move, much as the shape of clouds change as they move. However, these changes are very slow compared to the rate of movement past a stationary observer. Consequently, Taylor's hypothesis is believed to be valid for most practical applications. Note that only the component of $v(\mathbf{r})$ perpendicular to the line of sight causes temporal changes in the received light field.

Because turbulence is a random process, it must be described in terms of statistical quantities. In particular, we will require, in the sections that follow, the structure function[8]

$$D_{n_1}(\mathbf{r}_1, \mathbf{r}_2) = E[\{n_1(\mathbf{r}_1) - n_1(\mathbf{r}_2)\}^2] \tag{5.1.6}$$

and the correlation function of the index of refraction fluctuations

$$\Gamma_{n_1}(\mathbf{r}_1, \mathbf{r}_2) = E[n_1(\mathbf{r}_1)n_1(\mathbf{r}_2)] \tag{5.1.7}$$

The $E[\ ]$ notation denotes an average over the medium statistical ensemble. It is convenient to define new variables

$$\mathbf{R} = (\mathbf{r}_1 + \mathbf{r}_2)$$
$$\boldsymbol{\rho} = (\mathbf{r}_1 - \mathbf{r}_2) \tag{5.1.8}$$

The structure and correlation functions can then be written

$$\Gamma_{n_1}(\mathbf{r}_1, \mathbf{r}_2) = \Gamma_{n_1}(\mathbf{R}, \boldsymbol{\rho})$$
$$D_{n_1}(\mathbf{r}_1, \mathbf{r}_2) = D_{n_1}(\mathbf{R}, \boldsymbol{\rho}) \tag{5.1.9}$$

if the medium is homogeneous, the dependence on $\mathbf{R}$ can be suppressed and we have

$$\Gamma_{n_1}(\mathbf{r}_1, \mathbf{r}_2) = \Gamma_{n_1}(\boldsymbol{\rho})$$
$$D_{n_1}(\mathbf{r}_1, \mathbf{r}_2) = D_{n_1}(\boldsymbol{\rho}) \tag{5.1.10}$$

It can be shown that the structure and correlation functions are related by

$$D_{n_1}(\boldsymbol{\rho}) = 2[\mathrm{Var}\,(n_1) - \Gamma_{n_1}(\mathbf{R}, \boldsymbol{\rho})] \tag{5.1.11}$$

where Var (n_1) is the variance of n_1 and we have used $E[n_1] = 0$.

Using the Wiener-Khintchine theorem, we define the wavenumber spectrum of the index of refraction fluctuation $\Phi_n(\mathbf{K})$:

$$\Phi_n(\mathbf{K}) = 2\pi^{-3} \int_{-\infty}^{\infty} \Gamma_{n_1}(\boldsymbol{\rho}) \exp\{i\mathbf{K} \cdot \boldsymbol{\rho}\}\, d^3\boldsymbol{\rho} \tag{5.1.12}$$

Using Equations (5.1.11) and (5.1.12), we obtain the useful relationship

$$D_{n_1}(\boldsymbol{\rho}) = 2\pi^{-3} \int_{-\infty}^{\infty} \Phi_{n_1}(\mathbf{K})(1 - \exp\{i\mathbf{K} \cdot \boldsymbol{\rho}\})\, d^3\boldsymbol{\rho} \tag{5.1.13}$$

which allow one to calculate the index structure function given the wavenumber spectrum of n_1.

Historically, three different models for the function $\Phi_n(\mathbf{K})$ and $\Gamma_{n_1}(\boldsymbol{\rho})$ have been used. In early work, the Booker-Gordon and Gaussian models were used extensively since their simple form eased computational complexity. However, neither model described the characteristics of turbulence with sufficient accuracy to yield a complete description of all propagation phenenomena. We include them here only as a historical introduction to the Kolmogorov spectrum, which, because it was derived from considerations of the physical processes involved in fully developed turbulence, has become the standard statistical description of atmospheric turbulence.

The Booker-Gordon formula assumes that the function $\Gamma_{n_1}(\boldsymbol{\rho})$ is given by the exponential function[1]

$$\Gamma_{n_1}(\boldsymbol{\rho}) = \text{Var}\,(n_1)\exp\{-\boldsymbol{\rho}/l\} \tag{5.1.14}$$

The parameter l is referred to as the scale of the turbulence since it gives the $\exp\{-l\}$ correlation distance of the index of refraction fluctuations. Roughly speaking, l is the average size of the turbulent eddies. The Gaussian correlation function[1]

$$\Gamma_{n_1}(\boldsymbol{\rho}) = \text{Var}\,(n_1)\exp\{-(\boldsymbol{\rho}/l)^2\} \tag{5.1.15}$$

is usually assumed to be a better approximation of the behavior of a random medium.

The Kolmogorov theory, proposed in 1941, is based on the hypothesis that the kinetic energy associated with the larger eddies is redistributed without loss to eddies of decreasing size until it is finally dissipated by viscosity. The turbulent eddies are characterized by an outer scale L_0 and an inner scale l_0. Typically, l_0 is on the order of millimeters while L_0 can be a few meters. The turbulence is further categorized into three ranges which are related to the scale sizes L_0 and l_0. The input range is characterized by an eddy size greater than L_0. Here turbulent energy is introduced by wind shear and temperature gradients. For this range there is no single formula describing the turbulence. In the inertial subrange the eddy size is greater than l_0, but less than L_0. The turbulence is isotropic and the kinetic energy of the eddies overcomes the dissipation due to viscosity. The dissipation range is characterized by an eddy size less than l_0. In this case, energy loss from the eddies due to viscosity dominates and the spectrum is small.

Over the inertial subrange, the temperature structure function obeys an isotropic two-thirds power law[1]:

$$D_T(\boldsymbol{\rho})C_T^2 p^{2/3} \tag{5.1.16}$$

where C_T^2 is called the temperature structure parameter.

Equations (5.1.1) and (5.1.16) imply the following form for the index of refraction structure function

$$D_n(\boldsymbol{\rho}) = C_n^2 \boldsymbol{\rho}^{2/3} \tag{5.1.17}$$

where C_n^2, the index of refraction structure parameter, and C_T^2 are related by

$$C_n^2 = (dn/dt)^2 C_T^2 \tag{5.1.18}$$

Similarly,

$$\Phi_n(\mathbf{K}) = (dn/dt)^2 \Phi_T(\mathbf{K}) \tag{5.1.19}$$

It can be shown that the form of Equation (5.1.17) leads to the following expression for the Kolmogorov spectrum of the index of refraction fluctuations[1]:

$$\Phi_n(\mathbf{K}) = 0.033 C_n^2 k^{-11/3} \tag{5.1.20}$$

This spectrum is shown in Figure 5.1.

For some applications, it is necessary to know the form of the index fluctuation spectrum beyond the inertial subrange. While our present knowledge does not allow an exact specification of the spectrum for $L_0^{-1} \leq k^{-1} \leq l_0^{-1}$, certain modifications of Equation (5.1.20) have been proposed to effect a smooth transition to the input and dissipation ranges. For the dissipation range, it is clear that the spectrum must fall off more rapidly than $k^{-11/3}$.

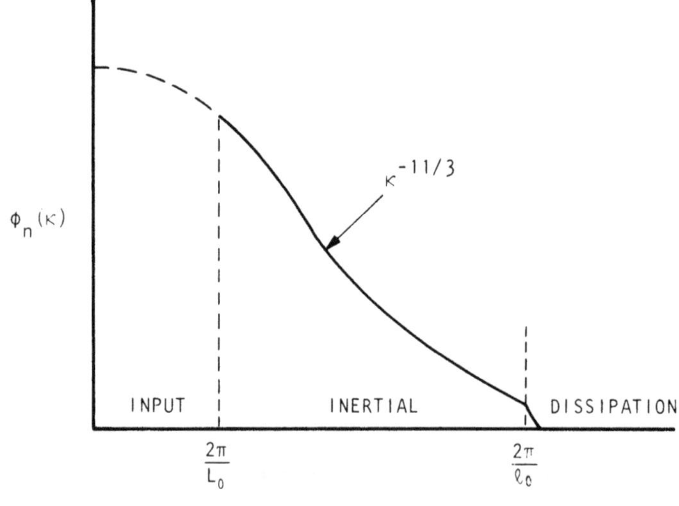

Figure 5.1. Kolmogorov spectrum.

Tatarski has proposed the following modification to Equation (5.1.20) which accomplishes this:

$$\Phi_n(\mathbf{K}) = 0.033C_n^2k^{-11/3}\exp\{-(k/k_m)^2\} \tag{5.1.21}$$

where $k_m = 5.92/l_0$. For the input range, the energy in the eddies larger than l_0 must be less than predicted by the Kolmogorov model. This can be approximated by a modified form of the Von Karman spectrum:

$$\Phi_n(\mathbf{K}) = 0.033C_n^2(k_l^2 + k^2)^{-11/6} \tag{5.1.22}$$

with $K_L = L_0^{-1}$. These spectra can, for mathematical convenience, be combined in the following expression:

$$\Phi_n(\mathbf{K}) = 0.033C_n^2k^{-11/3}\exp\{-(k/k_m)^2\}(k_l^2 + k^2)^{-11/6} \tag{5.1.23}$$

Again, we caution that this expression is only indicative of the behavior of the spectrum outside the inertial subrange. If the propagation phenomenon of interest depends strongly on L_0 of l_0, then one can assume that our present knowledge of the atmosphere will not allow its accurate quantification.

5.1.2. The Atmospheric Scattering Cross Section

The scattering cross section per unit volume, $\beta(\mathbf{o}, \mathbf{i})$, is defined through the equation

$$dE(\mathbf{r}) = H(\mathbf{r}')\beta(\mathbf{o}, \mathbf{i})\, dv'\, d\Omega \tag{5.1.24}$$

where, as show in Figure 5.2, $E(\mathbf{r})$ is the energy scattered in the far

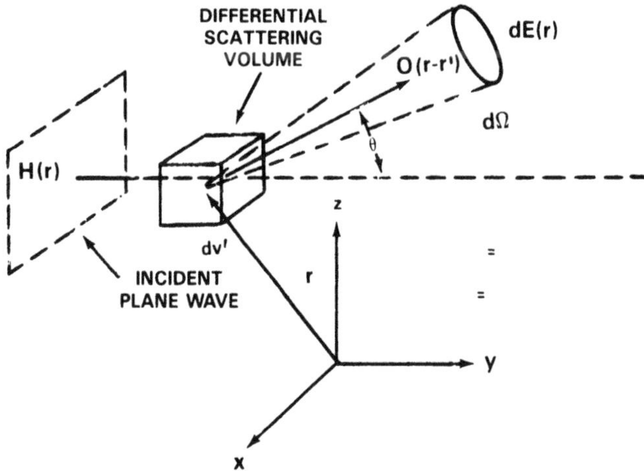

Figure 5.2. Illustrating the volume scattering function (scattering cross section per unit volume).

(Fraunhoffer) field in the direction $o(\mathbf{r}, \mathbf{r}')$ of the area located at $\mathbf{r}$ by the differential volume element dv' located at $\mathbf{r}'$, $H(\mathbf{r}')$ is the plane-wave irradiance, or energy per unit area, incident on the volume dv' in the direction $\mathbf{i}$, and $d\Omega$ is the solid angle subtended by dA at $\mathbf{r}'$, that is,

$$d\Omega = dA/|\mathbf{r} - \mathbf{r}'|^2 \tag{5.1.25}$$

To derive an expression which relates the volume scattering function to the random fluctuations in the atmospheric index of refraction, we first calculate the scattering amplitude in the Fraunhoffer diffraction zone of the scattering volume. We begin with the equation for the electromagnetic field in a random medium

$$\nabla^2 E(\mathbf{r}) + k^2[1 + n_1(\mathbf{r})]E(\mathbf{r}) = 0 \tag{5.1.26}$$

The solution of Equation (5.1.26) for the scattered field is

$$E_s(\mathbf{r}) = (k^2/2\pi)\sin(\chi)\int_{-\infty}^{\infty} G(\mathbf{r}, \mathbf{r}')n_1(\mathbf{r}')E(\mathbf{r}')\, d^3\mathbf{r}' \tag{5.1.27}$$

where $G(\mathbf{r}, \mathbf{r}')$ is the free-space Green's function. In the Born, or single-scattering, approximation the field at any point in the medium is given by the incident field at that point

$$E(\mathbf{r}') = E_0(\mathbf{r}') = H^{1/2}\exp\{i\mathbf{k}_i \cdot \mathbf{r}'\} \tag{5.1.28}$$

We are, therefore, neglecting the effects of the higher-order, multiply scattered fields. This is not a severe restriction for the present development, since we are concerned with the field emanating from a differential volume element where multiple scattering cannot play a significant role. In addition, since we are interested in the far-field approximation, we use the Fraunhoffer approximation to the free-space Green's function:

$$G(\mathbf{r}, \mathbf{r}') = \left(\frac{1}{|\mathbf{r} - \mathbf{r}'|}\right)\exp\{i\mathbf{k}_s \cdot \mathbf{r}'\} \tag{5.1.29}$$

where $\mathbf{k}_s = k\mathbf{s}$, with $\mathbf{s}$ a unit vector pointing from $\mathbf{r}'$ to $\mathbf{r}$ and $k = 2\pi/\lambda$. Substituting Equations (5.1.29) and (5.1.28) in Equation (5.1.27) we find

$$E_s(\mathbf{r}) = \left(\frac{H^{1/2}k^2\sin(X)}{2\pi}\right)\int_{-\infty}^{\infty} n_1(\mathbf{r}')\left(\frac{1}{|\mathbf{r} - \mathbf{r}'|}\right)\exp\{i\mathbf{K} \cdot \mathbf{r}'\}\, d^3\mathbf{r}' \tag{5.1.30a}$$

the vector **K** being given by

$$\mathbf{K} = \mathbf{k}_s - \mathbf{k}_i \tag{5.1.30b}$$

with modulus $K = 2k \sin \theta/2$ where θ is the angle between **i** and **s**. The irradiance, $H(\mathbf{r})$, at the point **r** is usually assumed to be equal to the mean square value of the electric field:

$$H(\mathbf{r}) = R[E(\mathbf{r}')E(\mathbf{r}'')]$$

$$= \left(\frac{Hk^4 \sin^2(X)}{16\pi^2}\right) \iint_{-\infty}^{\infty} \Gamma_{n_1}(\mathbf{R}, \boldsymbol{\rho})\left(\frac{1}{|\mathbf{r}-\mathbf{r}'||\mathbf{r}-\mathbf{r}''|}\right) \exp\{i\mathbf{K}_s \cdot \boldsymbol{\rho}\}\, d^3\mathbf{r}'\, d^3\mathbf{r}'' \tag{5.1.31}$$

where we have used Equation (5.1.7) and introduced the variables $(\mathbf{R}, \mathbf{p})$ defined by Equation (5.1.8). Using Equation (5.1.12), Equation (5.1.31) becomes

$$H(\mathbf{r}) = 2\pi Hk^2 \sin^2(X) \iint_{-\infty}^{\infty} \Phi_n(\mathbf{R}, \mathbf{K}_s)\left(\frac{1}{|\mathbf{r}-\mathbf{R}|^2}\right) d^3\mathbf{R} \tag{5.1.32}$$

where we have used Equation (5.1.8) and, as is customary in the Fraunhoffer diffraction zone, we have approximated $|\mathbf{r}-\mathbf{r}'||\mathbf{r}-\mathbf{r}''|$ by $|\mathbf{r}-\mathbf{R}|^2$ in the denominator. Writing Equation (5.1.32) in differential form we have

$$dH(\mathbf{r}) = 2\pi Hk^4 \sin^2(X)\Phi_n(\mathbf{R}, \mathbf{K}_s)\left(\frac{1}{|\mathbf{r}-\mathbf{R}|^2}\right) d^3\mathbf{R} \tag{5.1.33}$$

The differential energy, $dE(\mathbf{r})$, scattered in the direction of the differential area dA located at **r** from the volume dv' is, therefore,

$$dE(\mathbf{r}) = dH(\mathbf{r})\, DA$$

$$= 2\pi Hk^4 \sin^2(X)\Phi_n(\mathbf{R}, \mathbf{K}_s)\, d^3\mathbf{R}\, d\Omega \tag{5.1.34}$$

where

$$d\Omega = dA/|\mathbf{r}-\mathbf{R}|^2 \tag{5.1.35}$$

Comparing Equations (5.1.34) and (5.1.24) and noting that $dv = d^3\mathbf{R}$, we have

$$\beta(\mathbf{K}_s, \mathbf{R}) = 2\pi k^4 \sin^2(X)\Phi_n(\mathbf{R}, \mathbf{K}_s) \tag{5.1.36}$$

for the effective volume scattering function.

From Equation (5.1.36), it is apparent that the power scattered per steradian in the direction θ is determined by one spectral component of the function $\Phi_n(\mathbf{R}, \mathbf{k}_s)$ which corresponds to the spatial period

$$l = 2\pi/\mathbf{k}_s = \lambda/2 \sin(\theta/2) \tag{5.1.37}$$

Equation (5.1.37) is the well-known Bragg law for diffraction from objects which exhibit periodic spatial structure. Further insight into the physical significance of the function $\Phi_n(\mathbf{R}, \mathbf{k}_s)$ is provided by considering the special case in which a plane harmonic wave of amplitude A_o and wavevector k_i impinges on the spatial sinusoidal diffraction grating:

$$n_1(\mathbf{r}) = 1 + d \cos(\mathbf{k}_0 \cdot \mathbf{r}) \tag{5.1.38}$$

where

$$\mathbf{k}_0 = 2\pi\mathbf{a}/l \tag{5.1.39}$$

with l the spatial period in the direction $\mathbf{a}$. Substituting Equation (5.1.38) in Equation (5.1.30a), we have

$$E_s(\mathbf{r}) = \left[\frac{Ak^2 \sin(X)}{2\pi}\right] \int_{-\infty}^{\infty} \exp\{i\mathbf{k}_s \cdot \mathbf{r}\}[1 + d \cos(\mathbf{k}_0 \cdot \mathbf{r})]\, d^3\mathbf{r} \tag{5.1.40}$$

In computing Equation (5.1.40) we have neglected the finite size and shape of the boundary of the scattering medium by extending the limits of integration to infinity. The first term in Equation (5.1.40) is a forward scatter peak while the second and third terms represent the diffracted wavefronts. Note the there is a diffracted field when two conditions are met. First, if the vectors $\mathbf{k}_0$ and $\mathbf{k}_s$ are parallel, there is a specular term reflected from planes of constant $n_1(\mathbf{r})$ such that the incident and reflection angles are equal. The second condition is when

$$|\mathbf{k}_d| = |\mathbf{k}_s| \tag{5.1.41}$$

which implies

$$4\pi \sin(\theta/2)/\lambda = 2\pi/l \tag{5.1.42}$$

or

$$l = \lambda / 2 \sin(\theta/2) \qquad (5.1.43)$$

which, again, is Bragg's law. Thus, diffraction will be observed from the spatial sinusoidal diffraction grating when the conditions for specular reflection and Bragg's law are satisfied. The wavenumber spectrum of the index of refraction fluctuations can be considered to be a decomposition of the index of refraction fluctuations within the medium into a sum of sinusoidal spatial diffraction gratings of various periods and spatial orientations. A plane wave incident on a medium of infinite extent in the direction $\mathbf{i}$ is scattered in the direction $\mathbf{o}$ only by the particular grating which has its wavevector $\mathbf{k}_0$ parallel to $\mathbf{k}_s$. The remaining gratings have no effect on the scattering in the direction $\mathbf{o}$. If the medium is of finite extent with the characteristic linear dimension l, then each diffracted beam will have an angular extent on the order of λ / l due to diffraction by the boundaries of the medium. Thus, there will always be some overlap in the diffracted beams created by different spectral components of the function $\Phi_n(\mathbf{R}, \mathbf{k}_s)$. Consequently, scattering in a given direction is determined not only by the spatial diffraction gratings for which Bragg's condition is satisfied exactly but also by other gratings with different $\mathbf{k}_0$ values whose diffracted beams overlap due to their finite angular extent.

5.1.3. Models for the Wavenumber Spectrum Structure Parameter

In the peformance of systems calculations it is often necessary to integrate C_n^2 over the propagation path. It is usually assumed that C_n^2 is constant over planes of constant altitude; i.e., horizontal propagation. While, in general, C_n^2 and C_T^2 are complex functions of time of day, local winds, solar elevation angle, and type of terrain, for most applications it suffices to have an analytic relationship between C_n^2 and altitude. The atmospheric column can be divided into two parts, the free atmosphere which lies above the first temperature inversion layer and the atmosphere near the ground. The inversion layer may be over 1 km in altitude on a hot sunny afternoon to only a few tens of meters in the early morning.

Near the ground, thermal energy imparted to the air by the warm terrain causes convection currents which result in values of C_n^2 which are an order of magnitude higher than those in the free atmosphere. Furthermore, at the first inversion there is often a peak in the strength of C_n^2 which is an order of magnitude higher than that below the inversion, as shown in Figure 5.3. At night, when there is radiative cooling of the earth, a strong inversion layer develops at the ground and extends for several tens of meters.

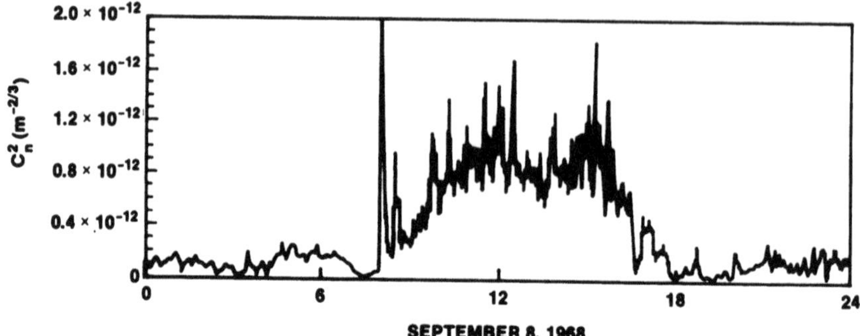

Figure 5.3. Refractive index structure function parameter versus time of day. The measurements were derived from temperature structure function measurements made with vertical spacings of 0.01 m at an elevation of 2 m.[9]

Convection is inhibited since this layer of cold air, created by heat conduction to the ground, is less dense than the warm air above. In this region C_n^2 may be assumed to be constant. A typical value is $2 \times 10^{-14}\ \mathrm{m}^{-2/3}$.

In the free atmosphere, the turbulence is even more complex. Figure 5.4 shows a measured profile of C_n^2 versus altitude. The considerable

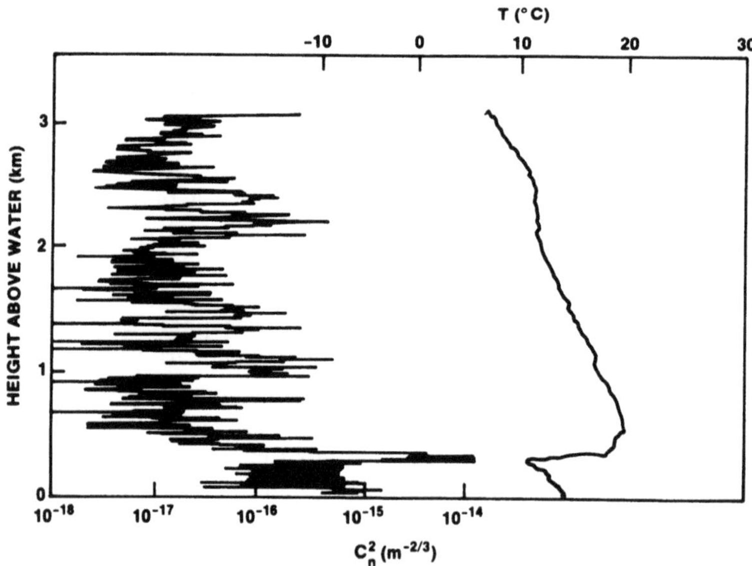

Figure 5.4. Refractive index structure function versus altitude for the first 3000 meters over the ocean. The local air temperature is shown on the right, with the scale in the upper right-hand corner.[9]

structure of greatly fluctuating values of C_n^2 is often associated with temperature inversions, strong wind shears, or atmospheric internal waves. We will present five relatively simple models for the altitude dependence of C_n^2. While these models do not reproduce the fine structure shown in Figure 5.4 nor the temporal dependence of C_n^2 shown in Figure 5.3, they may be regarded as giving values averaged over all realizations from the atmospheric statistical ensemble.

One of the earliest models for C_n^2 was given by Fried[7] who fitted an analytic function to data given by Hufnagel and Stanley[8]:

$$C_n^2(z) = K_0 z^{-1/3} \exp\{-z/z_0\} \tag{5.1.44a}$$

where K_0 is the turbulence strength parameter with units of meters$^{-2/3}$, and z is the altitude in meters. Typical values of K_0 for strong, moderately strong, and moderate turbulence are:

$$K_0 = \begin{cases} 6.7 \times 10^{-14} & \text{strong} \\ 8.0 \times 10^{-15} & \text{moderately strong} \\ 1.6 \times 10^{-15} & \text{moderate} \end{cases} \tag{5.1.44b}$$

In this prescription, $z_0 = 3200$ and may be regarded as the effective height of the turbulent atmosphere.

A very simple model given by Hufnagel and Stanley[11] is:

$$C_n^{-2}(z) = \begin{cases} 1.5 \times 10^{-13} z^{-1} & \text{for } z < 20\ \text{km} \\ 0 & z > 20\ \text{km} \end{cases} \tag{5.1.45}$$

Its simplicity eases some of the more complex path integrals which occur in propagation problems. Here, z is the altitude above local ground, assumed to be less than 2500 m. A more complex model for $C_n^2(z)$ has also been proposed by Hufnagel[9] and is given by

$$\begin{aligned} C_n^2(z) = {} & 8.2 \times 10^{-56} V^2 z^{10} \exp\{-z/100\} \\ & + 2.7 \times 10^{-16} \exp\{-z/1500\} \end{aligned} \tag{5.1.46}$$

where z is the altitude above sea level, and V is the rms wind speed averaged over the 5- to 50-km altitude interval and can be set to 27 m/s for a fixed model. Hufnagel suggests that a random model can be generated by allowing V to be a Gaussian random variable with a mean of 27 m/s and a 9-m/s standard deviation. These models are shown in Figure 5.5a. Another model for $C_n^2(z)$ consists of separate day and night estimates. It is also shown in Figure 5.5a but is depicted in greater detail in Figure 5.5b, where the functional form of its various straight-line elements is provided.

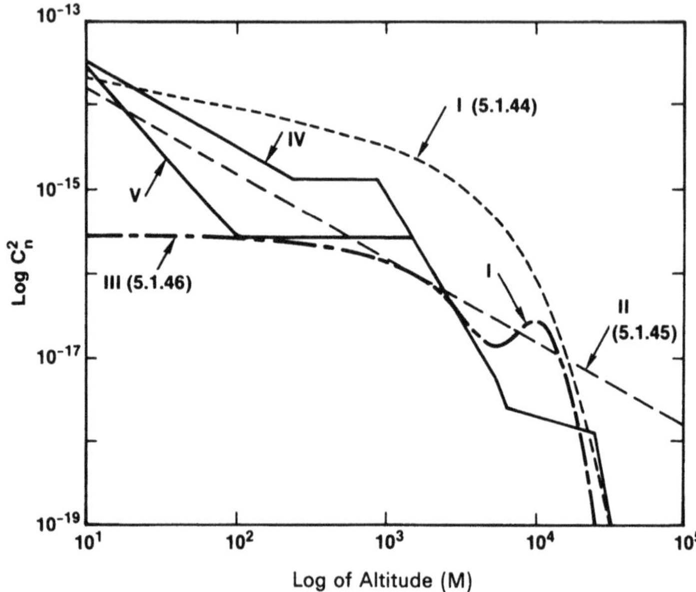

Figure 5.5a. Five models of the refractive index structure function as a function of altitude.

5.2. Optical Propagation through Turbulence

Having presented models for the index of refraction fluctuations, we will now proceed to obtain expressions for the scattered electric field under the assumption that the index fluctuations are weak. The early work in the theory of so-called "line of sight" propagation was developed by Tatarski[2,3] based on the Rytov approximation. In this section we will, beginning with Maxwell's wave equation, outline the development of this method. Our goal is the development of expressions which describe scintillation, beam spreading, spatial coherence propagation, angle of arrival fluctuations, and heterodyne detection. To accomplish this task, we will require expressions for the spectra of log amplitude and phase fluctuations. From these we can calculate the amplitude and intensity variance and the wave structure function. Subsequently, the wave structure function can be used to calculate the mutual coherence function (MCF), angle-of-arrival fluctuations, and beam spreading and to describe the effects of turbulence on heterodyne detection. This sequence of calculations, which can appear somewhat confusing and entangled at first reading, is displayed schematically in Figure 5.6. We will also discuss the limitations of the Rytov approach. In particular, we will find that while the expressions for the amplitude statistics of the field do not apply in the multiple-scatter regime because they do not predict the observed saturation of scintillation, the phase statistics can remain valid

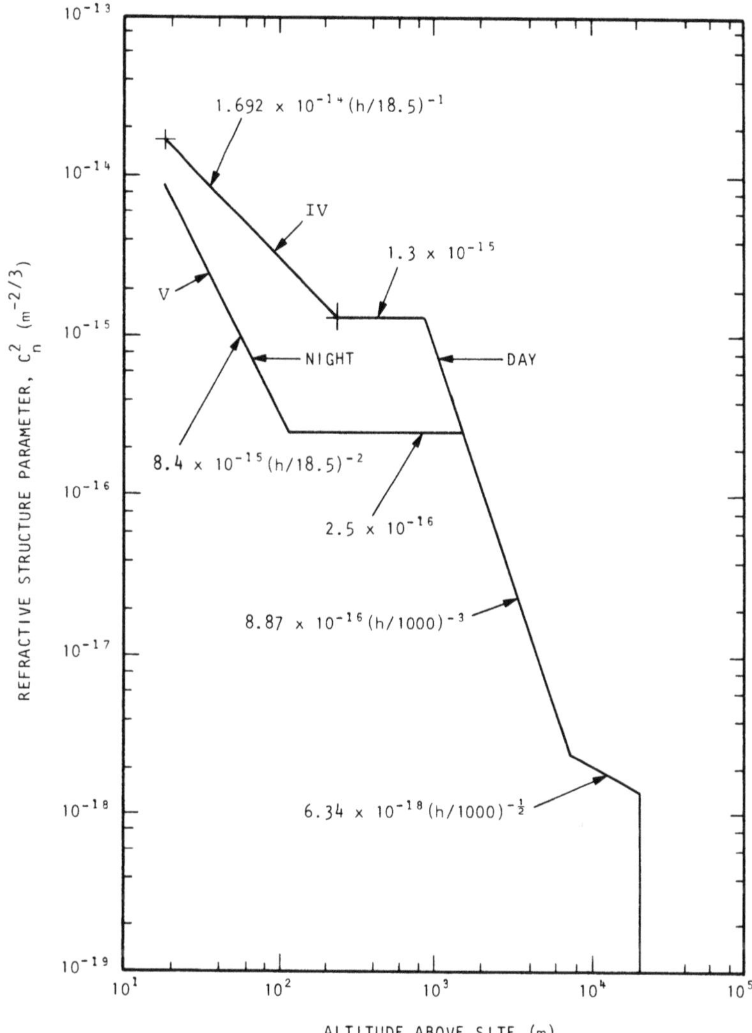

Figure 5.5b. Daytime and nighttime models for refractive index structure function versus altitude.

even in the presence of saturation. Our treatment of multiple-scattering theory will be presented during our discussion of the MCF where the parabolic differential equation for propagation of the MCF will be solved under the Markov approximation. Unfortunately, the resulting expressions cannot be evaluated for the Kolmogorov spectrum in terms of known mathematical functions; asymptotic approximations and numerical techniques are required.

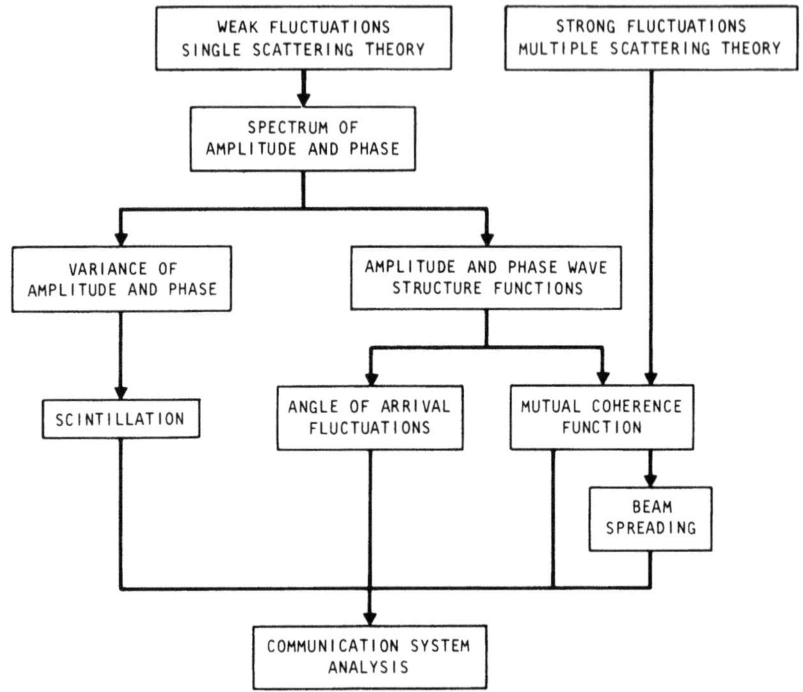

Figure 5.6. Illustrating sequence of theoretical developments.

5.2.1. The Rytov Method

The electric field of a narrowband field propagating in a random medium is described by the Maxwell wave equation[1,4]

$$\nabla^2 f(\mathbf{r}) + k^2[(1 + n_1(\mathbf{r})]^2 f(\mathbf{r}) - \nabla \cdot [\nabla \cdot f(\mathbf{r})] = 0 \tag{5.2.1}$$

with $k = 2\pi/\lambda$, and $\nabla = (d/dx)\mathbf{i} + (d/dy)\mathbf{j} + (d/dz)\mathbf{k}$, with $\mathbf{i}$, $\mathbf{j}$ and $\mathbf{k}$ unit vectors along the x, y, and z axes, respectively, is the gradient operator. The last term in Equation (5.2.1) can be shown to represent depolarization effects and is assumed to be negligible for optical waves in the atmosphere. It is, therefore, permissible to approximate Equation (5.2.1) by

$$\nabla^2 f(\mathbf{r}) + k^2[1 + n_1(\mathbf{r})]^2 f(\mathbf{r}) = 0 \tag{5.2.2}$$

To obtain the Rytov solution to Equation (5.2.2), we begin by writing the electric field in the form

$$f(\mathbf{r}) = \exp\{\Phi(\mathbf{r})\} \tag{5.2.3}$$

Substituting Equation (5.2.3) in Equation (5.2.2) we get the differential equation

$$\nabla^2\Phi + (\nabla\Phi)^2 + k^2(1 + n_1)^2 = 0 \tag{5.2.4}$$

We can also write the electric field in the form

$$f(\mathbf{r}) = f_0(\mathbf{r})T(\mathbf{r}) = \exp\{\Phi_0 + \Phi_1\} \tag{5.2.5}$$

where

$$\begin{aligned} T(\mathbf{r}) &= \exp\{\Phi_1\} \\ f_0(\mathbf{r}) &= \exp\{\Phi_0\} \end{aligned} \tag{5.2.6}$$

and $\Phi(\mathbf{r})$ satisfies the vacuum equation

$$\nabla^2\Phi_0 + (\nabla\Phi_0)^2 + k^2 = 0 \tag{5.2.7}$$

Note that Equation (5.2.5) implies that the field at any point in the medium can be written as the product of the free-space field and the complex amplitude transmittance $T(\mathbf{r})$. Thus, $\Phi_1(\mathbf{r})$ is a measure of the extent to which the wave field departs from its unperturbed free-space form. In particular, if we write

$$\begin{aligned} f(\mathbf{r}) &= A(\mathbf{r})\exp\{i\phi(\cdot r)\} \\ f_0(\mathbf{r}) &= A_0(\mathbf{r})\exp\{i\phi_0(\mathbf{r})\} \end{aligned} \tag{5.2.8}$$

then Equations (5.2.5) and (5.2.8) give

$$\Phi_1 = \log\{A(\mathbf{r})/A_0(\mathbf{r})\} + i[\varnothing(\mathbf{r}) - \phi_0(\mathbf{r})] = X + iS \tag{5.2.9}$$

The real part of Φ_1, X, is called the log-amplitude fluctuation and the imaginary part, S, is the phase fluctuation.

Substituting Equation (5.2.5) in Equation (5.2.2), assuming that $dn_1 \gg n_1^2$ and neglecting $|\nabla\Phi_1|$ in comparison to $|\nabla\Phi_0|$ we have

$$\nabla^2\Phi_1 + 2\nabla\Phi_0 \cdot \nabla\Phi_1 + 2k^2 n_1(\mathbf{r}) = 0 \tag{5.2.10}$$

which is the nonlinear Ricatti differential equation whose solution is[4]

$$\Phi_1(\mathbf{r}) = \left(\frac{k^2}{2\pi}\right)\int_{-\infty}^{\infty} \frac{n_1(\mathbf{r}')f_0(\mathbf{r}')\exp\{ik|\mathbf{r}-\mathbf{r}'|\}}{f_0(\mathbf{r})|\mathbf{r}-\mathbf{r}'|}\, d^3\mathbf{r}' \tag{5.2.11}$$

Equation (5.2.11) serves as the basis for our development of the spectra of amplitude and phase flctuations to be presented in Section 5.2.2.

5.2.2. The Spectrum of Log-Amplitude and Phase Fluctuations

To quantify turbulence-induced scintillation we will need expressions for the log-amplitude and phase spectra. We recall from Section 5.2.1 that

$$\begin{vmatrix} X \\ S \end{vmatrix} = \begin{vmatrix} \widetilde{\operatorname{Re}\Phi_1} \\ \widetilde{\operatorname{Im}\Phi_1} \end{vmatrix}$$

The spectra of log amplitude and phase can then be written compactly in terms of the random variables X and S:

$$\begin{bmatrix} \phi_X(\mathbf{k}) \\ \phi_S(\mathbf{k}) \end{bmatrix} = E\left[\begin{vmatrix} \tilde{X} \\ \tilde{S} \end{vmatrix}^2\right] = E\left[\begin{vmatrix} \widetilde{\operatorname{Re}\Phi_1} \\ \widetilde{\operatorname{Im}\Phi_1} \end{vmatrix}^2\right] \tag{5.2.12}$$

where $\sim$ denotes the two-dimensional spatial Fourier transform of a function. It is apparent from Equation (5.2.12) that we must find an expression for $\Phi_1(\mathbf{K}, z)$ whose functional form depends on the choice of the incident light field properties through the term $f_0(\mathbf{r}')$. Since we will be primarily concerned with the propagation of plane and spherical waves, it is convenient to begin with the spherical-wave case and later show how those results can be reduced to the plane-wave case. Furthermore, we will assume that the Fresnel diffraction approximation is valid. Under these conditions the spherical-wave field can be written in the quadratic form

$$f_0(\mathbf{r}) = \exp\{ik[x + (p^2/2x)]\}/x \tag{5.2.13}$$

and

$$\frac{\exp\{ik|\mathbf{r} - \mathbf{r}'|\}}{|\mathbf{r} - \mathbf{r}'|} = \frac{\exp\{ik[(x - x') + (\mathbf{p} - \mathbf{p}')^2]\}}{2(x - x')} \tag{5.2.14}$$

where $\mathbf{p} = y\hat{\mathbf{j}} + z\hat{\mathbf{k}}$. Substituting Equations (5.2.14) and (5.2.13) in Equation (5.2.11) and taking the Fourier transform, we get

$$\Phi_1(\mathbf{k}, x) = \left(\frac{k^2}{2\pi}\right) \int_{-\infty}^{\infty} \int_{-\infty}^{\infty} \int_0^L \times \frac{n_1(\mathbf{p}, x') \exp\left\{\dfrac{ik(\beta\mathbf{p} - \mathbf{p}')^2}{2\beta(L - x')} - \mathbf{k} \cdot \mathbf{p}\right\}}{2\beta(L - x')} d^2\mathbf{p}\, d^2\mathbf{p}'\, dx' \tag{5.2.15}$$

where $\beta = x'/L$ and $\mathbf{k} = \mathbf{k}_y\hat{\mathbf{j}} + \mathbf{k}_g\hat{\mathbf{k}}$. It can be shown that for $\beta = 1$, Equation (5.2.15) gives $\Phi_1(K, x)$ for a plane wave; i.e., $E_0(\mathbf{r}) = \exp\{ikx\}$. Performing the $\mathbf{p}$ and $\mathbf{p}'$ integrations in Equation (5.2.15) we find

$$\Phi_1(\mathbf{k}, x) = ik \int_0^L \frac{\tilde{n}_1\left(\dfrac{\mathbf{k}}{\beta}, x'\right) \exp\left\{\dfrac{ik^2(L - x')}{2k\beta}\right\}}{\beta} dx' \tag{5.3.16}$$

Substituting Equation (5.2.16) in Equation (5.2.12) we have

$$\begin{vmatrix} \phi_X(\mathbf{k}) \\ \phi_S(\mathbf{k}) \end{vmatrix} = 2\pi k^2 \int_0^L \int_0^L \Phi_{n_1}\left(\frac{\mathbf{k}}{\beta}, x', x''\right) \times \frac{\begin{matrix}\sin\\ \cos\end{matrix}\left|\dfrac{k^2(L - x')}{2\beta k}\right|}{\beta^2} \frac{\begin{matrix}\sin\\ \cos\end{matrix}\left|\dfrac{k^2(L - x'')}{2\beta k}\right|}{\beta^2} dx'\, dx'' \tag{5.2.17}$$

where

$$\Phi_{n_1}\left(\frac{\mathbf{k}}{\beta}, x', x''\right) = E\left[n_1\left(\frac{\mathbf{k}}{\beta}, x'\right) n_1\left(\frac{\mathbf{k}}{\beta}, x''\right)\right]$$

To make any significant analytical progress, it is necessary to assume that the wavenumber spectrum takes the form

$$\Phi_{n_1}\left(\frac{\mathbf{k}}{\beta}, x', x''\right) = 2\pi\Phi_{n_1}\left(\frac{\mathbf{k}}{\beta}\right)\delta(x' - x'') \tag{5.2.18a}$$

It follows directly from Equation (5.2.18a) that the index of refraction correlation function is delta correlated in the x-direction:

$$B_n(\boldsymbol{\rho}, x' - x'') = B_n(\boldsymbol{\rho})\delta(x' - x'') \tag{5.2.18b}$$

This implies that the turbulent eddies look like disks with their short axes aligned parallel to the x-axis, the correlation of n_1 in the transverse direction affecting only the transverse correlation of the field.

Substituting Equation (5.2.18) in Equation (5.2.17), we achieve our final result:

$$\begin{vmatrix} \phi_X(\mathbf{k}) \\ \phi_S(\mathbf{k}) \end{vmatrix} = 2\pi k^2 \int_0^L \Phi_{n_1}\left(\frac{\mathbf{k}}{\beta, x', x''}\right) \frac{\begin{matrix}\sin^2\\ \cos^2\end{matrix}\left|\dfrac{k^2(L - x')}{2\beta k}\right|}{\beta^2} dx' \tag{5.2.19}$$

Equation (5.2.19) is an important link in our analysis since, as we will show in the following sections, it is required for calculation of the wave structure function.

5.2.3. The Log-Amplitude, Phase, and Wave Structure Functions

The log-amplitude, phase, and wave structure functions are defined by the expressions:

$$D_w(\boldsymbol{\rho}) = D_\phi(\boldsymbol{\rho}) + D_X(\boldsymbol{\rho}) \tag{5.2.20a}$$

$$D_\phi(\boldsymbol{\rho}) = E[\{\phi(\mathbf{r}_1) - \phi(\mathbf{r}_2)\}^2] \tag{5.2.20b}$$

$$D_X(\boldsymbol{\rho}) = E[\{X(\mathbf{r}_1) - X(\mathbf{r}_2)\}^2] \tag{5.2.20c}$$

We will use the wave structure function in Section 5.2.5 to describe wavefront tilting and in Section 5.2.4 to characterize spatial coherence. We obtain the expression for $D_w(\boldsymbol{\rho})$ by inserting Equation (5.2.20a) in Equation (5.1.13):

$$D_w(\boldsymbol{\rho}) = \int_{-\infty}^{\infty} [\Phi_\phi(\boldsymbol{\rho}) + \Phi_X(\boldsymbol{\rho})][1 - \exp\{i\mathbf{k}\cdot\boldsymbol{\rho}\}]\, d^3\mathbf{k} \tag{5.2.21}$$

It is usually assumed that for most engineering applications $D_w(\boldsymbol{\rho}) \approx D_\phi(\boldsymbol{\rho})$,[10] so we eliminate the $\Phi_X(\boldsymbol{\rho})$ term from Equation (5.2.21). Substituting Equation (5.2.19) in Equation (5.2.21) and performing the integration, we find for homogeneous isotropic turbulence obeying Kolmogorov's spectrum:

$$D_W(\boldsymbol{\rho}) = 2.914k^2\rho^{5/3} \int_0^L c_n(x)\, dx \tag{5.2.22}$$

for plane waves and

$$D_w(\boldsymbol{\rho}) = 2.914k^2\rho^{5/3} \int_0^L C_n^2(x)\left(\frac{x}{L}\right)^{5/3} dx \tag{5.2.23}$$

for spherical waves. When $C_n^2(x)$ is uniform along the optical path, the above expressions reduce to

$$D_w(\boldsymbol{\rho}) = 2.914k^2\rho^{5/3}C_n^2 L \tag{5.2.24}$$

for plane waves and

$$D_w(\boldsymbol{\rho}) = 1.093k^2\rho^{5/3}C_n^2 L \tag{5.2.25}$$

for spherical waves.

The physical significance of the wave structure function can readily be appreciated by considering a plane wave with a tilt angle of θ, shown in Figure 5.7. The phase difference between two points separated by a distance p is given by $\Delta\phi = kp\theta_T$ for small θ. The phase structure function is then seen to be proportional to $\Delta\phi^2 = (kp\theta_T)^2$. Since $d_w(\mathbf{p}) = d_\phi(\mathbf{p})$, it is clear that a pure tilt of the wavefront is associated with a p^2 dependence in $D_w(\mathbf{p})$. In particular, the wave structure function is given by

$$D_w(\mathbf{p}) = k^2\rho^2 E[\theta_T^2]$$

where $E[\theta_T^2]$ is the mean square tilt angle. The wave structure functions presented above, with their $p^{5/3}$ dependence, imply that the average wavefront distortion is nearly a pure tilt.

5.2.4. Spatial Coherence

In this section we will show how the effects of turbulence decrease the spatial coherence of an initially coherent optical field as it propagates through the atmosphere. He will begin by deriving an expression for the mutual coherence function in a turbulent medium. These results will be used to show how the signal-to-noise ratio of an optical heterodyne receiver, which was derived in Chapter 3, is in fact limited by turbulence-induced coherence losss.

The mutual coherence function, which is discussed fully in Chapter 2, was defined by the expression

$$\Gamma(\mathbf{r}_1, \mathbf{r}_2) = E[f(\mathbf{r}_1)f(\mathbf{r}_2)] \tag{5.2.26}$$

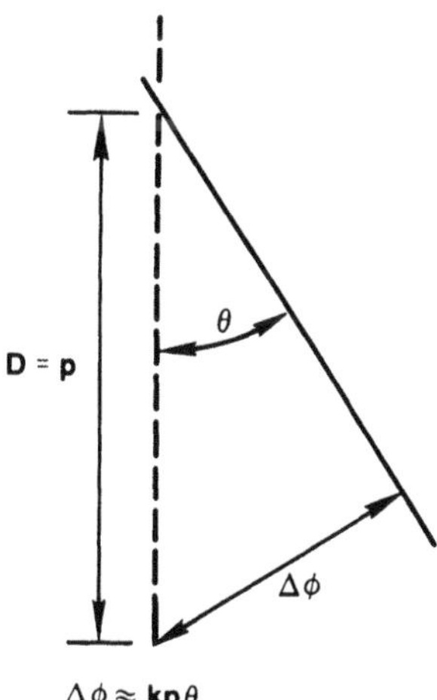

Figure 5.7. Illustrating wavefront tilt.

where $E[\ ]$ denotes, in this case, an average over the turbulent medium statistical ensemble. Writing $f(\mathbf{r})$ in the form of Equation (5.2.3), and using Equation (5.2.9) we have

$$\Gamma(\mathbf{r}_1, \mathbf{r}_2) = E[\exp\{\chi(\mathbf{r}_1) + \chi(\mathbf{r}_2) + i[S(\mathbf{r}_2) - S(\mathbf{r}_2)]\}] \tag{5.2.27}$$

To evaluate Equation (5.2.27) we will use the fact that χ and S are homogeneous, isotropic, independent Gaussian random variables. This allows us to use the equation

$$E[\exp\{ax + by\}] = \exp\{1/2(a^2\sigma_x^2 + b^2\sigma_x^2 + aE[x] + bE[y])\} \tag{5.2.28}$$

where $(x; y)$ are independent Gaussian random variables, $(a; b)$ are arbitrary constants, and σ_x^2 and σ_y^2 are, respectively, the variances of x and y. Applying Equation (5.2.28) to Equation (5.2.27) we have

$$\Gamma(\boldsymbol{\rho}) = \exp(1/2\{\sigma_\chi^2(\mathbf{r}_1, \mathbf{r}_2) + E[\chi] - D_\phi(\boldsymbol{\rho})\})\exp(1/2\{E[\chi(\mathbf{r}_1, \mathbf{r}_2)] - 2E[\chi^2] + 2E[\chi] - D_\phi(\boldsymbol{\rho})\}) \tag{5.2.29}$$

It is usually assumed that turbulence does not absorb energy, but only scatters it. Consequently, for infinite plane waves and spherical waves, it is usually assumed that $E[A] \approx A_0$. This implies that $E[\exp\{2\chi(\mathbf{r})\}] \approx 1$ and, using Equation (5.2.28), that $-\sigma_\chi^2 \approx E[\chi]$. This last equation can be written in the form

$$-E[\chi]^2 + E[\chi] = -E[\chi^2] \tag{5.2.30}$$

Substituting Equation (5.2.30) in Equation (5.2.29) and using the fact that $D_X(\boldsymbol{\rho}) = 2\{E[\chi^2] - E[\chi(\mathbf{r}_1)\chi(\mathbf{r}_2)]\}$, we achieve our final result:

$$\Gamma(\boldsymbol{\rho}) = \exp\{-1/2[D_\phi(\boldsymbol{\rho}) + D_X(\boldsymbol{\rho})]\} = \exp\{-1/2D_w(\boldsymbol{\rho})\} \tag{5.2.31}$$

This result will not hold for finite-diameter beams since turbulence will scatter energy out of the beam, $E[A] = A_0$, and the assumptions leading to Equation (5.2.31) are not valid. However, if it can safely be assumed that $D_w(\mathbf{p}) \approx D_\phi(\mathbf{p})$, then the range of validity of Equation (5.2.31) can be extended.

5.2.4.1. Spatial Coherence in Weak Turbulence

Using the weak-turbulence Rytov results, Equations (5.2.22) and (5.2.23), we can cast Equation (5.2.31) in the form:

$$\Gamma(\boldsymbol{\rho}) = A^2\exp\{-(\boldsymbol{\rho}/\boldsymbol{\rho}_0)^{5/3}\} \tag{5.2.32}$$

where

$$\boldsymbol{\rho}_0 = \left[1.45k^2 \int_0^L C_n^2(x)\,dx\right]^{-3/5} \tag{5.2.33}$$

for plane waves and

$$\boldsymbol{\rho}_0 = \left[1.45k^2 \int_0^L C_n^2(x)\left(\frac{x}{L}\right)^{5/3} dx\right]^{-3/5} \tag{5.2.34}$$

for spherical waves. For engineering use, it is convenient to write Equation (5.2.32) in the form

$$\Gamma(\boldsymbol{\rho}) = A^2 \exp\{-3.44(\boldsymbol{\rho}/\mathbf{r}_0)^{5/3}\} \tag{5.2.35}$$

where

$$\mathbf{r}_0 = (3.44)^{5/3}\boldsymbol{\rho}_0 \tag{5.2.36}$$

The apparently arbitrary selection of the constant in Equation (5.2.35) will be clarified when we discuss the signal-to-noise ratio of an optical heterodyne receiver.

We observe that in accordance with the discussion of field coherence in Chapter 2 which leads to Equation (2.7.20), both $\boldsymbol{\rho}_0$ and $\mathbf{r}_0$ may be conveniently regarded as field transverse coherence lengths. In particular, $\boldsymbol{\rho}_0$ gives the transverse distance at which the field coherence is reduced to e^{-1} and $A_c = \pi\boldsymbol{\rho}_0^2$ gives the area over which the fields remain correlated to within e^{-1}. Furthermore, the coherence solid angle defined by Equation (2.7.22) is $2(\lambda/\boldsymbol{\rho}_0)^2/\pi$ and the number of coherence areas within a receiver of radius r is given by $(r/\boldsymbol{\rho}_0)^2$.

Fried[10] has calculated $\mathbf{r}_0$ for a space-to-earth communication system (Figure 5.8), assuming (1) that the source is sufficiently high so that the field impinging on the atmosphere is effectively plane, (2) that the signal source is viewed at an angle θ by a receiver at altitude H, and (3) that the functional form of $c_n^2(z)$ is given by Equation (5.1.44a). The results are

$$r_0 = 0.05\lambda^{6/5}\cos^{3/5}(\theta)[\Gamma(2/3)/\Gamma(2/3, H/h_0)]^{3/5} \tag{5.2.37}$$

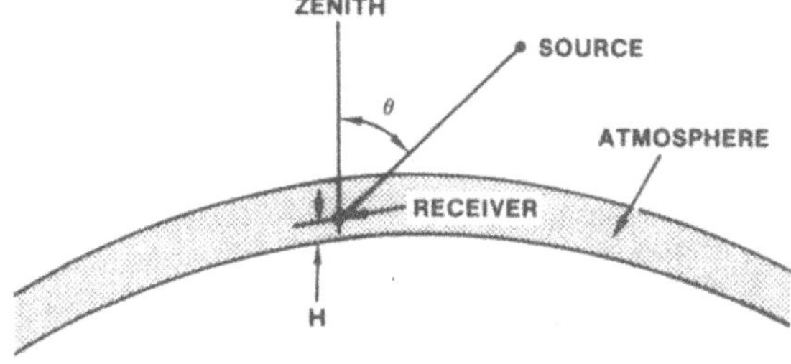

Figure 5.8. Geometry of a space-to-earth communication system.

for daytime and approximately a factor of two larger for nighttime conditions. A nomograph showing the composite dependence of r_0 on (λ, θ, H) is given in Figure 5.9. For horizontal propagation, the structure parameter is nearly constant along the path and Equation (5.2.34) gives

$$r_0 = [(1.2 \times 10^{-8})\lambda^{6/5} C_n^{-6/5}]x^{-3/5} \tag{5.2.38}$$

Figure 5.10 is a nomograph showing the dependence of r_0 on (λ, C_n^2, x).

For spherical waves in homogeneous turbulence, it can be shown that

$$\rho_0 = (0.45k^2 C_n^2 x)^{-3/5} \tag{5.2.39}$$

Figures 5.11 and 5.12 give the spherical-wave coherence length as functions of C_n^2 and propagation distance for $\lambda = 0.5\,\mu\text{m}$ and $10.6\,\mu\text{m}$ for homogeneous turbulence.

5.2.4.2. Spatial Coherence in Strong Turbulence

In Section 5.2.1 we used the Rytov method to develop equations for the scattered electric field when the fluctuations in the atmospheric index

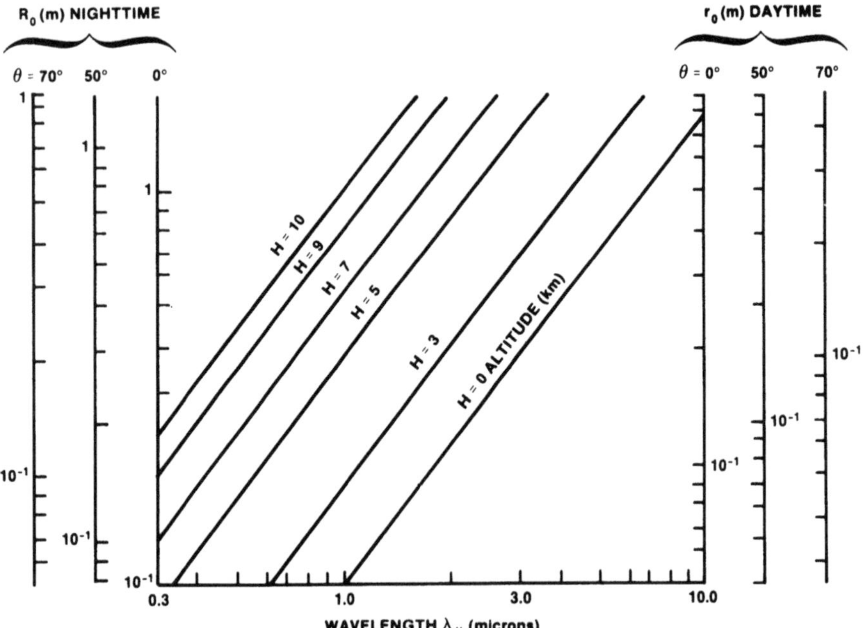

Figure 5.9. The dependence of r_0 upon wavelength λ, zenith angle θ, and altitude H for optical heterodyne reception of a space-to-earth signal. For daytime conditions, the three scales on the right should be used; the three on the left pertain to nighttime operation. Each of the three scales refers to a particular indicated angle.[10]

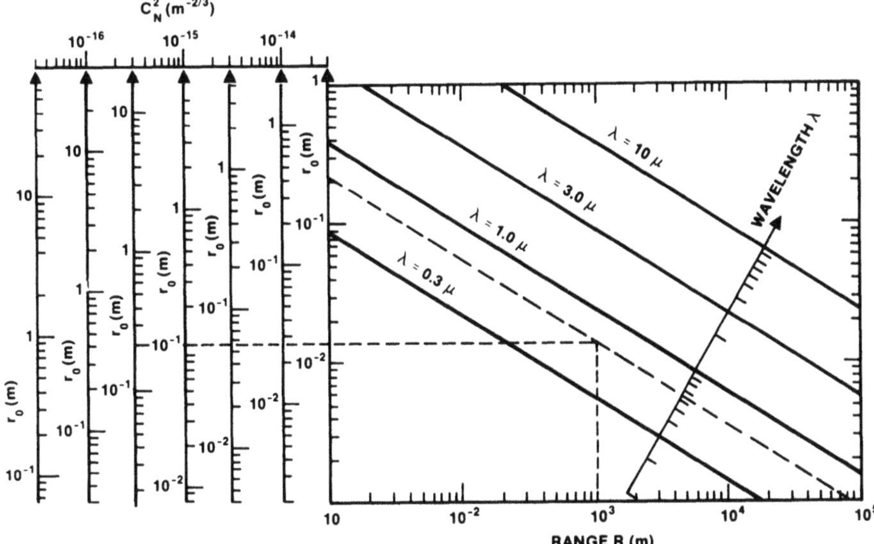

Figure 5.10. The dependence of r_0 upon wavelength λ, path length, and strength of turbulence as measured by the refractive index structure constant C_n^2, for horizontal propagation between transmitter and receiver. The C_n^2 scale on the upper left covers the range of values normally encountered within several tens of meters of the ground. To use the graph, a wavelength line is drawn parallel to the four heavy lines and passing through the wavelength scale at the appropriate point. Where the vertical line, drawn through the range of interest, intercepts the wavelength line, we pass horizontally over to one of the vertical r_0 scales on the left. Which of the r_0 scales we read from depends on the value of C_n^2 we wish to consider. As an example, we have drawn the wavelength line for 0.7 μm and considered a range of 1 km. For turbulence strength denoted by $C_n^2 = 3 \times 10^{-15}$, we see that r_0 falls between approximately 0.21 and 0.11 m.(10)

of refraction are weak. We can extend this to include the effects of multiple scattering for fields with small angular spread. Several techniques have been proposed to address this so-called strong-fluctuation case. These methods include the parabolic equation, transport theory(11), and the extended Huygens–Fresnel principle(12). It can be shown that each of these techniques gives equivalent results for the mutual coherence function. However, the situation with regard to the higher moments of the field is not presently clear. While the Huygens–Fresnel approach gives relatively simple results for the fourth-order moment, the equivalence between this and the other methods has not been demonstrated. Indeed, while the parabolic equations for all higher moments of the field have been obtained, they have not yet been solved. The starting point for our development will be the derivation of the parabolic differential equation for the complex field. Coupling this with certain crucial simplifying assumptions, we will derive the equations

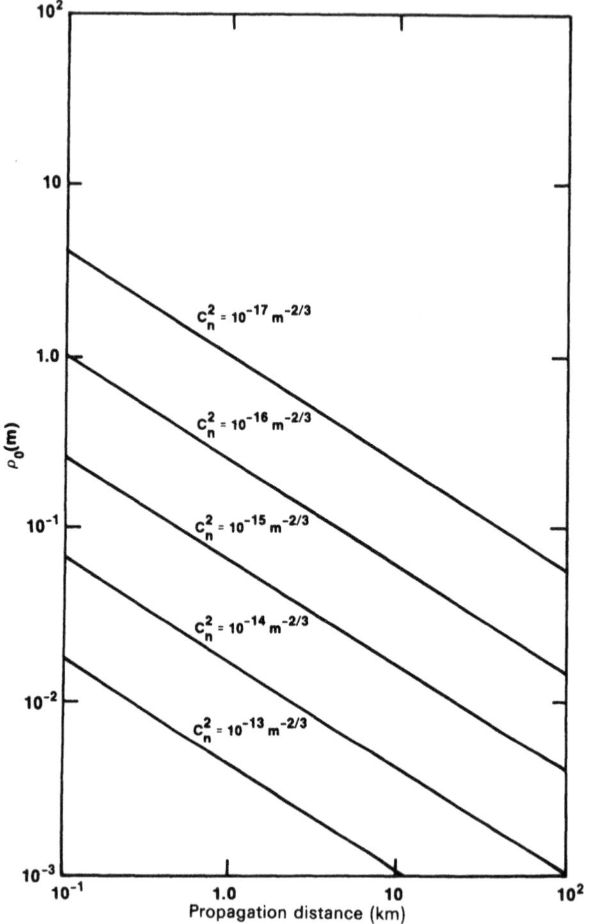

Figure 5.11. Spherical wave transverse coherence length for $\lambda = 0.5\ \mu$m.

which describe the propagation of the average field and the mutual coherence function. Using the parabolic equations as a foundation, we will then develop the transport theory and extended Huygens–Fresnel formulations of the problem, showing that each of these approaches leads to the same solution for the mutual coherence function. Unfortunately, we will find that the expression for the mutual coherence function cannot be evaluated for the Kolmogorov or Von Karman spectra using known mathematical functions; numerical methods must be employed. Using approximate expressions for the Von Karman spectrum coupled with numerical integration, several authors have obtained plots which display the behavior of the mutual

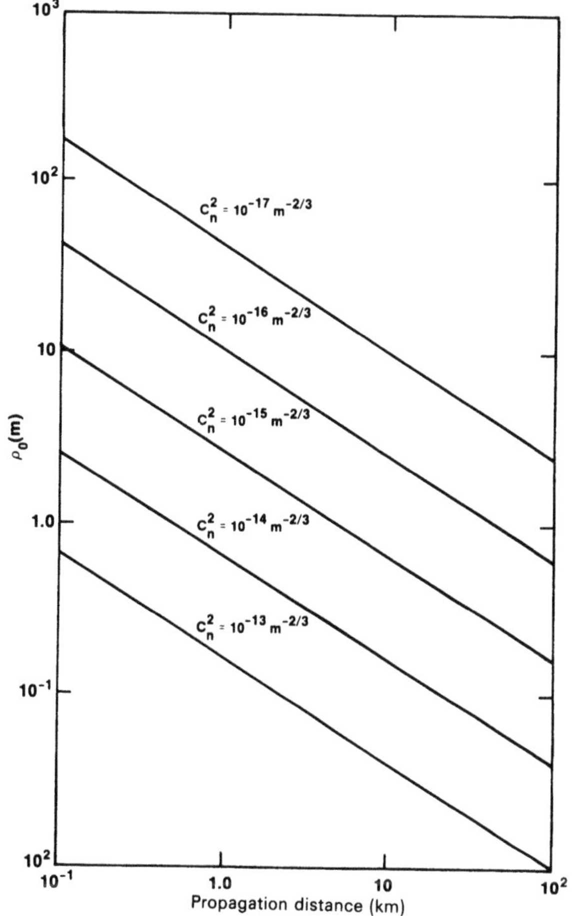

Figure 5.12. Spherical wave transverse coherence length for $\lambda = 10.6\ \mu$m.

coherence function as a function of propagation distance. We will discuss representative examples of these numerical efforts.

We begin by observing that for a light beam propagating in the x-direction with narrow angular spread, the phase of the electric field at any point is to a good approximation given by ikx. We may therefore write

$$E(\mathbf{r}) = f(x, y, z) \exp\{ikx\} \tag{5.2.40}$$

where $f(r)$ is a slowly varying function of x. Indeed, $U(r)$ will vary over distances on the order of the turbulence scale size, l_0, of the random medium.

Under these conditions

$$\left|\frac{k\,df}{dx}\right| \gg \frac{d^2f}{dx^2} \tag{5.2.41}$$

if $l_0 \gg \lambda$. Substituting Equation (5.2.40) in Equation (5.2.2) we have

$$2ik\frac{df}{dx} + \nabla^2 f - k^2 f + k^2 f + k^2(1 + n_1)^2 f = 0 \tag{5.2.42}$$

By virtue of Equation (5.2.41), we may neglect d^2f/dx^2 with respect to $k(df/dx)$. If, in addition, we approximate $2n_1 + n_1^2$ by $2n_1$, Equation (5.2.42) becomes

$$2ik\frac{df}{dx} + \frac{d^2f}{dy^2} + \frac{d^2f}{dz^2} + 2k^2 n_1 f = 0 \tag{5.2.43}$$

Equation (5.2.43) is known as the parabolic wave equation and is, by virtue of the assumptions expressed in Equations (5.2.40) and (5.2.41), applicable to the propagation of light beams with narrow angular spread. It can be shown that the parabolic approximations are equivalent to the Fresnel approximation of scalar diffraction theory. Unfortunately, this simplified and more restrictive form of the propagation equation cannot be solved for the complex field. To make any analytical progress in its solution, early workers in the field were forced to make two additional simplifying assumptions, collectively known as the Markov approximations.

The Markov approximations assume that (1) n_1 obeys Gaussian statistics and (2) the fluctuation n_1 is delta function correlated in the direction of propagation. The second assumption is the same one required in our development of the single-scattering Rytov theory and expressed in Equations (5.2.18a) and (5.2.18b). In the context of the present development, it implies that the index of refraction correlation function takes the form

$$E[n_1(x, \mathbf{r})n_1(x', \mathbf{r}')] = \delta(x - x')A(x, \boldsymbol{\rho}) \tag{5.2.44}$$

and that the turbulent eddies are disklike with their short axes aligned along the direction of propagation. Substituting (5.2.44) in (5.1.12), the wavenumber spectrum of the index of refraction fluctuations is then given by

$$\Phi_n(\mathbf{k}) = 2\pi^{-3}\int_{-\infty}^{\infty} A(x, \mathbf{p}) \exp\{-i\mathbf{k}\cdot\boldsymbol{\rho}_t\}\, d^2\boldsymbol{\rho} \tag{5.2.45}$$

where we have used

$$\mathbf{K} = \mathbf{k} + K_x\hat{\mathbf{i}} \tag{5.2.46}$$

To obtain a differential equation for the average field, we take the ensemble average of the parabolic equation (5.2.43):

$$2ik\frac{dE[f]}{dx}+\frac{d^2E[f]}{dy^2}+\frac{d^2E[f]}{dz^2}+2k^2E[n_1f]=0 \qquad (5.2.47)$$

Unfortunately, this is not a differential equation for the average field alone but involves the correlation function $E[n_1 f]$. Since n_1 and U are not, in general, independent random variables, the average $E[n_1 f]$ cannot be factored into the form $E[n_1]E[f]$. Indeed, the field $f(\mathbf{r})$ is a functional of $n_1(\mathbf{r})$. However, if the correlation function $E[n_1 f]$ can be written in the form

$$E[n_1(\mathbf{r})f(\mathbf{r})]=g(\mathbf{r})E[f(\mathbf{r})] \qquad (5.2.48)$$

we use the Novikov–Furutsu formula, which states that if n_1 is a Gaussian random variable and $f(n_1)$ is an arbitrary functional of n_1, then

$$E[n_1(\mathbf{r})f\{n_1(\mathbf{r})\}]=\int_{-\infty}^{\infty}E[n_1(\mathbf{r})n_1(\mathbf{r}')]E\left[\frac{\delta f\{n_1(\mathbf{r})\}}{\delta n_1(\mathbf{r})}\right]d^3\mathbf{r} \qquad (5.2.49)$$

where $\delta f/\delta n_1$ is a variational derivative. Substituting Equation (5.2.44) in Equation (5.2.49), we obtain

$$E[n_1(\mathbf{r})f\{n,\mathbf{r}\}]=\int_{-\infty}^{\infty}A(x,\mathbf{p}-\mathbf{p}')E\left[\frac{\delta f(x,\mathbf{p})}{\delta n_1(x',\mathbf{p}')}\right]d\mathbf{p}' \qquad (5.2.50)$$

It can be shown that if the field at $(x,\mathbf{p})$ depends on $n_1(x'\,\mathbf{p}')$ for $x'<x$, i.e., backscattering from the inhomogeneities in the region $x'>x$ are neglected, then

$$E\left[\frac{\delta f(x,\mathbf{p})}{\delta n_1(x',\mathbf{p}')}\right]=ik\delta(\mathbf{p}-\mathbf{p}')E[f(x,\mathbf{p})]/4 \qquad (5.2.51)$$

Substituting Equation (5.2.51) in Equation (5.2.50), we have

$$E[n_1(\mathbf{r})f(\mathbf{r})]=ikA(x,0)/4 \qquad (5.2.52)$$

Inserting Equation (5.2.52) in Equation (5.2.47) then gives

$$2ik\frac{dE[f]}{dx}+\frac{d^2E[f]}{dy^2}+\frac{d^2E[f]}{dz^2}+ik^3A(x,0)E[f]/4=0 \qquad (5.2.53)$$

which is subject to the boundary condition $f(x=0,\mathbf{p})=f_0(\mathbf{p})$. To solve Equation (5.2.53), we assume a solution of the form

$$E[f(x,\mathbf{p})]=g(x,\mathbf{p})\exp\left\{-\int_0^x s(x')\,dx'\right\} \tag{5.2.54}$$

Substituting Equation (5.2.54) in Equation (5.2.53), we have

$$2ik\frac{dg}{dx}+\frac{d^2g}{dy^2}+\frac{d^2g}{dz^2}+ik^3A(x,0)g/4+2ikx=0 \tag{5.2.55}$$

If we make the identification

$$s(x)=k^2A(x,0)/8 \tag{5.2.56}$$

Equation (5.2.55) becomes

$$\left[2ik\frac{d}{dx}+\frac{d^2}{dy^2}+\frac{d^2}{dz^2}\right]g=0 \tag{5.2.57}$$

Since Equation (5.2.57) is the parabolic differential equation for propagation in vacuum ($n_1=0$) we have

$$g(x,\mathbf{p})=f(x,\mathbf{p}) \tag{5.2.58}$$

Substituting Equations (5.2.58) and (5.2.56) in Equation (5.2.54), and using the boundary condition, we have

$$\begin{aligned}E[f(x\cdot\mathbf{p})]&=f_0(\mathbf{p})k^2\exp\left\{-\int_0^x A(x',0)\,dx'\right\}\Big/8\\&=f_0(\mathbf{p})\exp\left\{-\int_0^x b(x')\,dx'\right\}\end{aligned} \tag{5.2.59}$$

which is the general expression for the average field. The coherent intensity is given by

$$|E[f(x),\mathbf{p})]|^2=I(\mathbf{p})\exp\left\{-2\int_0^x b(x')\,dx'\right\} \tag{5.2.60}$$

where $|f_0(\mathbf{p})|^2=I(\mathbf{p})$ and we have defined the total scattering cross section per unit volume:

$$b(x')=k^2A(x',0)/4 \tag{5.2.61}$$

This relationship can also be derived directly from the definition:

$$2\pi k_0^4 \Phi_n(x, \mathbf{k}) = \beta(x, \mathbf{k}) \tag{5.2.62}$$

Integrating both sides over solid angle, $d^2 r = d^2\mathbf{k}/k_0^2$, we have

$$\begin{aligned} b(x') &= \int_{-\infty}^{\infty} \beta(x, \mathbf{k})\, d^2\mathbf{k}/k_0^2 \\ &= 2\pi k_0^2 \int_{-\infty}^{\infty} \Phi_n(x', \mathbf{k})\, d^2\mathbf{k} = k^2 A(x, 0)/4 \end{aligned} \tag{5.2.63}$$

If $\Phi_n(x, \mathbf{k})$ is independent of x, then the volume scattering coefficient is independent of x and Equation (5.2.60) reduces to

$$I(x, \mathbf{p}) = I_0(\mathbf{p})^2 \exp\{-bx) \tag{5.2.64}$$

which is the usual expression displaying the exponential decay for power transmission per unit area over a path of length x in a scattering medium.

We will now derive the parabolic differential equation for the mutual coherence function which was defined in Equation (5.2.26) and is repeated below:

$$\Gamma(z, \mathbf{p}_1, \mathbf{p}_2) = E[f(z, p_1) f^*(z, p_2)] \tag{5.2.65}$$

We begin with the parabolic equation

$$\left\{2ik\frac{d}{dz} + \nabla_1^2 + k^2 n_1(z, \mathbf{p}_1)\right\} U(z, \mathbf{p}_1) = 0 \tag{5.2.66a}$$

where ∇_1 is the Laplacian with respect to the variable $\mathbf{p}_1$. Multiplying Equation (5.2.66a) by $U^*(z, \mathbf{p}_2)$, we obtain

$$\begin{aligned} &2ikU(\mathbf{p}_2, z)\, dU(\mathbf{p}_1, z)/dz + \nabla_1 U(\mathbf{p}_1, z) U^*(\mathbf{p}_2, z) \\ &+ k^2 n_1(t, \mathbf{p}_1) U(\mathbf{p}_1, z) U^*(\mathbf{p}_2, z) = 0 \end{aligned} \tag{5.2.67a}$$

Equation (5.2.66a) can also be written in the form

$$\left\{2ik\frac{d}{dz} + \nabla_1^2 + k^2 n_1(z, \mathbf{p}_2)\right\} U(z, \mathbf{p}_2) = 0 \tag{5.2.66b}$$

Multiplying both sides of Equation (5.2.66b) by $U(z, r_1)$ we have

$$2ikU(\mathbf{p}_1, z)\, dU(\mathbf{p}_2, z)/dz + \nabla_1 U^*(\mathbf{p}_2, z) U(\mathbf{p}_1, z) + k^2 n_1(z, \mathbf{p}_1) U(\mathbf{p}_1, z) U^*(\mathbf{p}_2, z) = 0 \tag{5.2.67b}$$

Subtracting Equation (5.2.67b) from Equation (5.2.67a) and ensemble averaging the resulting equation, we have

$$\left\{2ik\frac{d}{dz} + \nabla_1 - \nabla_2\right\}\Gamma(\mathbf{p}_1, \mathbf{p}_2, z) + k^2 E[\{n_1(\mathbf{p}_1, z) - n_1(\mathbf{p}_2, z)\} U(\mathbf{p}_1, z) U^*(\mathbf{p}_2, z)] \tag{5.2.68}$$

It can be shown that[1]:

$$E[\{n_1(\mathbf{p}_1, z) - n_1(\mathbf{p}_2, z)\} U(\mathbf{p}_1, z) U^*(\mathbf{p}_2, z)] = \frac{ik}{2}\{A(z, 0) - A(z, \mathbf{p}_1 - \mathbf{p}_2)\}\Gamma(\mathbf{p}_1, \mathbf{p}_2, z) \tag{5.2.69}$$

Substituting Equation (5.2.69) in Equation (5.2.68), we obtain

$$\left[2ik\frac{d}{dz} + \nabla_1 - \nabla_2 + \frac{ik^3}{2}\{A(z, 0) - A(z, \mathbf{p}_1 - \mathbf{p}_2)\}\right]\Gamma(\mathbf{p}_1, \mathbf{p}_2, z) = 0 \tag{5.2.70}$$

which is the parabolic differential equation for the mutual intensity function. We now introduce the variables $(\mathbf{R}, \boldsymbol{\rho})$ defined by Equation (5.1.8). It can be shown that

$$\nabla_1 - \nabla_2 = 2\nabla_{\mathbf{R}}\nabla_{\boldsymbol{\rho}} \tag{5.2.71}$$

where $\nabla_{\mathbf{R}}$ and $\nabla_{\boldsymbol{\rho}}$ are, respectively, the gradient operators with respect to the variables $\mathbf{R}$ and $\boldsymbol{\rho}$. Substituting Equation (5.2.71) in Equation (5.2.70) gives

$$\left\{2ik\frac{d}{dz} + 2\nabla_{\mathbf{R}}\nabla_{\boldsymbol{\rho}} + \frac{ik^3}{2}H(z, \boldsymbol{\rho})\right\}\Gamma(\mathbf{R}, \boldsymbol{\rho}, z) = 0 \tag{5.2.72}$$

where

$$H(z, \boldsymbol{\rho}) = \{A(z, 0) - A(z, \boldsymbol{\rho})] \tag{5.2.73}$$

In Appendix B we show that there is a direct link between the parabolic wave optics equation for propagation of the mutual coherence function and the small-angle transport equation for propagation of the generalized radiance. In particular, we derive the small-angle transport equation

$$2ik_0 \frac{\partial J(\mathbf{R}, Z, \boldsymbol{\rho})}{\partial z} + 2\nabla_{\boldsymbol{\rho}} \cdot \nabla_{\mathbf{R}} J(\mathbf{R}, Z, \boldsymbol{\rho}) + \frac{ik_0^3}{2} H(Z, \boldsymbol{\rho}) J(\mathbf{R}, Z, \boldsymbol{\rho}) = 0 \tag{5.2.74}$$

from the parabolic equation, Equation (5.2.72), thereby establishing their equivalence. Central to this derivation is the fact that the generalized radiance and the mutual coherence function are a two-dimensional spatial Fourier transform pair:

$$N_G(\mathbf{R}, z, \mathbf{k}) = (k_0/2\pi)^2 \int_{-\infty}^{\infty} \Gamma(\mathbf{R}, \boldsymbol{\rho}, z) \exp\{\mathbf{k} \cdot \boldsymbol{\rho}\}\, d^2\boldsymbol{\rho}$$

$$\Gamma(\mathbf{R}, \boldsymbol{\rho}, z) = k_0^{-2} \int_{-\infty}^{\infty} N_G(\mathbf{R}, z, \mathbf{k}) \exp\{-\mathbf{k} \cdot \boldsymbol{\rho}\}\, d^2\mathbf{k} \tag{5.2.75}$$

Until recently, it was widely believed that the generalized radiance, $N_G(\mathbf{R}, z, \mathbf{k})$ defined above, and the traditional radiance, $N_T(\mathbf{R}, z, \mathbf{k})$ defined in Chapter 1, were identical. This led many researchers to two incorrect conclusions. The first of these was that the traditional radiance contains a complete specification of the first-order correlation properties of the complex field. The second was that the traditional radiance transport equation and the parabolic wave optics equation are entirely equivalent formulations which lead to identical descriptions of the first-order correlation properties of the complex field. These conclusions are inconsistent with the fact that the traditional radiance transport equation neglects interference effects. Indeed, it can be derived solely on the basis of energy balance considerations without regard to the complex field quantities. It has been shown that (1) the generalized and traditional radiance functions are not, in general, identical, (2) the small-angle transport equation correctly describes the propagation of the generalized radiance, and (3) the generalized radiance transport equation reduces to a transport equation for the traditional radiance if and only if the generalized and traditional radiance functions are identical throughout the scattering medium. These issues are discussed further in Appendix B.

In Appendix C, we use Equations (5.2.74) and (5.2.75) to derive the general solution for the mutual coherence function. The result is:

$$\Gamma(\mathbf{R}, z, \boldsymbol{\rho}) = \exp\{-cz\}F(\mathbf{q}, \boldsymbol{\rho} + \mathbf{q}z\} \exp\left[-i\mathbf{q}\cdot\mathbf{R} + \int_0^z \beta(\boldsymbol{\rho} + \mathbf{q}n)\, d\eta\right] d^2\mathbf{q}/4\pi^2 \tag{5.2.76}$$

where

$$F(\mathbf{q}, \boldsymbol{\rho}) = \Gamma(\mathbf{R}, z = 0, \boldsymbol{\rho}) \exp\{i\mathbf{q}\cdot\mathbf{R}\}\, d^2\mathbf{R} \tag{5.2.77}$$

and

$$\beta(\boldsymbol{\rho}) = \int_{-\infty}^{\infty} \beta(\mathbf{n}) \exp\{i\boldsymbol{\rho}\cdot\mathbf{n}\}\, d^2\mathbf{n} \tag{5.2.78}$$

The solution can also be expressed in the following form which explicitly shows the dependence on the wavenumber spectrum of the index of refraction fluctuations, $\Phi_n(\mathbf{k})$:

$$\Gamma(\mathbf{R}, z, \boldsymbol{\rho}) = k^2 \iint_{-\infty}^{\infty} \Gamma(\mathbf{R}', z, \boldsymbol{\rho}') \exp\{i[k(\boldsymbol{\rho} - \boldsymbol{\rho}') \times (\mathbf{R} - \mathbf{R}')/z - H]\}/4\pi^2 z^2\, d^2\boldsymbol{\rho}'\, d^2\mathbf{R}' \tag{5.2.79a}$$

where

$$\begin{aligned} H &= k^2 \int_0^Z \{A(0) - A[\boldsymbol{\rho}' + (\boldsymbol{\rho} - \boldsymbol{\rho}')z/Z]\}\, dz/4 \\ &= 4\pi^2 k^2 \int_0^Z \int_{-\infty}^{\infty} [1 - J_0(kp)\Phi_n(k)k]\, dk\, dz \end{aligned}$$

with

$$\rho = \boldsymbol{\rho}' + (\boldsymbol{\rho} - \boldsymbol{\rho}')z/Z \tag{5.2.79b}$$

This expression has the advantage that all of the effects due to propagation in the turbulent medium are conveniently lumped into the parameter H. When H is zero, as would be the case when $\Phi_{n_1}(\mathbf{k}) = 0$ and, consequently, the index of refraction fluctuation $n_1 = 0$, Equation (5.2.79a) reduces to the Fresnel formula for free-space propagation of the mutual coherence function.

Another method of describing the propagation of electromagnetic fields in turbulent media has been developed by Lutomirski and Yura and is called the extended Huygens–Fresnel principle[12]. Because it treats the random turbulence channel as a linear system and employs linear systems notation, it has enjoyed considerable popularity among optical designers. In the following development, we will begin by writing expressions for the complex field in terms of the random medium impulse response, introducing the important concept of point-to-point reciprocity. This formalism will then be used to describe propagation of the mutual coherence function using a quantity called the two-point-source spherical wave structure function. The reader should review Chapter 2, and Section 2.7 in particular, for a more detailed discussion of linear random channels.

To formalize this technique, we begin by writing the random complex field in terms of the random turbulent medium spatial impulse response, or the Green's function $g(\mathbf{r}_r, \mathbf{r}_s)$:

$$F(\mathbf{r}_r) = \int_{-\infty}^{\infty} F(\mathbf{r}_s) g(\mathbf{r}_r, \mathbf{r}_s)\, d^2\mathbf{r}_s \tag{5.2.80}$$

where

$$g(\mathbf{r}_r, \mathbf{r}_s) = \exp\{ik(\mathbf{r}_s - \mathbf{r}_r)^2/2x + \phi(\mathbf{r}_s, \mathbf{r}_r)\} \tag{5.2.81}$$

and $\phi(\mathbf{r}_s, \mathbf{r}_r)$ is the random part of the complex phase of a spherical wave propagating in the turbulent medium from $\mathbf{r}_s$ to $\mathbf{r}_r$. Equation (5.2.80) is an extension of the Huygens–Fresnel principle to a random medium and reduces to the Fresnel approximation of the Huygens–Fresnel integral for free-space propagation when $\phi(\mathbf{r}_s, \mathbf{r}_r) = 0$.

It can be shown[6] that the Green's function $g(\mathbf{r}_r, \mathbf{r}_s)$ is reciprocal, that is,

$$g(\mathbf{r}_s, \mathbf{r}_r) = g(\mathbf{r}_r, \mathbf{r}_s) \tag{5.2.82}$$

This simply means that the complex field at $\mathbf{r}_r$ due to a point source at $\mathbf{r}_s$ is identical to the complex field at $\mathbf{r}_s$ due to a point source at $\mathbf{r}_r$. Indeed, it can be shown[13] that point-to-point reciprocity as expressed in Equation (5.2.82) is a necessary condition for deriving a superposition integral of the form of Equation (5.2.80) from the scalar wave equation. Furthermore, since Equation (5.2.82) implies that reciprocity holds for instantaneous values of the electromagnetic field, one would suspect that reciprocity should also hold for a statistic of the complex field. That this is indeed the case can be verified by examination, for example, of our expression for the variance of the log-amplitude flcutations for plane and spherical waves. If one interchanges the source and measurement points in Equations (5.2.103) and (5.2.104) by making the change of variable $x = L - x$, then the equations are found to be invariant under this transformation. Reciprocity plays an important role in the analysis of adaptive optical system performance in the turbulent atmosphere and can be used to simplify some systems calculations.

Using the definition of the mutual coherence function and Equation (5.2.82), we obtain the following expression for the propagation of coherence in the turbulent medium:

$$\Gamma(\mathbf{r}_{r1}, \mathbf{r}_{r2}) = \iint_{-\infty}^{\infty} \Gamma(\mathbf{r}_{s1}, \mathbf{r}_{s2}) E[g(\mathbf{r}_{s1}, \mathbf{r}_{r1}) g^*(\mathbf{r}_{s2}, \mathbf{r}_{r1})] \, d^2\mathbf{r}_{s1} \, d^2\mathbf{r}_{s2} \quad (5.2.83)$$

with

$$E[g(\mathbf{r}_{s1}, \mathbf{r}_{r1}) g^*(\mathbf{r}_{s2}, \mathbf{r}_{r1})] = (\lambda L)^{-2} \exp\{ik(|\mathbf{r}_{r1} - \mathbf{r}_{s1}|^2 - |\mathbf{r}_{r2} - \mathbf{r}_{s2}|^2)/2L - D(\boldsymbol{\rho}_r - \boldsymbol{\rho}_s)$$

where

$$\boldsymbol{\rho}_r = (\mathbf{r}_{r1} - \mathbf{r}_{r2})$$
$$\boldsymbol{\rho}_s = (\mathbf{r}_{s1} - \mathbf{r}_{s2}) \quad (5.2.84)$$

and $D(\boldsymbol{\rho}_r - \boldsymbol{\rho}_s)$ is the two-source spherical wave structure function given by

$$D(\boldsymbol{\rho}_r - \boldsymbol{\rho}_s) = 2.91 k^2 \int_0^L C_n^2(x) \left[\frac{|\boldsymbol{\rho}_r x + \boldsymbol{\rho}_s(L - x)|}{L}\right]^{5/3} dx \quad (5.2.85)$$

for the Kolmogorov spectrum. We should emphasize that the extended Huygens–Fresnel formalism contains no new information beyond that which is contained in the parabolic and transport equation approaches. Its primary asset lies in the convenience associated with the linear systems representation.

While Equations (5.2.74) and (5.2.76) give the complete solution for the propagation of the mutual coherence function in a turbulent medium, they are not easily evaluated for either the Kolmogorov of Van Karman spectra. In general, a closed-form expression for the mutual coherence function in terms of known mathematical functions does not exist. Although some closed-form approximate expressions under special limiting conditions do exist and can be useful, a numerical solution of Equation (5.2.79a) is usually required. We will now present both representative numerical results which characterize general propagation trends and approximate closed-form results for the propagation of a beam wave which is characterized by an initial field given by

$$f(z = 0, \mathbf{r}) = f_0 \exp\{-(W_0^{-2} + 2ikr^2 R_0^{-1})\} \quad (5.2.86)$$

where W_0 is the effective beam cross section and R_0 gives the wavefront radius of curvature. In the literature one often finds the beam cross section expressed as effective beam areas, s, where $s = \pi W_0^2/2$. If $R_0 < 0$, then the beam is focused at $z = -R_0$ in vacuum, while $R_0 = \infty$ represents a collimated beam with divergence angle $\pi^{1/2}/ks$. The infinite plane-wave case is obtained by setting $W_0 = \infty$, $R_0 = -\infty$.

Ishimaru[1] has evaluated Equation (5.2.79a) numerically using the following approximate expressions for $A(0) - A(\boldsymbol{\rho})$:

$$A(0) - A(\boldsymbol{\rho}) = \begin{cases} 6.56 C_n^2 \rho^2 / l_0^{1/3} \\ 5.83 C_n^2 \rho^{5/3} \\ 3.127 C_n^2 L_0^{5/3} \end{cases} \tag{5.2.87}$$

His results appear in Figures 5.13, 5.14, and 5.15. Figure 5.13 gives the normalized average on-axis intensity for collimated and focused beams as a function of propagation distance for $W_0 = 5$ cm. Figure 5.14 gives the normalized average intensity as a function of the distance R and the normalized coherence function for $R = 0$ as a function of p for 0.5-km and 5-km propagation distances in the collimated beam case, while Figure 5.15 gives the same results for the focused beam case.

Several general trends can be deduced from these figures. First, the average intensity of a focused beam approaches that of a collimated beam after a relatively short distance. Second, at the short distance of 500 m, while beam spreading is negligible, compared to the free-space result, the transverse coherence length is reduced to approximately one-fourth of the free-space value. At 5 km, the beam has further reduced coherence and has spread out considerably. Third, the focused beam case shows that at $z =$ 500 m, the beam has spread slightly, but there is not much degradation of the spatial coherence. At $z = 5$ km, the beam spread and coherence length are almost the same as in the collimated beam case.

Using approximate expressions for $\beta(\mathbf{p})$, Fante[14] has solved Equation (5.2.76) numerically using the Von Karman spectrum for $z < 4/k^2 C_n^2 l_0^{5/3}$; the results appear in Figures 5.16–5.19. Figure 5.16 and 5.17 plot the modulus of the normalized mutual coherence function for optical thickness values of $\tau = 3$ and 20, respectively, as a function of ρ/l_0 for various values of the beam area parameter s. Figure 5.18 plots $\Gamma(\rho)$ as a function of p/L_0 for τ varying from 3 to 300, given the fixed value $s = 196$ cm^2. The expected decrease in coherence length for increasing values of τ is clearly displayed. Figure 5.19 shows the effect of varying s on the e^{-1} coherence length. Notice that the correlation length is a minimum for $ks/\pi z = 1$, and for sufficiently large distances is nearly independent of s.

Using the extended Huygens–Fresnel principle, Yura[15] has obtained an expression for the MCF which can be shown to be a good approximation to the exact results given by Equation (5.2.76). His result, which gives

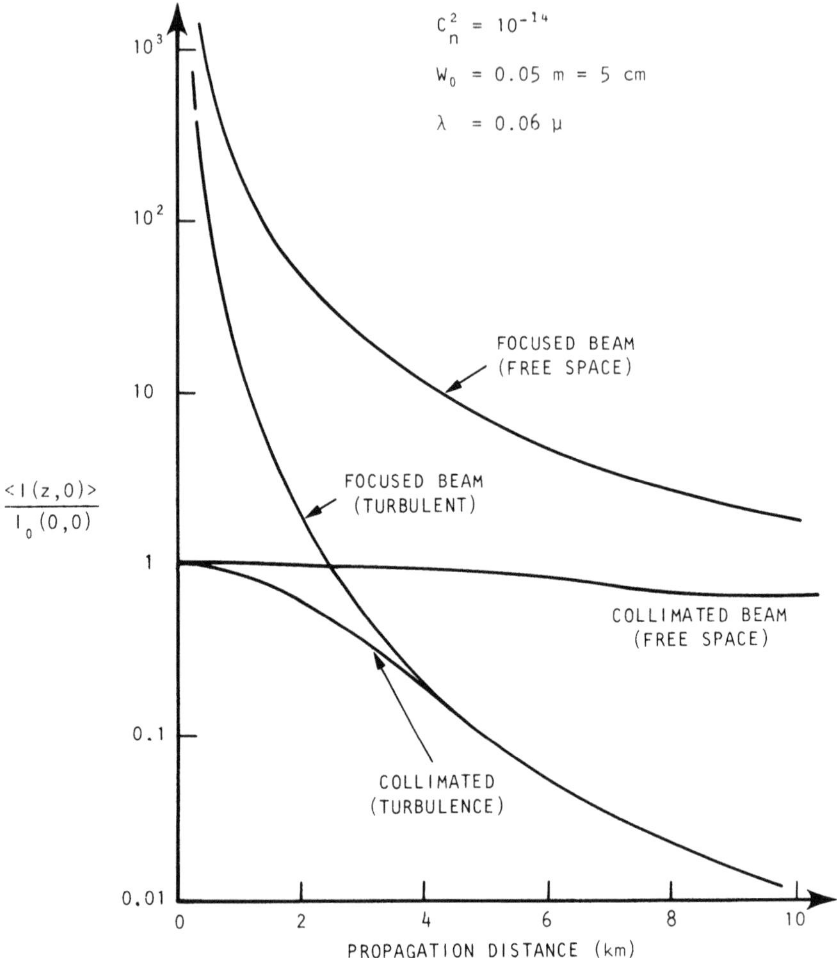

Figure 5.13. Average intensity of collimated and focused beams on the beam axis as a function of distance.[1]

the coherence relative to the undisplaced center of the beam wave, is

$$\Gamma(z, \boldsymbol{\rho}) = \exp\{-(\rho/\rho_c)^2\} \tag{5.2.88}$$

where p_c is the beam lateral coherence length and is given by:

$$\begin{aligned} p_c &= p_0(\{[1-(z/f)]^2 + \Omega^2[1+(\delta^2/3)]\}\{(1-(13/3)(z/f)+(11/3)(z/f)^2 \\ &\quad + (\Omega^2/3)[1+(\delta^2/4)]\}^{-1})^{1/2} \\ &= p_0(\{[1-(z/f)]^2 + (z/z_b)^2[1+(\delta^2/3)][1/(1+\delta^2)]\}\{1-(13/3)(z/f) \\ &\quad + (11/3)(z/f)^2 + [(z/z_b)^2/3][1+(\delta^2/4)][1/(1+\delta^2)]\}^{-1})^{1/2} \end{aligned} \tag{5.2.89}$$

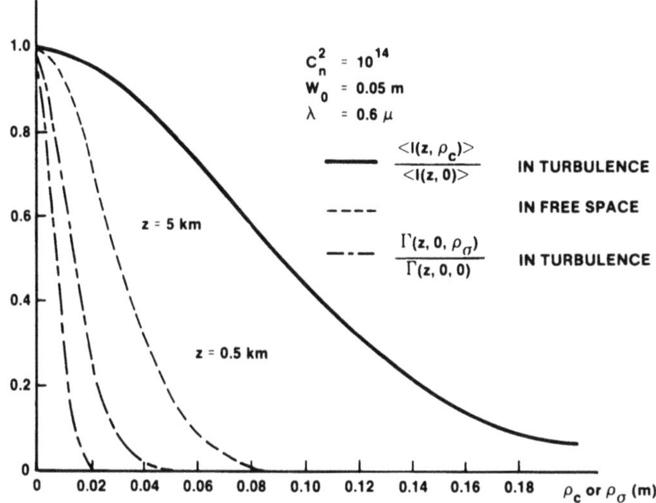

Figure 5.14. Collimated beam case: normalized average intensity as functions of transverse distance at $z = 0.5$ km and $z = 5$ km and normalized mutual coherence function on beam axis as functions of separations.[1]

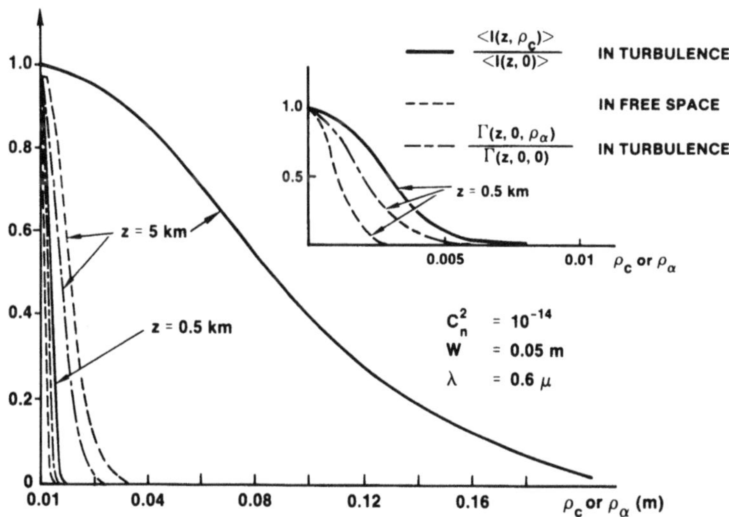

Figure 5.15. Focused beam case: normalized average intensity as functions of transverse distance at $z = 0.5$ km and $z = 5$ km and normalized mutual coherence function on beam axis as functions of separations.[1]

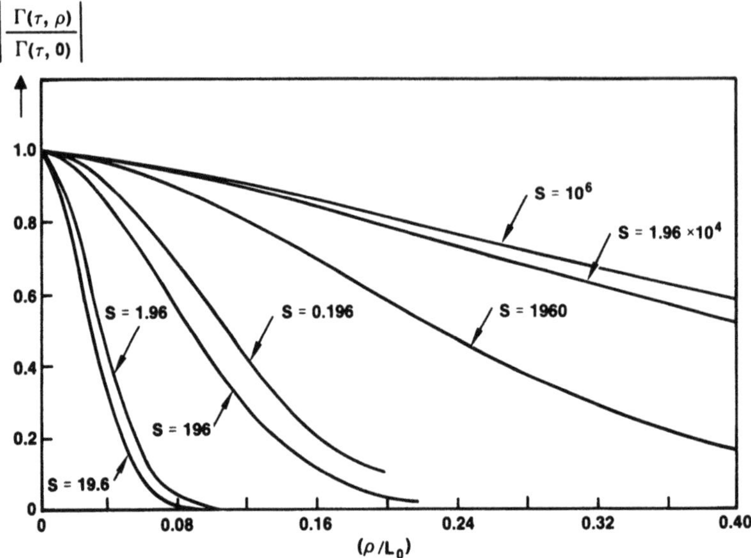

Figure 5.16. Normalized MCF for $\tau = 3$, $F = -\infty$, $k = 6280\ \text{cm}^{-1}$, and C_n^2 $3 \times 10^{-15}\ \text{cm}^{-1}$. Values of s are in $\text{cm}^2\lambda$.(14)

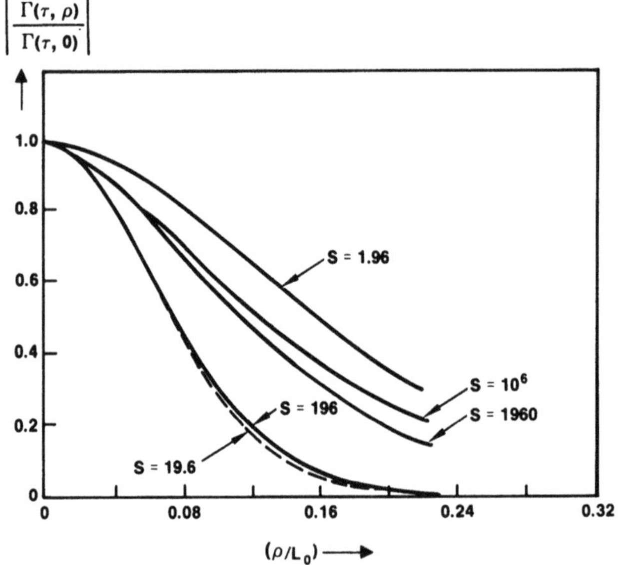

Figure 5.17. Normalized MCT for $\tau = 20$, $F = -\infty$, $k = 6280\ \text{cm}^{-1}$, and $C_n^2 = 3 \times 10^{-15}\ \text{cm}^{-1}$. Values of s are in cm^2.(14)

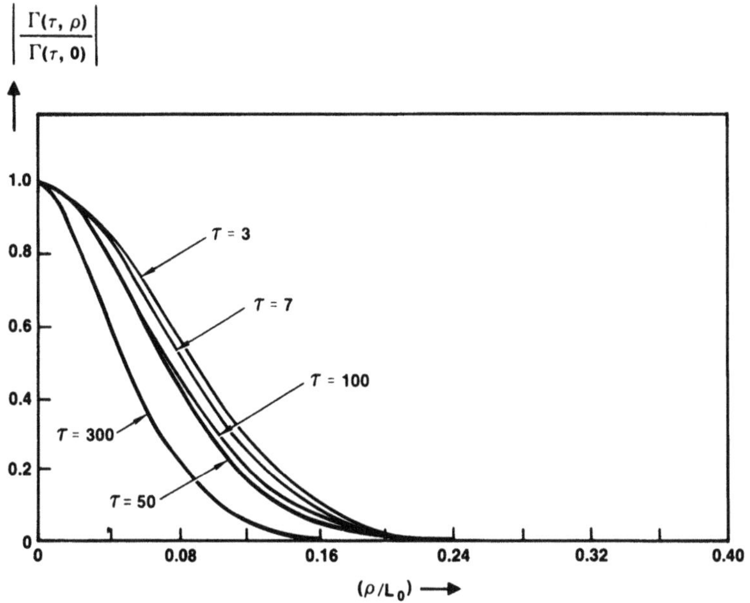

Figure 5.18. Normalized MCF for $s = 196\ \text{cm}^2$, $F = -\infty$, $k = 6280\ \text{cm}^{-1}$, and $C_n^2 = 3 \times 10^{-15}\ \text{cm}^{-1}$.[14]

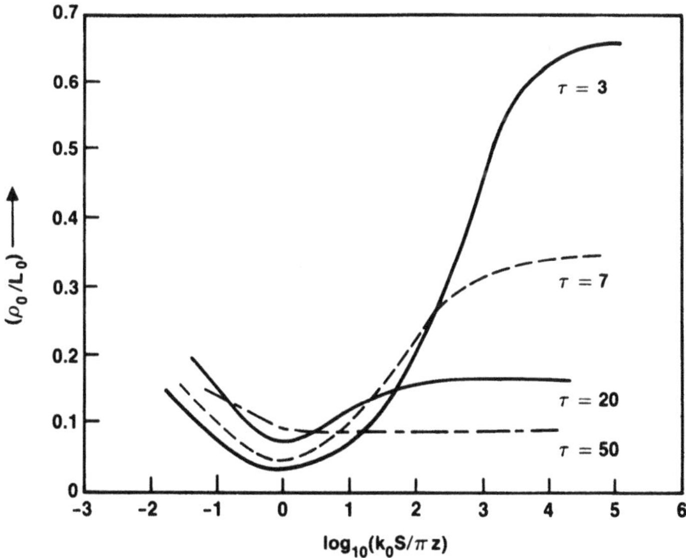

Figure 5.19. Effects of varying s on the e^{-1} correlation length for $F = -\infty$, $k = 6280\ \text{cm}^{-1}$, and $C_n^2 = 3 \times 10^{-15}\ \text{cm}^{-1}$.[14]

where ρ_0 is the plane-wave coherence length given by

$$\rho_0 = \left[1.46k^2 z \int_0^1 C_n^2(zy)\, dy\right]^{-3/5} \tag{5.2.90}$$

and

$$\begin{aligned} \Omega &= z/kw_0^2 \\ \delta &= 2w_0/p_0 \\ z/z_b &= \Omega(1+\delta^2)^{1/2} \\ f &= R_0/4 \end{aligned} \tag{5.2.91}$$

Because of the quadratic dependence on the variable ρ, the normalized coherence function exhibits a Gaussian-like behavior. Equation (5.2.84) reduces to the plane-wave coherence length, i.e., $\rho_c = \rho_0$, when the beam diameter w_0 and radius of curvature f go to infinity. In the spherical-wave limit, $w_0 = 0$ and the coherence length reduces to $p_c = (3\rho_0)^{1/2}$. In addition, we observe that when z is much less than z_b, the degree of coherence is to a good approximation given by the plane-wave result. When z is much greater than z_b, the degree of coherence is well approximated by the spherical-wave result. In general, the degree of coherence of a beam wave lies between that of plane and spherical waves.

In addition to spatial coherence, we can also estimate the temporal spectrum of a plane wave undergoing turbulence. We begin by invoking the frozen-flow hypothesis, assuming that the turbulence is transported with a mean wind velocity transverse to the propagation path. The mutual coherence function then satisfies the parabolic differential equation

$$\left\{\frac{2ik}{dz} + \nabla_{t1}^2 + \nabla_{t2}^2 + ik^3[A(0) - A(\mathbf{r}_1 - \mathbf{r}_2 - V\tau)]\right\}\Gamma(\mathbf{r}_1, \mathbf{r}_2, Z, \tau) = 0 \tag{5.2.92}$$

The solution of this equation is identical to Equation (5.2.70) except that $A(\boldsymbol{\rho})$ is replaced by $A(\boldsymbol{\rho} - V\tau)$. Using this solution, Fante[14] has calculated the temporal spectrum for two cases of interest.

For distances such that $\tau \gg (L_0/l_0)^{5/3}$, the temporal power spectrum has the form

$$S(\tau, \omega) = L_0 \exp\{-(\omega - \omega_0)^2 L_0^2/V^2 4\beta\tau\}/2V(\pi\beta\tau)^{1/2} \tag{5.2.93}$$

which has an e^{-1} spectrum width given by

$$\Delta\omega = 3.56\, VkC_n Z^{1/2}/l_0^{1/6} \tag{5.2.94}$$

For $1 \ll \tau \ll (L_0/l_0)^{5/3}$ a closed-form solution does not exist, but numerical evaluation shows that the spectral width is

$$\Delta\omega = 4.1\, Vk^{6/5}(C_n^2)^{3/5} Z^{3/5} \tag{5.2.95}$$

Thus, for $1 \ll \tau \ll (L_0/l_0)^{5/3}$ the spectrum width increases as $Z^{3/5}$ while for $\tau \gg (L_0/l_0)^{5/3}$ it increases as $Z^{1/2}$.

5.2.5. Optical Heterodyne Detection in Atmospheric Turbulence

The heterodyne receiver for the free-space channel has been discussed in detail in Chapter 3 (Sections 3.2 and 3.3). In this section we reinvestigate heterodyne detection for the atmospheric channels. In the following analysis we will assume that (1) the local oscillator power is much larger than the signal power, (2) the effects of background noise are negligible, and (3) the local oscillator and signal fields are properly matched over the photodetector surface. It can then be shown that the expression (3.3.18) for the signal-to-noise ratio takes the form

$$\mathrm{SNR} = \{8\alpha/B_n D^2\} \int_0^D p k_0(p, D)\Gamma(p)\, dp \tag{5.2.96}$$

where D is the diameter of the receiver aperture and $k_0(p, D)$ is the well-known circular aperture OTF given by Equation (2.8.12):

$$k_0(p, D) = \begin{cases} (D^2/2)\cos^{-1}\{(p/D) - (p/D)[1-(p/D)^2] & \text{for } p < D \\ 0 & \text{for } p < D \end{cases} \tag{5.2.97}$$

To evaluate the dependence of the signal-to-noise ratio on the aperture diameter in the presence of atmospheric turbulence, we define the function

$$F(D/r_0) = (32/\pi r_0^2 D^2) \int_0^D p k_0(p, D)\Gamma(p)\, dp \tag{5.2.98a}$$

Thus,

$$\mathrm{SNR} = (\pi r_0^2/2h\nu Df)F(D/r_0) \tag{5.2.98b}$$

Defining the variables $U = D/r_0$ and $u = p/D$ and using Equation (5.2.97) and (5.2.35), we have

$$F(U) = (16U^2/\pi) \int_0^1 u[\cos^{-1}(u) - u(1-u^2)^{1/2}] \times \exp\{-3.44(Uu)^{5/3}\}\, du \tag{5.2.99}$$

Fried [16] has evaluated this integral numerically. The results are plotted in Figure 5.20, which shows the dependence of F on the normalized receiver diameter D/r_0. There is a distinct knee in the curve at $D/r_0 = 1$, which implies that there is very little improvement in SNR if the aperture diameter exceeds the field transverse coherence length r_0. This explains the apparently arbitrary choice of the constant in Equation (5.2.35); it makes the knee occur at $D = r_0$.

Qualitatively speaking, the turbulence breaks the optical field up to such an extent that increasing the receiver diameter beyond r_0 results in the collection of very little additional coherent signal. Examination of Equation (5.2.99) reveals that the asymptotic behavior of $\Gamma(p)$ for large p determines the amount of coherent signal energy collected by increasing D beyond r_0. If $\Gamma(p)$ approaches a low-level constant as p increases, then increasing D will increase the SNR. This behavior is apparent in Figure 5.20, where $F(D/r_0)$ does increase slightly beyond r_0.

5.2.6. Scintillation

When a wavefront propagates through a turbulent medium, refraction by the turbulent eddies causes different portions of the wavefront to experience different phase changes, as shown in Figure 5.21. This results in warping of the wavefront, that is, deviation of the local wave normals from that associated with the unperturbed wave so that rays which emanate from different portions of the wavefront will ultimately interfere in the receiver plane. This random interference phenomenon leads to fluctuations,

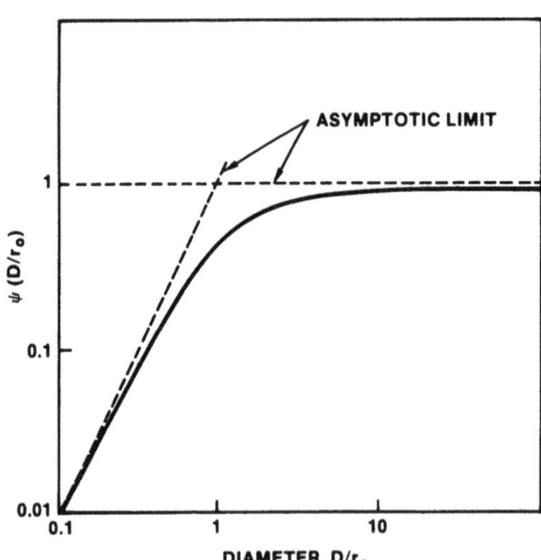

Figure 5.20. Dependence of the normalized signal-to-noise ratio, $F(D/r_0)$, upon the normalized diameter D/r_0. When D is very much less than r_0, the signal-to-noise ratio increases almost as the square of the diameter, as it should for a diffraction-limited system. Atmospheric optical effects are of no significance in this range of diameter. For diameters much larger than r_0, the consequences of atmospheric optical effects are so dominant that the signal-to-noise ratio cannot be improved by increasing the collector diameter.(10)

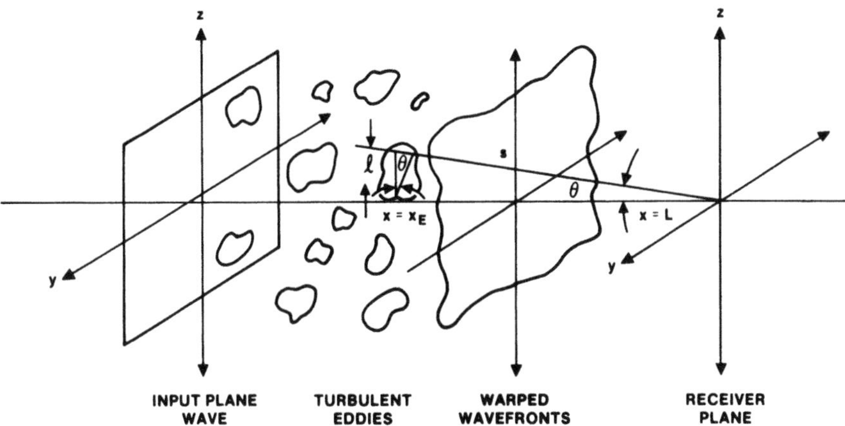

Figure 5.21. Schematic of interaction of a plane wave with atmospheric turbulence, showing geometry for determining eddy size most effective for producing scintillation in the receiver plane. The most effective eddy is equal in size to the first Fresnel zone as viewed from the receiver.

or scintillations, in the log amplitude and hence intensity. It is easily shown that the eddy size most effective for producing scintillations is roughly equal to the first Fresnel zone as viewed from the receiver plane. Considering the geometry shown in Figure 5.21, it is clear that to effectively produce scintillation at $x = L$, the phase shift across the eddy must be at least π. This implies that $s - (L - x_E) = \lambda/2$ or

$$\lambda/2l = \sin\theta = \frac{\ell}{s} \approx \ell/(L - X_E)$$

and we have for the minimum eddy size

$$l = \lambda(L - X_E)^{1/2}$$

which is the size of the first Fresnel zone for the path length from X_E to L. Eddies much larger than this will not produce scintillation at $x = L$ because they do not diffract light through a sufficiently large angle. An eddy of $\ell > \ell_0$ will scatter light through an angle $\theta = \lambda/l$. So to reach the point $x = L$, the diffraction angle must be $\theta = l_0(L - x_E)$. By assumption, $l_0 > [\lambda(L - x_E)]^{1/2}$, so we have $\lambda/l_0 < l_0/(L - x_E)^{1/2}$ and very little diffracted energy reaches the receiver. Eddies smaller than l_0 contribute less at $x = L$ because the scattered amplitude is less, due to the weaker refractivity fluctuations associated with them in the Kolmogorov spectrum.

5.2.6.1. Log-Amplitude and Phase Fluctuations for Kolmogorov Turbulence

The following two equations for the spectrum of the log-amplitude and phase fluctuations, assuming that the turbulence obeys Kolmogorov's law, l_0 is negligibly small, and L_0 is large, can be derived by substituting Equation (5.1.20) in Equation (5.2.20) and making the appropriate choice for β. For plane waves we have

$$\begin{bmatrix} \phi_X(\mathbf{k}) \\ \phi_S(\mathbf{k}) \end{bmatrix} = 2\pi k^2 0.0333 k^{-11/3} \int_0^L C_n^2(x') \cos^2 |k^2(L - x')/2k| \, dx' \quad (5.2.100)$$

while for spherical waves

$$\begin{bmatrix} \phi_X(\mathbf{k}) \\ \phi_S(\mathbf{k}) \end{bmatrix} = 2\pi k^2 0.0333 k^{-11/3} \int_0^L C_n^2(x')(x/L)^{5/3} \cos^2 |k^2(L - x')/2k\gamma| dx' \quad (5.2.101)$$

The variance of X, σ_{X^2}, for the isotropic turbulence is given by

$$\sigma_{X^2} = \int_0^\infty 2\pi k \phi_X(k) \, dk \quad (5.2.102)$$

Substituting Equations (5.2.101) and (5.2.100) in Equation (5.2.102) and performing the integration yields the following results:

$$\sigma_{X^2} = 0.56 k^{7/6} \int_0^L C_n^2(x)(L - x)^{5/6} \, dx \quad (5.2.103)$$

for plane waves and

$$\sigma_{X^2} = 0.56 k^{7/6} \int_0^L C_n^2(x)(x/L)^{5/6}(L - x)^{5/6} \, dx \quad (5.2.104)$$

for spherical waves.

For plane-wave propagation, the turbulence is weighted by the factor $(L - x)^{5/6}$, which attains its maximum value near the source and approaches zero at the receiver. This implies that turbulence near the source is more effective in producing scintillation than turbulence near the receiver. In contrast to the plane-wave case, the spherical-wave weighting factor $(x/L)^{5/6}(L - x)^{5/6}$ has its maximum value at the midpoint of the path and approaches zero at the source and the receiver. Thus, if the turbulence is concentrated near the transmitter, e.g., upward propagation through the atmosphere, the scintillation of a spherical wave will be less than that of a plane wave.

We can use these results to obtain the irradiance covariance function defined by Equation (3.1.13):

$$C_I(x, \mathbf{r}_1, \mathbf{r}_2) = \{E[I(x, \mathbf{r}_1)I(x, \mathbf{r}_2)] - E[I(x, \mathbf{r}_1)]E[I(x, \mathbf{r}_2)]\}/E[I(x, \mathbf{r}_1)]E[I(x, \mathbf{r}_2)] \quad (5.2.105)$$

In particular, under the assumption of lognormal statistics for X it can be shown that the log-amplitude covariance is related to the intensity covariance by

$$C_I(x, \mathbf{r}_1, \mathbf{r}_2) = \exp\{4C_X(x, \mathbf{r}_1, \mathbf{r}_2)\} - 1 \quad (5.2.106)$$

The variance of the irradiance fluctuations,

$$\sigma_{I^2} = E[I^2] - E[I]^2 \quad (5.2.107)$$

is then given by

$$\sigma_{I^2} = \exp\{4C_X(x, 0, 0)\} - 1 = \exp\{4\sigma_X^2\} - 1 \quad (5.2.108)$$

Equations (5.2.100)-(5.2.104) are based on the Rytov approximation and predict that ϕ_X and hence σ_{X^2} increase without limit as C_n^2 or the path length increases. However, it is found experimentally that with increasing C_n^2 or path length, σ_{X^2} reaches a peak value of approximately 0.3 and then slowly decreases as shown in Figure 5.22. This saturation effect can occur for visible light with horizontal paths on the order of a kilometer or less.

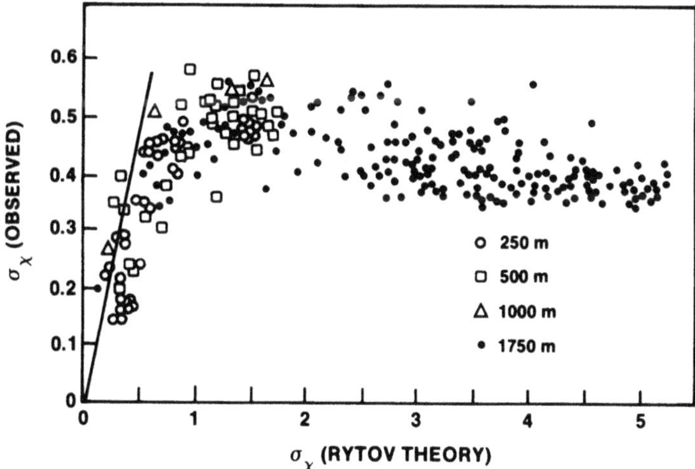

Figure 5.22. Observed strength of scintillation versus the scintillation predicted by Rytov theory for various horizontal path lengths with evidence of saturation.[9]

We recall that the Rytov approximation requires that the magnitude of the scattered field phase be small compated to the unperturbed phase gradient. For this reason, the Rytov technique does not correctly account for multiple-scattering effects. Qualitatively, saturation occurs because multiple scattering can cause the incident wave to become increasingly incoherent as it propagates into the medium. A single source of light can, therefore, appear as extended multiple sources scintillating with random phase. When these multiple apparent source fields are added, the resultant intensity scintillation is limited.

Based on some experimental work in strong turbulence, the effects of saturation on the wavefront phase are relatively minor. Thus, the Rytov statistics for phase, and other related quantities such as angle of arrival fluctuations, will have a greater range of validity than the log amplitude statistics, and for most engineering purposes they can be used even in the presence of saturation.[11]

5.2.6.2. Scintillation Statistics

We have shown in the previous sections that the random spatial and temporal fluctuations in the refractive index of the atmosphere generate scintillations in the irradiance received by an optical communication system. It is not possible, in general, to determine the probability density function for the intensity scintillations under arbitrary atmospheric conditions and beam parameters. The statistics of scintillation continue to be the subject of numerous experimental and theoretical investigations whose results we summarize below.

Heuristically, one can view a turbulent medium as consisting of a series of independent slabs, each of which is thicker than the outer scale L_0, so that fluctuations between slabs are independent. Using the notation of Section 5.2.1, the field after N slabs is given by:

$$f_N(\mathbf{r}) = \exp\{\Theta_0(\mathbf{r}) + \Theta_1(\mathbf{r})\} = \exp\{\Theta_0(\mathbf{r}) + \sum_{i=1}^{N} \Theta_{1i}(\mathbf{r})\} \qquad (5.2.109a)$$

where $T_i(\mathbf{r}) = \exp(\Theta_{1i}(\mathbf{r})$ is the complex amplitude transmittance for the ith slab. Since the slab thickness exceeds L_0, the $\Theta_{1i}(\mathbf{r})$, $i = 1, \ldots, N$, are independent random variables. Because the number of slabs can be large, the received field is the result of many multiplicative effects. Alternatively, the field can also be written as a random phasor sum of N complex fields $\mathbf{a}_i$:

$$f(\mathbf{r}) = \sum_{i=1}^{N} \mathbf{a}_i = \sum_{i=1}^{N} a_i \exp\{i\theta_i\} \qquad (5.2.109b)$$

where the phasor step lengths a_i are generated by scattering from the refractive index inhomogeneities and θ_i are the associated phase shifts. In conjunction with certain additional assumptions, we will now show these models for the scattered field can be used to evaluate the statistics of intensity scintillations.

Considering the representation (5.2.109a), because the number of slabs can be large, the received field is the result of many multiplicative effects. Thus, by the central limit theorem, the sum:

$$\Theta_1(\mathbf{r}) = \sum_{i=1}^{N} \Theta_{1i}(\mathbf{r}) \tag{5.2.110}$$

and its real part, $X = \text{Re}\,\Theta_1(\mathbf{r})$ will be normally distributed:

$$P(X) = \exp\{X - E[X])^2\}/(2\pi\sigma_X^2)^{1/2} \tag{5.2.111}$$

There have been claims that scintillations continue to obey lognormal statistics even into the saturation regime. However, recent evidence suggests that this is not the case and, in fact, departure from lognormal statistics increases as the turbulence strength or the optical path increases [5.23, 5.24].

Several different mathematical models have been proposed to describe this departure from lognormal statistics. Their ranges of validity are usually expressed in terms of the normalized intensity variance, σ_{NI}^2, defined as

$$\sigma_{NI}^2 = \sigma_I^2/E[I]^2 = E[I^2]/E[I]^2 - 1 \tag{5.2.112}$$

In particular, we can identify three regions for σ_{NI}^2. For $\sigma_{NI}^2 < 0.75$, or for path lengths less than 100 m, the lognormal distribution is valid. For $\sigma_{NI}^2 = 1$ or path lengths of several kilometers, the number of independent scatterings in the representation (5.2.109b) is a large fixed number, the phase θ_i will be uniformly distributed over the interval $[-\pi, \pi]$ and, as a consequence of the central limit theorem, the scattered field will be a zero mean, circular complex Gaussian process. This is the fully developed speckle result which leads to the Rayleigh distribution for the field amplitude. Rayleigh statistics for the amplitude imply negative exponential statistics for the intensity, the probability density given by [5.3]

$$P(I) = \exp\{-I/b\}/b \tag{5.2.113}$$

where $b = E[I]$. For $3 < \sigma_{NI}^2 < 4$, or intermediate path lengths, the number of independent sources comprising the scattered amplitude may not be large enough to justify use of the central limit theorem. One set of models used to describe the intensity statistics in this nonGaussian transition region between lognormal and negative exponential statistics are the so-called K-related distributions.[1-6] We will now focus our attention on the underlying assumptions which lead to a K-statistical description.

In our previous statistical descriptions, it has been assumed that N is a fixed asymptotically large number. In the transition region it is certainly plausible that N may be small and could potentially fluctuate. In our development of the lognormal statistics, we identified N with the number of correlation cells which are characterized by the outer scale size of the atmosphere. Within these large cells many small cells will contribute to the scattered field in a correlated manner. This modulation of small scale inhomogeneities over distances on the order of the outer scale size of the atmosphere is referred to as a clustering effect. The underlying hypothesis of the K-statistical model is that the number of these correlated small scale cells can fluctuate. One way to account for this effect in the representation (5.2.109b) is to adopt a number-fluctuation model for N, where N is now interpreted as the number of small scale inhomogeneities. It has been shown that the choice of a Poisson model for N does not lead to new results in the limit of large N, but that when the negative binomial distribution for N (5.2.114a) is used then the K-statistics (5.2.114b) are obtained [5.22]:

$$P(N) = \binom{N+\alpha-1}{N}(E[N]/\alpha)^N/(1+E[N]/\alpha)^{N+\alpha} \tag{5.2.114a}$$

$$P_K(I) = 2E(I)^{-1}y^{(y-1)}/2x^{(y-1)}/2K_{y-1}(2(xy)^{1/2}))/\Gamma(y)$$

$$N_K^{[m]} = m!\Gamma(m+y)/y^m\Gamma(y)$$

$$x = I/E[I], y = 2/(\sigma_{NI}^2 - 1) \tag{5.2.114b}$$

where (a/b) represents the standard binomial coefficient, α is a parameter characterizing the clustering in the negative binomial distribution, K_v are modified Bessel functions, Γ is the Gamma function, and $n_K^{[m]}$ is the mth moment of the intensity. The negative binomial distribution is an exact stationary solution for the population in a birth–death–immigration process, the parameter α given by the ratio v/λ, where v is the immigration rate and λ is the birth rate.[7] For atmospheric turbulence it is assumed that it describes the evolution of the turbulent eddies. The selection of the negative binomial distribution for $P(N)$ appears to be a reasonable one based on the success of the K-distribution in fitting experimental data.

Recently, a universal statistical model has been proposed by Phillips and Andrews.[18] They break the field into two parts, a specular, or line-of-sight, component, $A_S \exp\{i\theta\}$, and a diffuse component, $A_D \exp\{i\phi\}$, generated by multiple scattering. The total intensity is then given by

$$I = |A_T|^2 = A_S^2 + A_D^2 + 2A_D A_S \cos(\theta - \phi) \tag{5.2.115}$$

The PDF (probability density function) for the total intensity can then be expressed in the form:

$$P(I) = (1/2)\left\{\int_0^\infty zJ_0[(Iz)^{1/2}]E[J_0(A_S z)]E[J_0(A_D z)]\right\} \tag{5.2.116}$$

where

$$E[J_0(A_S)] = \int_0^\infty P(A_S)J_0(A_S)\, dA_S \tag{5.2.117}$$

and

$$E[J_0(A_D)] = \int_0^\infty P(A_D)J_0(A_D)\, dA_D \tag{5.2.118}$$

$J_0(\ )$ is the Bessel function of the first kind of zero, and $P(A_D)$ and $P(A_S)$ are the PDFs for the diffuse and specular components, respectively. To account for the transition to a large number of scatterings as the diffuse component becomes prominent, the diffuse PDF is assumed to be a generalized Rayleigh density which becomes Rayleigh distributed as the number of scatterings becomes large. A density which exhibits this behavior is the m distribution given by

$$P(A_D) = 2m^m A_D^{2m-1} \exp\{-mA_D^2/b\}/\Gamma(m)b^m, \qquad A_D > 0 \tag{5.2.119}$$

where $b = E[A_D^2]$, m is related to $\sigma_{A_D}^{-2}$, and Γ is the gamma function. It is also assumed that the specular amplitude is described by the M distribution:

$$P(A_S) = 2M^M A_S^{2M-1} \exp\{-MA_S^2/c\}/\Gamma(M)c^M, \qquad A_S > 0 \tag{5.2.120}$$

where $C = E[A_S^2]$. Substituting Equations (5.2.120) and (5.2.119) in Equation (5.2.16) through Equations (5.2.117) and (5.2.118), it can be shown that the universal PDF is

$$P(I) = m\Gamma(m) \exp\{-mI/b\} \sum_{jk=0}^{\infty}\sum (=1)^k \begin{vmatrix} k \\ j \end{vmatrix} L_k(mI/b)\Gamma(M+j)$$
$$\times (rm/M)/\Gamma(m-k+j)/b\Gamma(M) \tag{5.2.121}$$

where

$$\begin{vmatrix} k \\ j \end{vmatrix} = k!/j!(k-j)! \tag{5.2.122}$$

is the binomial coefficient and

$$r = c/b \tag{5.2.123}$$

The parameter r defined by Equation (5.2.123) is the power ratio of the specular-to-diffuse components of the field. For conditions of relatively weak turbulence, where most of the energy of the field is concentrated in the specular component, the numerical value of r is relatively large, whereas it monotonically decreases to zero as the strength of turbulence increases. It can be shown that as r gets large, $p(I)$ transitions to the negative exponential PDF.

5.2.6.3. Aperture Averaging of Scintillation

It is well known that the deleterious effects of scintillation can be reduced by increasing the size of the receiver aperture in an optical communication system. Qualitatively, we may regard the aperture as consisting of an array of nonoverlapping regions; within each region the scintillations are perfectly correlated while between regions there is no significant correlation. If the aperture is sufficiently large, then uncorrelated scintillations from many regions will be averaged together, thereby reducing scintillation-induced signal variance. The mechanism here is analogous to the one which results in the saturation effect described previously. Fried has analyzed this phenomenon for the case of an infinite plane wave[(24)]. We will now develop the mathematical formalism required to describe this effect and present numerical results.

We consider a circular receiver aperture of diameter D defined by the function:

$$P(\mathbf{r}, D) = \begin{cases} 1 & \text{for } |\mathbf{r}| < D/2 \\ 0 & \text{for } |\mathbf{r}| > D/2 \end{cases} \tag{5.2.124}$$

The signal collected by the aperture can then be written

$$S(D) = \int_{-\infty}^{\infty} P(\mathbf{r}, D) I(\mathbf{r})\, d^2\mathbf{r} \tag{5.2.125}$$

where $I(\mathbf{r})$ is the irradiance impinging on the aperture. The signal variance is defined by

$$\sigma^{2S} = E[S^2] - E[S]^2 \tag{5.2.126}$$

Substituting Equation (5.2.125) in Equation (5.2.126) and using the fact that $E[S] = \pi D^2 I_0/4$, where $I_0 = E[I(\mathbf{r})]$, we obtain

$$\sigma_S^2 = \int_0^D pk(p, D)C_I(p)\,dp \tag{5.2.127}$$

where we have used the transformation of variables in Equation (5.1.8), $C_I(p)$ is the intensity covariance function defined for homogeneous, isotropic statistics by

$$C_I(p) = E[\{I(\mathbf{R} + \mathbf{p}/2) - I_0\}\{I(\mathbf{R} - \mathbf{p}/2)\}] \tag{5.2.128}$$

and $k(p, D)$ is given by

$$k(p, D) = \int_{-\infty}^{\infty} P(R + \mathbf{p}/2, D)P(R - \mathbf{p}/2, D) \tag{5.2.129}$$

and gives the area of overlap of two circles of diameter D whose centers are displaced a distance p, and is identical to the circular aperture OTF given by Equation (5.2.97) and repeated below:

$$k_0(p, D) = \begin{cases} (D^2/2)\cos^{-1}\{(p/D) - (p/D)[1 - (p/D)^2]\} & \text{for } p < D \\ 0 & \text{for } p > D \end{cases} \tag{5.2.130}$$

It is convenient to define the aperture-averaging factor, Θ, by dividing Equation (5.2.127) by the signal variance that would be observed if the scintillation were perfectly correlated over the entire aperture, that is, $(\pi D^2/4)^2 C_I(0)$,

$$\Theta = \sigma_S^2/(\pi D^2/4)^2 C_I(0) \tag{5.2.131}$$

Combining Equations (5.2.131), (5.2.127), and (5.2.106) we achieve our final result:

$$\Theta = (16/\pi D^2)\int_0^D \frac{[\exp\{4C_X(p)\} - 1]}{[\exp\{4C_X(0)\} - 1]} k(p, D)\,p\,dp \tag{5.2.132}$$

Visualization of the role of Θ in the context of the heuristic explanation of aperture averaging presented at the beginning of this section is aided by expressing Θ in the form:

$$\Theta = (\pi d_0^2/4)/(\pi D^2/4) \tag{5.2.133}$$

with

$$d_0 = D\Theta^{1/2} \tag{5.2.134}$$

Then d_0 may be regarded as the scintillation transverse coherence length, and the number of coherence areas within the aperture is just Θ^{-1}. The numerical results presented in this section verify the $1/D^2$ dependence of Θ predicted by Equation (5.2.133) when $d_0 < D/2$.

Using results of a rather complex derivation for $C_X(p)$, Fried[21] carried out calculations of Θ for an infinite plane wave for various values of $C_X(0)$ assuming the atmospheric model given by Equation (5.1.44). His results are presented in Figures 5.23 and 5.24 and Tables 5.1 and 5.2.

For small values of $C_I(0) = \sigma_I^2$, the aperture-averaging factor is given approximately by

$$\Theta = [1 + (D/2p_I)^2]^{-1} \tag{5.2.135}$$

where $p_I = (x/k)^{1/2}$ and $[h_0 \sec\theta/k]^{1/2}$ for homogeneous and inhomogeneous propagation paths, respectively. The value $(x/k)^{1/2}$ gives the transverse correlation length of the intensity fluctuations. We emphasize the fact that Fried's aperture-averaging results do not correctly account for multiple-scattering effects. For values of the normalized intensity variance

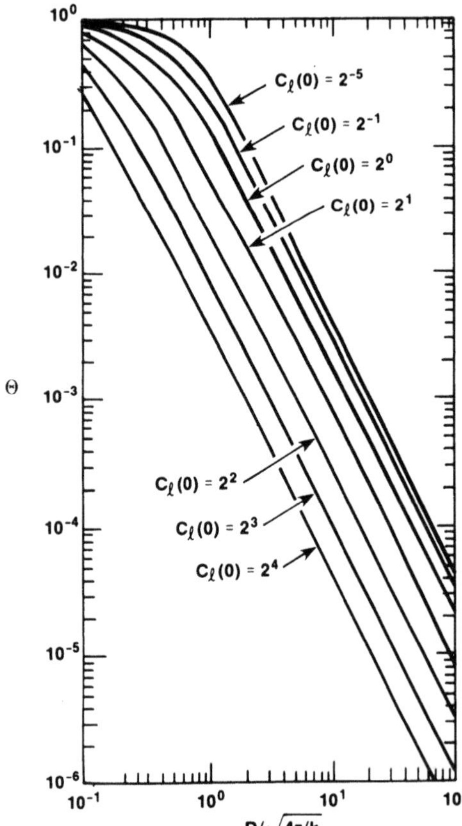

Figure 5.23. Dependence of the aperture-averaging factor, Θ, upon the normalized aperture diameter $D/(4z/k)^{1/2}$. The calculated curves, shown for various values of $C_l(0)$, the log-amplitude variance, are based upon the statistics of propagation of an infinite plane wave, with wave number k, traveling a horizontal path of length z.[24]

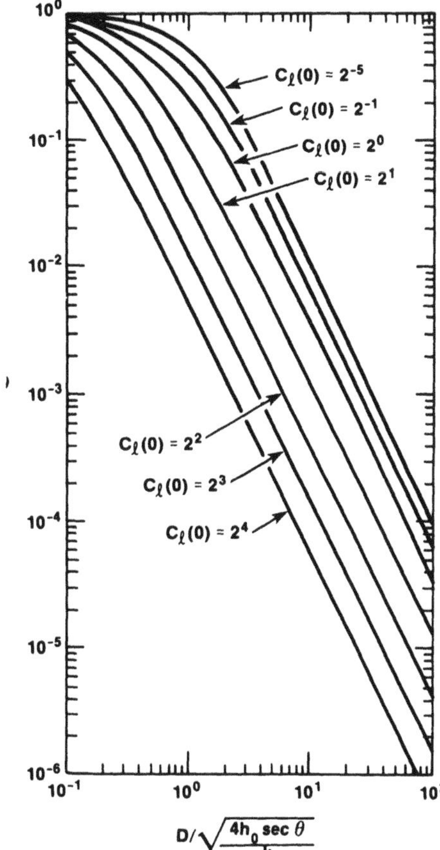

Figure 5.24. Dependence of the aperture-averaging factor, Θ, upon the normalized aperture diameter $D/(rh_0 \sec \theta/k)^{1/2}$. The calculated curves, shown for various values of $C_l(0)$, the log-amplitude variance, are based upon the statistics of propagation of an infinite plane wave, with wave number k, traveling from space to the ground and arriving with a zenith angle $\theta(h_0 + 3200 \text{ m})$.[24]

$\sigma_{NI}^2 = \sigma_I^2/E[I]^2 \geq 1$, or values of the log-amplitude variance $\sigma_{\chi^2} \geq 0.3$, the accuracy of the results can be expected to degrade considerably. The reason for this is that when one encounters the multiple-scattering saturation regime associated with long propagation paths or large values of C_n^2, $(x/k)^{1/2}$ is, by virtue of the single-scattering assumptions employed in the theory used to develop it, an overestimate of the correlation length P_I. We now propose a modification of Equation (5.2.135) which attempts to correct this deficiency.

As a first-order correction, one might expect the field coherence length, p_0, as given by the Rytov theory, to be a better estimate of p_I since it remains valid even under conditions of saturation. Thus, deep in the multiple-scattering regime we write:

$$\Theta = [1 + (D/2p_0)^2]^{-1} \tag{5.2.136}$$

Table 5.1. Aperture-Averaging Factor, Θ, for Values of the Normalized Collector Diameter $D/(4x/k)^{1/2}$ and Values of the Log-Amplitude Variance $C_l(0)$ [a] [(24)]

$D/(4x/k)^{1/2}$	Θ^b for $C_l(0) =$									
	2^{-5}	2^{-4}	2^{-3}	2^{-2}	2^{-1}	2^{0}	2^{1}	2^{2}	2^{3}	2^{4}
0.1	9.67(−1)	9.65(−1)	9.61(−1)	9.52(−1)	9.31(−1)	8.82(−1)	7.87(−1)	6.35(−1)	4.40(−1)	2,54(−1)
0.2	9.09(−1)	9.04(−1)	8.94(−1)	8.72(−1)	8.23(−1)	7.19(−1)	5.47(−1)	3.45(−1)	1.81(−1)	8.44(−2)
0.4	7.72(−1)	7.63(−1)	7.42(−1)	7.00(−1)	6.15(−1)	4.62(−1)	2.71(−1)	1.30(−1)	5.65(−2)	2.39(−2)
0.7	5.70(−1)	5.58(−1)	5.33(−1)	4.84(−1)	3.93(−1)	2.53(−1)	1.20(−1)	4.95(−2)	2.01(−2)	8.20(−3)
1.0	4.03(−1)	3.92(−1)	3.71(−1)	3.30(−1)	2.57(−1)	1.54(−1)	6.62(−2)	2.57(−2)	1.02(−2)	4.10(−3)
2.0	1.24(−1)	1.21(−1)	1.15(−1)	1.03(−1)	8.01(−2)	4.66(−2)	1.87(−2)	6.85(−3)	2.64(−3)	1.05(−3)
4.0	2.95(−2)	2.90(−2)	2.79(−2)	2.55(−2)	2.06(−2)	1.24(−2)	4.94(−3)	1.77(−3)	6.71(−4)	2.65(−4)
7.0	9.22(−3)	9.10(−3)	8.82(−3)	8.19(−3)	6.74(−3)	4.13(−3)	1.65(−3)	5.84(−4)	2.21(−4)	8.68(−5)
10.0	4.42(−3)	4.37(−3)	4.26(−3)	3.98(−3)	3.40(−3)	2.04(−3)	8.15(−4)	2.88(−4)	1.09(−4)	4.26(−5)
20.0	1.08(−3)	1.07(−3)	1.05(−3)	9.83(−4)	8.25(−4)	5.14(−4)	2.06(−4)	7.24(−5)	2.72(−5)	1.07(−5)
40.0	2.66(−4)	2.64(−4)	2.59(−4)	2.44(−4)	2.06(−4)	1.29(−4)	5.17(−5)	1.81(−5)	6.82(−6)	2.67(−6)
70.0	8.63(−5)	8.57(−5)	8.42(−5)	7.96(−5)	6.73(−5)	4.22(−5)	1.69(−5)	5.93(−6)	2.23(−6)	8.73(−7)
100.0	4.22(−5)	4.19(−5)	4.12(−5)	3.90(−5)	3.30(−5)	2.07(−5)	8.30(−6)	2.91(−6)	1.09(−6)	4.28(−7)

[a] Ref. 20.

[b] The negative number in parentheses is the power of 10 by which the three-digit number is to be multiplied.

Table 5.2. Aperture-Averaging Factor, Θ, for Values of the Normalized Collector Diameter $D/(4x \sec \theta/k)^{1/2}$ and Values of the Log-Amplitude Variance $C_l(0)$ [a] (24)

$D/(4x \sec \theta/k)^{1/2}$	Θ^b for $C_l(0) =$ 2^{-5}	2^{-4}	2^{-3}	2^{-2}	2^{-1}	2^{0}	2^{1}	2^{2}	2^{3}	2^{4}
0.1	9.74(−1)	9.72(−1)	9.69(−1)	9.61(−1)	9.44(−1)	9.05(−1)	8.25(−1)	6.91(−1)	5.05(−1)	3.09(−1)
0.2	9.29(−1)	9.25(−1)	9.16(−1)	8.98(−1)	8.57(−1)	7.69(−1)	6.14(−1)	4.13(−1)	2.29(−1)	1.09(−1)
0.4	8.24(−1)	8.15(−1)	7.98(−1)	7.61(−1)	6.85(−1)	5.39(−1)	3.40(−1)	1.71(−1)	7.54(−2)	3.17(−2)
0.7	6.67(−1)	6.55(−1)	6.31(−1)	5.82(−1)	4.86(−1)	3.29(−1)	1.65(−1)	6.84(−2)	2.73(−2)	1.10(−2)
1.0	5.32(−1)	5.19(−1)	4.94(−1)	4.44(−1)	3.53(−1)	2.16(−1)	9.47(−2)	3.61(−2)	1.39(−2)	5.51(−3)
2.0	2.47(−1)	2.39(−1)	2.24(−1)	1.94(−1)	1.44(−1)	7.72(−2)	2.83(−2)	9.80(−3)	3.64(−3)	1.41(−3)
4.0	7.04(−2)	6.82(−2)	6.40(−2)	5.58(−2)	4.14(−2)	2.19(−2)	7.66(−3)	2.55(−3)	9.31(−4)	3.58(−4)
7.0	2.29(−2)	2.13(−2)	2.01(−2)	1.78(−2)	1.35(−2)	7.34(−3)	2.58(−3)	8.45(−4)	3.07(−4)	1.18(−4)
10.0	1.04(−2)	1.01(−2)	9.59(−3)	8.55(−3)	6.57(−3)	3.62(−3)	1.28(−3)	4.17(−4)	1.51(−4)	5.77(−5)
20.0	2.47(−3)	2.42(−3)	2.31(−3)	2.08(−3)	1.62(−3)	9.12(−4)	3.24(−4)	1.05(−4)	3.79(−5)	1.45(−5)
40.0	6.03(−4)	5.91(−4)	5.66(−4)	6.13(−4)	4.04(−4)	2.29(−4)	8.15(−5)	2.63(−5)	9.49(−6)	3.62(−6)
70.0	1.95(−4)	1.91(−4)	1.83(−4)	1.66(−4)	1.31(−4)	7.49(−5)	2.67(−5)	8.62(−6)	3.10(−6)	1.18(−6)
100.0	9.50(−5)	9.32(−5)	8.94(−5)	8.13(−5)	6.44(−5)	3.67(−5)	1.31(−5)	4.22(−6)	1.52(−6)	5.80(−7)

[a] Ref. 20.

[b] The negative number in parentheses is the power of 10 by which the three-digit number is to be multiplied.

A smooth transition between Equations (5.2.135) and (5.2.136) is achieved with the following expression:

$$\Theta = [1 + d^2(p_I^{-2} + p_0^{-2})]^{-1} \qquad (5.2.137)$$

The aperture-averaging factor can be expressed simply in terms of σ_X^2. To achieve this form, we recall that

$$\sigma_X^2 = \begin{cases} 0.3075k^{7/6}C_n^2 x^{11/6} \\ 0.4963k^{7/6}k_0 h_0^{3/2} \sec^{11/6}\theta \end{cases} \qquad (5.2.138)$$

and

$$p_0^{-5/3} = \begin{cases} 0.423k^2 C_n^2 x \\ 0.375k^2 k_0 h^{2/3} \sec\theta \end{cases} \qquad (5.2.139)$$

for homogeneous horizontal and nonhomogeneous slant paths, respectively. Using Equations (5.2.138) and (5.2.139), we find the following expressions for p_0^{-2}:

$$p_0^{-2} = \begin{cases} 1.47k\sigma_X^{1.2}/x \\ 0.71k\sigma_X^{1.2}/h_0 \sec\theta \end{cases} \qquad (5.2.140)$$

Combining Equations (5.2.137) and (5.2.140) we obtain the following expressions for Θ:

$$\Theta = \begin{cases} [1 + D^2k(1 + 1.47\sigma_X^{1.2})/4x]^{-1} \\ [1 + D^2k(1 + 0.71\sigma_X^{1.2})/4h_0 \sec\theta]^{-1} \end{cases} \qquad (5.2.141)$$

for horizontal homogeneous and inhomogeneous slant paths, respectively.

Our expressions for Θ allow us to make some very general statements about how the turbulence strength impacts aperture averaging. In weak ($\sigma_{NI} \ll 1$) turbulence, C_n^2 dictates the value σ_{NI^2} but has very little impact on Θ since Θ scales with $(x/k)^{1/2}$. When σ_{NI^2} approaches one and beyond, the magnitude of C_n^2 has negligible effect on σ_{NI^2} and significantly impacts Θ, thereby enhancing the scintillation reductions achieved with aperture averaging.

5.2.7. Angle-of-Arrival Fluctuations

When an electromagnetic wave propagates through a random medium, different portions of the wave experience different phase shifts due to the spatial fluctuations of the medium index of refraction. This leads to fluctuations in the angle of arrival, θ, of the wavefront and can, in sufficiently strong turbulence, lead to beam breakup. In this section we will derive the mean square angle of arrival at the receiver aperture and the angle-of-arrival temporal spectrum for an optical field propagating through turbulence.

These results are important because they establish the performance requirements for adaptive optical systems.

We consider a receiving aperture of diameter D. The phase difference, ϕ, across this aperture is given approximately by

$$\phi = kD \sin\theta \approx k\, d\theta \tag{5.2.142}$$

for small θ. The mean square angle-of-arrival fluctuations can then be written

$$E[\theta^2] = E[\phi^2]/(kD)^2 = D_\phi(x = L, p = D)/(kD)^2 \tag{5.2.143}$$

The beam diameter is usually much larger than the diameter of the receiver aperture, and the phase structure function can be approximated by its plane-wave limit:

$$D_\phi(\mathbf{p}) \approx \begin{cases} D_1(\mathbf{p})/2 & \text{for } l_0 \ll p \leq (\lambda x)^{1/2} \\ D_1(\mathbf{p}) & \text{for } (\lambda x)^{1/2} \ll p \end{cases} \tag{5.2.144}$$

where $D_1(\mathbf{p})$ is the plane-wave structure function given by Equation (5.2.22). Combining Equations (5.2.144) and (5.2.143), we have

$$E[\theta^2] \approx \begin{Bmatrix} 1.146 \\ 2.92 \end{Bmatrix} D^{-1/3} \int_0^L C_n^2(x)\, dx \qquad \begin{matrix} \text{for } l_0 \ll D \ll (\lambda x)^{1/2} \\ \text{for } L_0 > D \gg (\lambda x)^{1/2} \end{matrix} \tag{5.2.145}$$

Equation (5.2.145) is strictly valid only for weak turbulence. It can be shown that the above expression can also be applied to the strong-turbulence case, the results being accurate to within a numerical coefficient of order unity.[(4)]

The spectrum of the angle-of-arrival fluctuations, $S_\phi(\omega)$, is given by

$$S_\phi(\omega) = (RD)^{-2} \int_0^\infty C_\phi(\tau) \cos(\omega\tau)\, d\tau \tag{5.2.146}$$

where

$$\phi(t) = \phi(x, \mathbf{r}_1, t) - \phi(x, \mathbf{r}_2, t) \tag{5.2.147}$$

and

$$C_\phi(\tau) = E(\phi(t)\phi(t+\tau)] \tag{5.2.148}$$

If we invoke the frozen-flow hypothesis, then

$$\begin{aligned} C_\phi(\tau) = E[\{&\phi(x, 0, t) - \phi(x, \mathbf{r}, t)\} \\ &\times \{\phi(x, -\mathbf{v}t, t) - \phi(x, \mathbf{r} - \mathbf{v}t, t)\}] \end{aligned} \tag{5.2.149}$$

where $\mathbf{v}$ is the flow velocity of the turbulence transverse to the direction of propagation. Using Equation (5.2.149) in Equation (5.2.146), we find that

$$S_\phi(\omega) = \left(\frac{1}{2(kD)^2}\right)[D_\phi(x, \mathbf{r} - \mathbf{v}t) + D_\phi(x, \mathbf{r} + \mathbf{v}t) - 2D_\phi(x, \mathbf{v}t)] \cos(\omega\tau)\, d\tau \tag{5.2.150}$$

It can be shown that for a plane wave in homogeneous turbulence[(4)]

$$S_\phi(\omega) = \begin{Bmatrix} 0.0326 \\ 0.0652 \end{Bmatrix} C_n^2 x v^{-5/3} D^{-1}(2\pi/\omega)^{8/3} \times [1 - \cos(\omega D/v)]/[1 + (1.07v/\omega L_0)]^{5/3} \qquad \begin{matrix} l_0 \ll D \ll (\lambda x)^{1/2} \\ D \gg (\lambda x)^{1/2} \end{matrix} \tag{5.2.150a}$$

For a spherical wave the result is[(4)]:

$$S_\phi(\omega) = \begin{Bmatrix} 0.0326 \\ 0.0652 \end{Bmatrix} C_n^2 x v^{-5/3} D^{-1}(2\pi/\omega)^{8/3}[1 - \sin(\omega D/v)(\omega D/v)^{-1}] \times [1 + (1.07v/\omega L_0)]^{-4/3} \qquad \begin{matrix} l_0 \ll D \ll (\lambda x)^{1/2} \\ D \gg (\lambda x)^{1/2} \end{matrix} \tag{5.2.150b}$$

The normalized angle-of-arrival spectrum for a spherical wave is plotted in Figure 5.25.

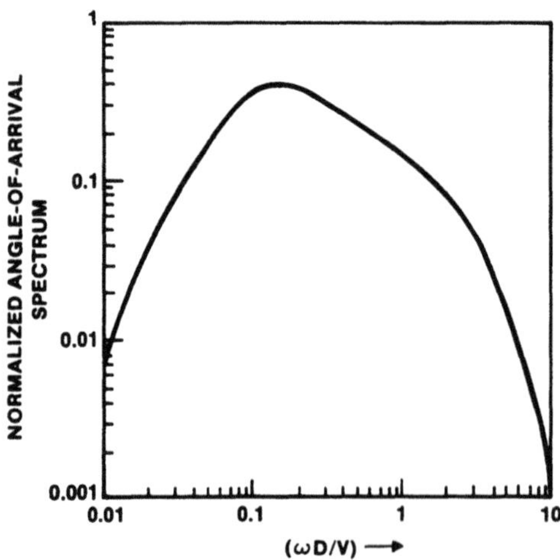

Figure 5.25. Angle-of-arrival spectrum for a spherical wave received by an aperture with $D—L_0 = 0.1$.[(4)]

5.2.8. Beam Spreading

In the absence of atmospheric turbulence, an optical beam exiting from an aperture of size D will have an angular spread due to diffraction on the order of λ/D, where λ is the wavelength. However, in the presence of turbulence, the beam is scattered by turbulent eddies, thereby inducing beam spread much greater than that associated with diffraction alone. In sufficiently strong turbulence, the beam can be broken up to such an extent that it appears as an ensemble of individual beams stretched out in stringlike fashion. Furthermore, since the turbulent eddies are in motion, we should expect the beam structure to be time dependent.

To describe these effects in greater detail, let us first assume that the medium is frozen so that the turbulent eddies have zero velocity. In general, those eddies which are large compared to the aperture diameter will tend to deflect the beam but not broaden it. The effect here is analogous to that which would be achieved by passing the beam through a glass wedge. The eddies which are smaller than the beam diameter tend to broaden the beam, but not deflect it significantly. This is loosely analogous to passing a beam through ground glass. These effects are depicted pictorially in Figures 5.26 and 5.27. Let us now set the medium into motion with a transverse flow velocity V. At any instant in time, we will observe two effects: broadening of the beam by the small eddies and deflection of the broadened beam by the large eddies. In particular, the broadened beam will be deflected in different directions in time intervals of order $T_w \approx D/V$. In addition to the aforementioned beam wander, there will also be a temporal change in the

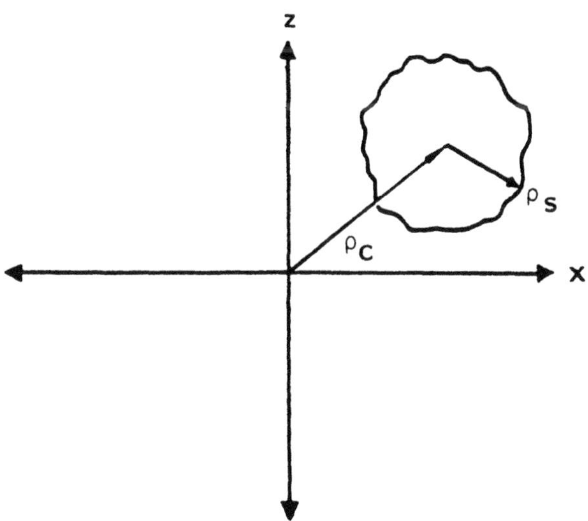

Figure 5.26a. Short exposure of laser spot illustrating beam wander.

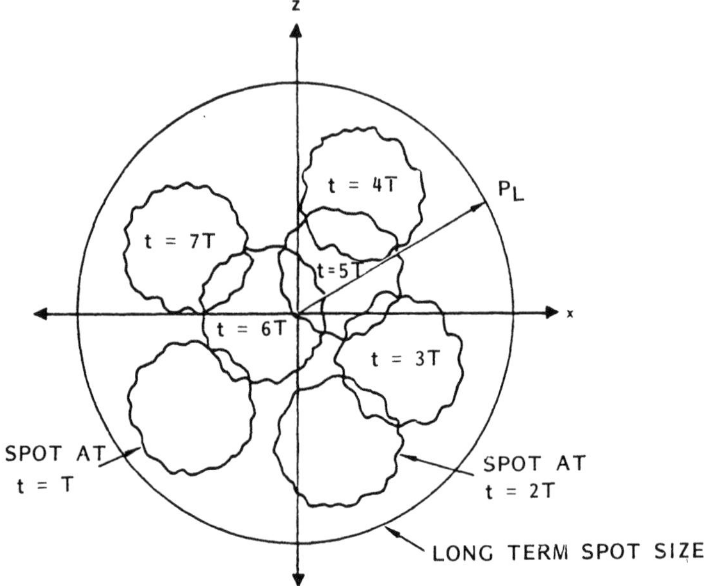

Figure 5.26b. Long exposure of laser beam illustrating beam wander and breathing.

specific structure of the broadened beam due to the velocity of the smaller eddies. This so-called beam breathing effect is due to the random interference pattern generated by eddies which lies within a transverse coherence length of the optical radiation. The characteristic time scale for beam breathing is $T_B \approx p_0/V$, where p_0 is the transverse coherence parameter discussed in Section 5.2.4.1. For the following conditions

$$D = 1\ \text{m}$$

$$p_0 = 10\ \text{cm}$$

$$V = 5\ \text{m/s}$$

values of T_w and T_B are

$$T_w = 0.2\ \text{s}, \quad T_B = 0.02\ \text{s}$$

while the characteristic frequencies are

$$\nu_w = 5\ \text{Hz}, \quad \nu_B = 50\ \text{Hz}$$

With reference to Figures 5.28 and 5.29, suppose that we now place a photographic plate in the receiver plane. It is clear that if the exposure time

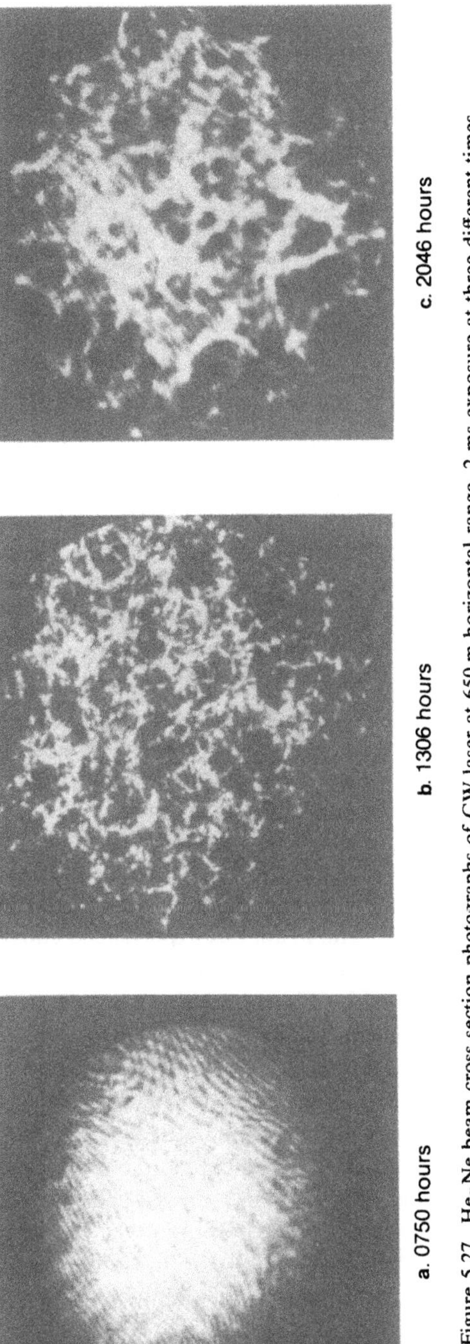

Figure 5.27. He–Ne beam cross section photographs of CW laser at 650 m horizontal range, 2-ms exposure at three different times.

is much larger than T_w, a broadened spot would be observed with a mean square radius given by[4]

$$E[p_L^2] = E[p_S^2] + E[p_C^2] + p_D^2 \tag{5.2.151}$$

where $E[p_S^2]$ is the short-term mean square radius due to beam breathing, $E[p_C^2]$ is the mean square radius described by the beam centroid wander, p_D^2 is the square of the spreading due to diffraction, and $E[p_L^2]$ is the square of the long-term beam spread. The description above is valid when the turbulence is relatively weak; when the turbulence is strong or the path length is long, the beam will break up as previously described.

The physical appearance of a laser beam cross section under varying conditions of turbulence is illustrated by the sequence of photographs in Figure 5.27. Dietz and Wright photographed a CW He–Ne laser beam cross section over a horizontal path at a range of 650 meters using a 2 millisecond exposure at three different times of day. The times at which this imagery was gathered corresponded to three meteorological conditions: inversion (before 0800 and after 1700 hours), neutral (0800 and 1700 hours), and lapse (0800 to 1700 hours). Figure 5.27a was taken at 0750 h during a neutral temperature gradient, at which time the atmosphere least affected the transmitted beam. The value of C_n was less than 10^{-8}. The presence of a low-contrast fringe pattern, which was caused by an interference filter in the optical system, is indicative of a large spatial coherence length made observable by a long temporal coherence length. Figure 5.27b was taken at 1306 h during high lapse conditions near the solar zenith. At this time the solar flux to the earth's surface was a maximum, producing the greatest heat transfer to the atmosphere at the air–ground interface, C_n reaching a peak value of 1.8×10^{-7}. The beam has experienced considerable break-up, the spread has increased, and the coherence effects which were prominent in 5.27a are not observable. Figure 5.27c was taken at 2046 h during temperature inversion conditions. C_n was less than 10^{-8}. Here again beam break-up is evident. In comparison with 5.27b, the scale size of the irradiance variations is larger and the edges of the stringlike cells are not as sharp. Spatial coherence effects are also apparent.

The following qualitative argument can be used to derive approximate expressions for beam spread.[26] As mentioned previously, the angular spreading due to a coherent radiator of diameter D is approximately $2/kD$. The effects of the turbulent medium can be viewed as being equivalent to a reduction of the coherent aperture to a diameter of p_0, the spherical wave coherence length. The angular spread is, therefore, given by $\theta_s \approx 2/kp_0$. For $p_0 \approx 10$ cm, $\lambda = 0.5\,\mu$m, $\theta_s = 5\,\mu$rad. The total beam spread due to turbulence can therefore be written as the product of the propagation distances and the spread angle due to turbulence:

$$E[p_C^2] + E[p_S^2] \approx 4x/k\rho_{0LT} \quad \text{long-term average}$$
$$E[p_S^2] \approx 4x/k\rho_{0ST} \quad \text{Short-term average} \tag{5.2.152}$$

Thus, the same formula is valid for short- and long-term exposures depending on the choice of long-or short-term coherence lengths, ρ_{0LT} or ρ_{0ST}. For the beam wave described by the Gaussian form

$$F(r) = \exp\{-p^2(2/D + ik/2F)\} \tag{5.2.153}$$

where D is the initial diameter and F is the radius of curvature of the wavefront, it can be shown that the spreading due to diffraction is

$$P_D^2 = 4x^2(kD)^2 + \{D(1-F)\}^2/4 \tag{5.2.154}$$

The total beam spread is obtained by combining Equations (5.2.154) and (5.2.151):

$$E[p_T^2]_{LT,ST} \approx 4x^2/(kD)^2 + \{D(1-F)\}^2/4 + 4x/k\rho_{0LT,ST} \tag{5.2.155}$$

We must now find expressions for ρ_{0LT} and ρ_{0ST} which account for the difference in the underlying mechanisms which produce beam wander and beam breathing. To accomplish this, we must examine the long- and short-term spherical-wave mutual coherence functions.

We have already derived the long-term mutual coherence function in Section 5.2.4. The value of ρ_0 corresponding to this case is given by Equation (5.2.34) and is repeated below:

$$\boldsymbol{\rho}_{0LT} = \left[1.45k^2 \int_0^L C_n^2(x)(1 - x/L)^{5/3}\, dx\right]^{-3/5} \tag{5.2.156}$$

which is valid for $L < (k^2 C_n^2 l_0^{5/3})^{-1}$. For $x \gg (k^2 C_n^2 l_0^{5/3})^{-1}$ Fante gives[4]

$$\boldsymbol{\rho}_{0LT} = \left[6.6k^2/4l_0^{1/3} \int_0^L C_n^2(x)(1 - x/L)^{5/3}\, dx\right]^{-3/5} \tag{5.2.157}$$

An expression for the short-term spherical-wave mutual coherence function can be derived using arguments similar to those given by Fried.[27] We recall that the mutual coherence function is given by

$$\Gamma(\boldsymbol{\rho}) = \exp\{-1/2D_w(\boldsymbol{\rho})\} \tag{5.2.158}$$

where $D_w(\boldsymbol{\rho})$ is the wave structure function. Our task, therefore, is reduced to finding the wave structure function associated with short-term effects. We have observed that the long-term spreading is due to beam wander which is associated with an overall tilt of the wavefront. Furthermore, we have shown in Section 5.2.4 that the wave structure function associated with a pure tilt is

$$D_w(\boldsymbol{\rho}) = n\rho^2 \tag{5.2.159}$$

where $n = h^2 E[\theta_T^2]$. The effects of tilt, therefore, can be removed from the long-term wave structure function by subtraction:

$$D_{w\mathrm{ST}}(\boldsymbol{\rho}) = D_{w\mathrm{LT}}(\boldsymbol{\rho}) - n\rho^2 \tag{5.2.160}$$

Substituting Equation (5.2.160) in Equation (5.2.158), we obtain the expression for the short-term mutual coherence function:

$$\Gamma_{\mathrm{ST}}(\boldsymbol{\rho}) = \Gamma_{\mathrm{LT}}(\boldsymbol{\rho}) \exp\{n\rho^2\} \tag{5.2.161}$$

Using the above results, it can be shown that[22]

$$\rho_{\mathrm{OST}} = \rho_{\mathrm{OLT}}[1 + 0.37(\rho_{\mathrm{OLT}}/D)^{1/3}] \tag{5.2.162}$$

Equations (5.2.156) and (5.2.157) together with (5.2.162) and (5.2.155) constitute our results for long- and short-term beam spread for $p \ll D$. An expression for the mean square radius of the beam centroid can be calculated from Equation (5.2.151). The results for $p \ll D$ are[4]:

$$E[p_c^2] = 2.97L^2/k^2\rho_{\mathrm{OLT}}^{5/3}D^{1/3} \tag{5.2.163}$$

From Equations (5.2.162) and (5.2.152), it is apparent that the long-term spread is greater than the short-term spread. This is a direct consequence of the fact that $\rho_{\mathrm{OST}} > \rho_{\mathrm{OLT}}$.

For $p_0 \geqslant D$, there are no simple expressions for ρ_{OST}. In this case, one must obtain a numerical result for the requirement that $\Gamma(\rho_{\mathrm{OST}}) = e^{-1}$. Figure 5.28[4] presents such results in terms of the parameters

$$\begin{aligned} \beta^2 &= [(kd)^2/4L]^2(1 - L/f)^2 \\ \mu^2 &= E[p_s^2]/E[p_1^2] \end{aligned} \tag{5.2.164}$$

For $\rho_0 \gg D$, there is very little beam wander and the long- and short-term spreads are approximately equal and given by Equation (5.2.155) with $p_0 = \rho_{\mathrm{OLT}}$.

The expressions for long- and short-term beam spreading which we have developed in this section can be used to aid in the specification of receiver aperture diameters and the short-term average irradiance within that aperture. Specifically, since one wishes to capture as much of the transmitted power as possible, the aperture diameter must be at least as large as the long-term beam spread area A_L. In addition, once absorption

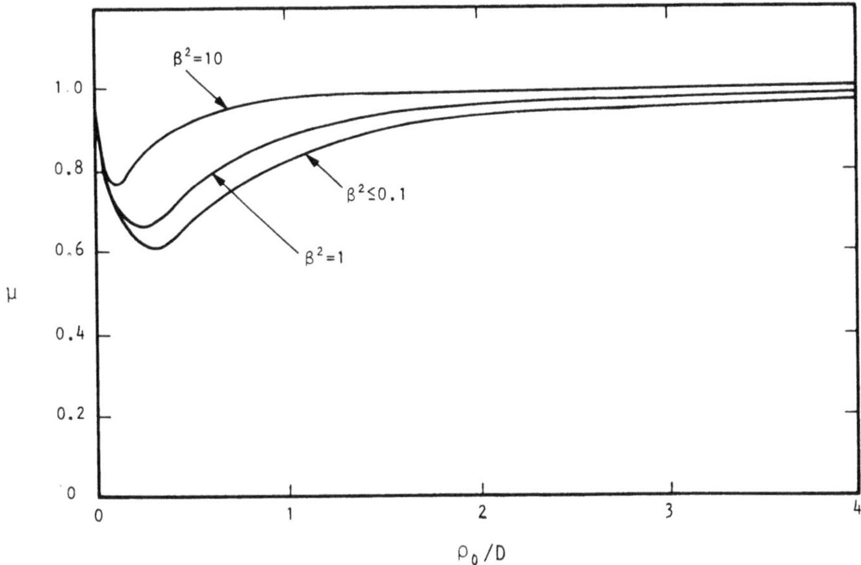

Figure 5.28. Ratio of the short- to long-term averaged beam spread.[4]

and scattering by particulates are accounted for, as described in Chapter 6, then the short-term power within the aperture is confined to the area A_s. We should emphasize that the physical mechanisms responsible for beam spreading are essentially the same as those responsible for the scintillation effects discussed in Section 5.2.6. Indeed, if the receiver aperture is smaller than the long-term beam spread, then the moving short-term spread area, A_s, will wander outside the receiver aperture, causing fades, or scintillation, in the received power. The depth of the fading induced by long-term beam wander will, of course, be inversely proportional to the receiver aperture diameter, approaching zero as D approaches A_L. At this point, we should caution that our notion of long- and short-term beam spreading breaks down in the region of saturated scintillation, where multiple-scattering effects become important. Under those conditions, the beam will have an appearance similar to that shown in Figure 5.27b and c, i.e., a somewhat circular area, whose diameter is nearly time invariant, which contains a time-varying speckle pattern. In this case one must employ the full multiple-scattering treatment discussed in Section 5.2.4.2.

5.3. Optical Propagation in Marine Turbulence

While our treatment thus far has been developed within the content of atmospheric turbulence, the mathematical formalism developed in Section 5.2 can, in principle, be applied to any turbulent medium; one need only specify the appropriate wavenumber spectrum of index of refraction fluctuations. In this section, the wavenumber spectrum of index of refraction fluctuations for marine turbulence is derived. Using a solution of the radiative transport equation, an expression for the scattered radiance is also derived.

5.3.1. $\Phi_n(k)$ for Marine Turbulence

The index of refraction for seawater depends on several variables, most notably, salinity, temperature, and pressure. Figure 5.29 gives a typical temperature–salinity diagram for ocean water density and refractive index. Generally, the influence of pressure variation on these latter two parameters is negligible; hence, we will ignore it in the analysis to come. In addition, without loss of generality, we will neglect the contributions of salinity to the index of refraction variations and only focus on the effects of temperature. Following Hodara,[24] we can write the rms refractive index fluctuations as

$$E[\Delta n^2]^{1/2} \approx (\partial n\ \partial t)E(\Delta T^2]^{1/2} \tag{5.3.1}$$

where $(\partial n/\partial t)$ is the refractive index gradient with respect to temperature and $E[\Delta T^2]^{1/2}$ is the rms variation of the temperature. In his development,

$$1.5 \times 10^{-5}/°\text{C} < (\partial n/\partial t) < 9 \times 10^{-4}/°\text{C} \tag{5.3.2}$$

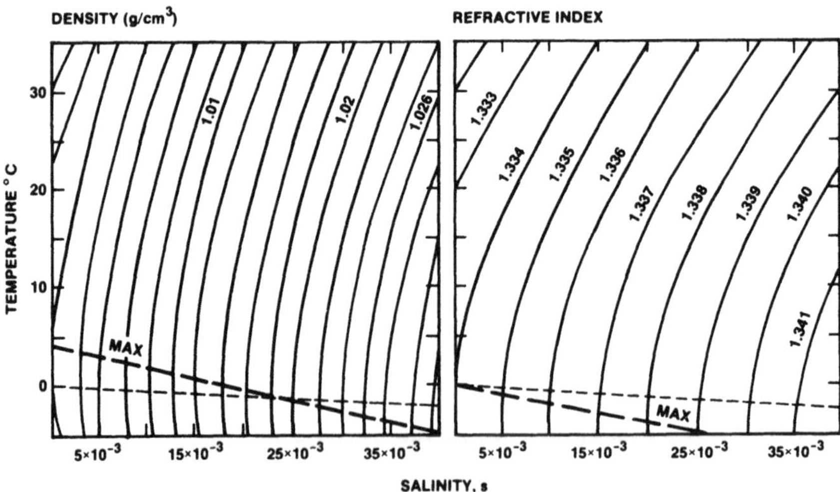

Figure 5.29. Diagrams for seawater density and refractive index.

Figures 5.30 and 5.31 show typical temperature power spectra for the waters off the Bahamas, as measured by Merterns.[24] It is apparent that $E(\Delta T^2]^{1/2}$ is between 10^{-3} and 10^{-2} °C for these waters and is depth dependent. Using Equation (5.3.1), this translates into rms refractive index fluctuations in the range

$$1.5 \times 10^{-8} < E[\Delta n^2]^{1/2} < 9 \times 10^{-5} \tag{5.3.3}$$

Figure 5.32 gives the correlation function for the temperature fluctuations averaged over several waters.[29] It is apparent that this function possesses an essentially exponential nature. In particular,

$$C_T(p) \approx \exp\{-\rho/L\} \tag{5.3.4}$$

where $L = 60$ cm. This dependence is in good agreement with what Mertens found for the Bahamas waters described above. Because of the proportionality relationship between temperature and refractive index, we expect the correlation function for the index of refraction variations to be given by

$$C_n(p) = \exp\{-\rho/L\} \tag{5.3.5}$$

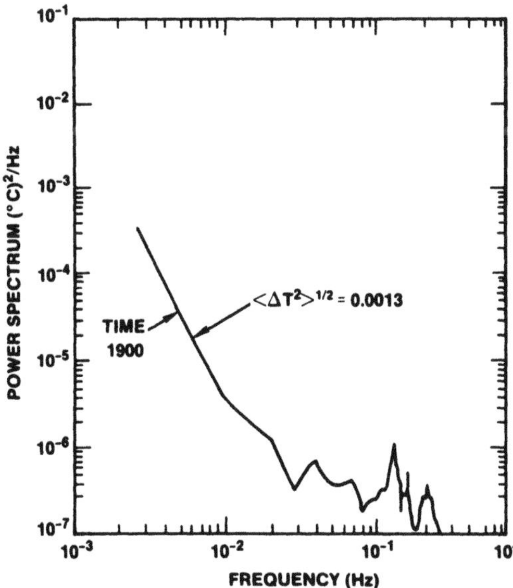

Figure 5.30. Temperature power spectrum in Bahamas waters.

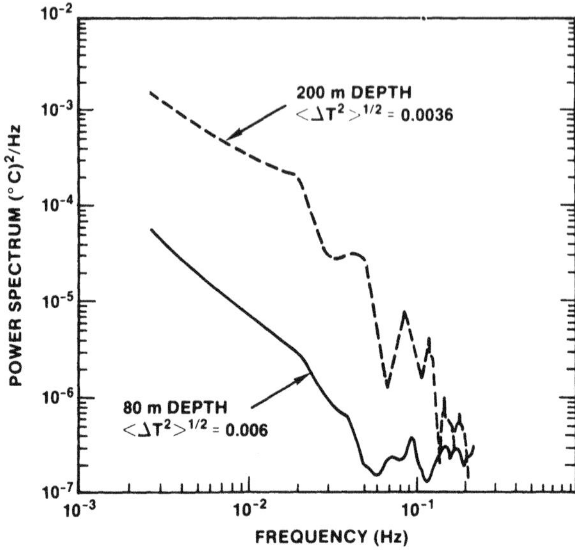

Figure 5.31. Temperature power spectrum at two different depths in Bahamas waters.

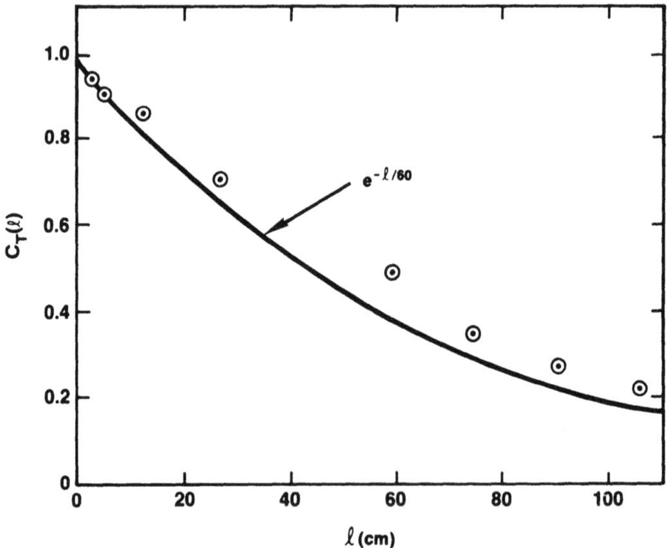

Figure 5.32. *In situ* temperature correlation function averaged over several waters.

Hence, the autocorrelation for the refractive index fluctuations is equal to

$$B_n(p) = E[\Delta n^2] \exp\{-p/L\} \tag{5.3.6}$$

and the wavenumber spectrum of the index of refraction fluctuations is then

$$\begin{aligned}\Phi_n(k) &= (2\pi)^2 \int_0^\infty B_n(p) \sin(kp) p^2/kp\, dp \\ &= E[\Delta n^2]/\pi^2 L[L^2 + k^2]^{-2}\end{aligned} \tag{5.3.7}$$

for seawater. Using Equation (5.3.7) in conjunction with Equation (5.1.35), the scattering cross section per unit volume is given by

$$\beta_g(\gamma) = 2E[\Delta n^2]/\pi L[(kL)^2 + \gamma^2]^{-2} \tag{5.3.8}$$

which yields a Fourier transform of the form

$$\boldsymbol{\beta}_g(v) = E[\Delta n^2] v K_1(v/k_0 L)/\pi\lambda_0 \tag{5.3.9}$$

where $K_1(x)$ is a modified Bessel function of the second kind.

5.3.2. The Scattered Radiance

In this section we shall develop an expression for the scattered radiance produced by marine turbulences using the method suggested by Fante.[15] Let us begin with a review of his approach. In deriving the transfer equation, one exploits the fact that the scattering cross section for most real atmospheres and oceans is strongly forward-biased in nature.[31] Fante noted that this would also allow one to expand the radiance within the integrand of (5.2.74) into a Taylor series about $\gamma = \gamma'$, and to retain only those terms less than order $b(\gamma_0^2)^2\nabla_\gamma^4 I$, where

$$\gamma_0^2 = b^{-1} \iint_{\infty}^{\infty} \sigma_g(\mathbf{g})\gamma^2\, d^2\boldsymbol{\gamma} \tag{5.3.10}$$

By separating the radiance I into an unscattered portion, I_u, plus a scattered portion, I_s, he rewrote the radiative transport equation in the form of the following set of differential equations:

$$dI_u/dz + \boldsymbol{\gamma} \cdot dI_u/d\tau \approx -cI_u \tag{5.3.11}$$

and

$$dI_s/dz + \boldsymbol{\gamma} \cdot dI_s/d\mathbf{r} - b\gamma_0^2\nabla^2 I_s/4 \approx \iint_{-\infty}^{\infty} \sigma_g(|\boldsymbol{\gamma} - \boldsymbol{\gamma}'|) I_u(\mathbf{r}, z, \boldsymbol{\gamma})\, d^2\boldsymbol{\gamma} \tag{5.3.12}$$

Equations (5.3.11) and (5.3.12) can be easily solved using Fourier transforms. In particular, their solutions are

$$(2\pi)^{-4} \iiiint_{-\infty}^{\infty} I_{u,s}(z, \mathbf{k}, \mathbf{v}) \exp\{-i(\mathbf{k} \cdot \tau + \mathbf{v} \cdot \boldsymbol{\gamma})\}\, d^2\mathbf{k}\, d^2\mathbf{v} \quad (5.3.13)$$

where

$$I_{u,s}(z, \mathbf{k}, \mathbf{v}) = \exp\{-\tau\} I_0(\mathbf{k}, \mathbf{v} + \mathbf{k}z) \quad (5.3.14)$$

and

$$I_s(z, \mathbf{k}, \mathbf{v}) \approx I_u(z, \mathbf{k}, \mathbf{v}) \int^{z_0} \exp\left\{bn - b\gamma_0^2 \int_0^n |\mathbf{v} + \mathbf{k}\mu|\, d\mu\right\} \sigma_g(\mathbf{v} + \mathbf{k}n)\, dn \quad (5.3.15)$$

With this behind us, we will now turn to the characterization of optical propagation in the marine channel. Substituting Equation (5.3.9) in (5.3.15) we have

$$I_s(z, \mathbf{k}, \mathbf{v}) = E[\Delta n^2] \exp\{-\tau\} I_0(\mathbf{k}, \mathbf{v} + \mathbf{k}z) \int_0^z \exp\left\{bn - b\gamma_0^2 \int_0^n |\mathbf{v} + \mathbf{k}\mu|\, d\mu\right\}$$
$$\times |v + \mathbf{k}n| K_1(|v + \mathbf{k}z|/k_0 L)/\pi\lambda_0\, dn \quad (5.3.16)$$

We need only Fourier invert this expression to obtain the scattered radiance.

As an example of the utility of Equation (5.3.16), we shall now calculate the power received by a Gaussian planar aperture located a distance z into the turbulent medium: specifically, the quantity

$$p(z) = \iiiint_{-\infty}^{\infty} A(\mathbf{r}) I(z, \mathbf{r}, \boldsymbol{\gamma})\, d^2\mathbf{r}\, d^2\boldsymbol{\gamma} \quad (5.3.17)$$

with receiver aperture function

$$A(\mathbf{r}) = \exp\{-r^2/r_1^2\} \quad (5.3.18)$$

Using Equations (5.3.13) and (5.3.18), we can rewrite Equation (5.3.17) as

$$P(z) = A_1 \exp\{-\tau)(2\pi)^{-2} \iint_{-\infty}^{\infty} \exp\{-k^2 r_1^2/4\} I_0(\mathbf{k}, \mathbf{k}z) + A_1 E[\Delta n^2]$$
$$\times \exp\{-\tau\}(2\pi)^{-2}(\pi\gamma_0)^{-1} \int_0^z n \exp\{bn\} \iint_{-\infty}^{\infty} k K_1(kn/k_0 L)$$
$$\times \exp\{-(k^2/4)(r_1^2 + b\gamma_0^2 n^3/4\} I_0(\mathbf{k}, \mathbf{k}z)\, dn \quad (5.3.19)$$

where

$$A_1 = \pi r_1^2 \tag{5.3.20}$$

If we assume the initial radiance to be Gaussian distributed according to

$$I_0(\mathbf{r}, \boldsymbol{\gamma}) = P_0(\pi^2 r_0 \gamma_0)^{-2} \exp(\{-(r/r_0)^2 - (\gamma/\gamma_0)^2\} \tag{5.3.21}$$

then

$$I_0(\mathbf{k}, \mathbf{v} + \mathbf{k}z) = P_0 \exp\{-[(kr_0)^2 - (\gamma_0|\mathbf{v} + \mathbf{k}z|)^2]/4\} \tag{5.3.22}$$

and we have

$$\begin{aligned} P(z) &= P_0 A_1 \exp\{-\tau\}\{\pi[r_0^2 + r_1^2 + (\gamma_0 z)^2]\}^{-1} \\ &\quad + P_0 A_1 E[\Delta n^2] L \exp\{-\tau\}[2\pi^2\gamma_0]^{-1} \int_0^z n \exp\{bn\} \\ &\quad \times \int_0^\infty K_1(kn/k_0 L) \exp\{-(ka_0)^2/4\} k^2\, dk\, dn \\ &= P_0 A_1 \exp\{-\tau\}\{\pi[r_0^2 + r_1^2 + (\gamma_0 z)^2]\}^{-1} \\ &\quad + P_0 A_1 E[\Delta n^2] L \exp(-\tau\}[2\pi^2\gamma_0]^{-1} \int_0^z n a_0^{-2} \exp\{bn\} \\ &\quad \times \exp\{(n/k_0 L a_0)^2/8\} W_{-1,1/2}[n^2/(2k_0 L a_0)^2]\, dn \end{aligned} \tag{5.3.23}$$

where $W_{\alpha,\beta}(x)$ is Whittaker's function[32] and

$$a_0^2 = [(r_0^2 + r_1^2 + (\gamma_0 z)^2 + b(\gamma_n)^3/3] \tag{5.3.24}$$

Using Ref. 12, Equation (5.3.23) can be written

$$\begin{aligned} P(z) &= P_0 A_1 \exp\{-\tau\}\{\pi[r_0^2 + r_1^2 + (\gamma_0 z)^2]\}^{-1} \\ &\quad + P_0 A_1 E[\Delta n^2] L \exp\{-\tau\}[2\pi\gamma_0^2]^{-1} \int_0^z n a_0^{-2} \exp\{bn\} \\ &\quad \times [1 - b_0 n^2 \exp\{b_0 n^2\} E_1(b_0 n^2)]\, dn \end{aligned} \tag{5.3.25}$$

where $E_1(x)$ is the exponential integral and

$$b_0 = (2k_0 L a_0)^{-2} \tag{5.3.26}$$

Unfortunately, there are no experimental data with which to compare the above to establish the validity range. However, Fante has postulated that the above is valid if $E[\Delta n^2] \ll 1$ and the photon mean free path $(1/b)$ is large compared to L. That is, we require $E[\Delta n^2] \ll 1$ with $(kL)^2 E[\Delta n^2] \ll 1$.

5.4. An Example–Multimode Heterodyne Detection

In this section we illustrate the use of the equations and concepts presented in this chapter by considering the possibility that the turbulence-induced limit on the SNR of an optical heterodyne detection system can be improved by performing "multimode" heterodyne detection. Our goal here is to obtain first-level estimates of SNR improvement which might be achieved using this technique and to identify critical issues which would have to be addressed by a higher-level design effort.

In Section 5.2.5 we found that there is very little improvement in the SNR of an optical heterodyne receiver if the aperture is increased beyond the scaled received field transverse coherence length, $r_0 = (3.44)^{5/3}\rho_0$, given by Equation (5.2.36). We also observed in Section 5.2.4 that the number of coherence areas within a receiver aperture of diameter D is given approximately by

$$N_c \approx (D/2\rho_0)^2 \tag{5.4.1}$$

Each of these coherence areas can be associated with a particular spatial mode in the angular spectrum decomposition of the electromagnetic field and forms a single-mode diffraction pattern in the detector plane. This effect is schematically illustrated in Figure 2.2. The phase associated with a particular modal pattern is spatially uniform but the value of that phase varies in a random fashion among the various modes.† While the centers of the patterns associated with different modes are spatially separated, there is some overlap in the associated patterns because of their finite spatial extent. This modal overlapping coupled with the random relative phase between patterns produces random interference which results in the generation of a speckle pattern in the detector plane. One can think of speckle as a two-dimensional noise pattern. The physical appearance of the speckle will be similar to that shown in Figure 5.27. The average size of the detector plane speckle over which the amplitude and phase remain correlated is given approximately by the size of a single-mode diffraction pattern for an aperture of diameter D which determines the spatial bandwidth. Here again the relative phase between speckles is a random variable. In single-mode

†As shown in Ref. 7 these modes are independent eigenmodes.

heterodyne detection as described in Section 5.2.5, a single detector with an aperture diameter which is less than or equal to the speckle correlation length is aligned with a particular speckle. The aperture effectively limits the field of view of the detector system to a value less than or equal to $\theta_s = \lambda / p_s$, where λ is the wavelength of the radiation and p_s is speckle correlation length. If the detector aperture is larger than this maximum value, then more than one speckle will, on the average, be detected. Because of the random phase between the signal fields associated with separate speckles, this leads to a corresponding reduction in the SNR if no phase correction is made. The SNR will therefore be a monotonically decreasing function of detector field of view once θ_s has been exceeded.

In multimode heterodyne detection one could employ a detector array to sample more than just one mode. After appropriate processing, the heterodyne signals from the separate detectors are combined to increase the SNR. The precombination processing consists of phase locking the heterodyne beat signals from the various detectors to compensate for the random phase between modes and applying weighting factors. The multimode SNR_m can therefore be expressed in terms of the single-mode SNR_s using Equation (5.4.1):

$$\mathrm{SNR}_m \approx N_c \times \mathrm{SNR}_s \tag{5.4.2}$$

where SNR_s is given by Equation (5.2.98). It is apparent from this expression that increases in the receiver diameter will lead to corresponding increases in the heterodyne composite SNR as long as the various spatial modes are properly detected and combined.[(33)] The limitations imposed on heterodyne detection by atmospheric turbulence will have therefore been greatly reduced.

In practice, the performance of a multimode heterodyne detection system could be degraded by a combination of environmental and system effects. We will now briefly discuss some of these effects to illustrate how one could use the concepts presented in the text to identify critical issues which would have to be addressed in a higher-level system design. We begin by observing that the speckle pattern in the detector plane is a complex random spatial structure which must be properly sampled to achieve the multimode heterodyning advantage. In particular, for optimum performance each detector in the array must be positioned so that it intercepts a region of the pattern over which the amplitude and phase remain correlated.[†] If it is positioned so that it intercepts the boundary between speckles, then two or more randomly phased signals will be detected and summed within

[†]This is equivalent to sampling a time function with a narrow gate. The sampling time should be short relative to variations in the function.

a single detector element, leading to a corresponding reduction in the instantaneous SNR for that element in the array with time scales associated with beam wandering and breathing, i.e., 0.2 and 0.02 s, respectively, as described in Section 5.2.8. One would therefore expect the SNR_s for each detector element in the array to be time dependent. These deleterious effects could potentially be minimized by increasing the number, N_d, of fixed detector elements in the array and weighting their individual contributions in the summed output in proportion to their respective SNRs. This would effectively reduce the temporal variations in SNR_m. One would also expect the performance of such a system to improve as N_d is increased. This is because the number of high-SNR_s elements in the array at any given time will be an increasing function of the number of detector elements. In effect, we are increasing the spatial diversity of the system with a corresponding improvement in SNR_m. There will, of course, be limits to the improvements in SNR_m that can be obtained by increasing the number of detector elements. For example, as N_d becomes larger, the detector element spacing must necessarily decrease. This could impose limitations on the detector size with corresponding reductions of SNR_s because less signal power is detected by each element in the array. If SNR_s is reduced below the value required to perform phase-locking of the array elements, then the multimode detection advantage would be lost. In all cases, $N_d \leqslant N_c$.

From a mathematical point of view, we have from Equations (3.3.18) (3.3.19), and (3.3.20) that

$$\mathrm{SNR} = \frac{(2\alpha a_r^2)A_c}{(1 + \alpha N_{\mathrm{ob}})2B_n} \tag{5.4.3}$$

where A_c is the coherence area defined as

$$A_c = \int_{A_d}\int_{A_d} \Gamma_{f_s}(\mathbf{r}_1, \mathbf{r}_2) f_L(\mathbf{r}_1) f_L^*(\mathbf{r}_2)\, d\mathbf{r}_1\, d\mathbf{r}_2 \tag{5.4.4}$$

with $f_L(\mathbf{r})$ the normalized diffraction pattern. Using Mercer's theorem,[7] we can expand the coherence function $\Gamma_{f_s}(\mathbf{r}_1, \mathbf{r}_2)$ as

$$\Gamma_{f_s}(\mathbf{r}_1, \mathbf{r}_2) = \sum \gamma_i \phi_i(\mathbf{r}_1) \phi_i^*(\mathbf{r}_2) \tag{5.4.5}$$

where $\phi_i(\mathbf{r})$ is an orthonormal set of eigenfunctions which are a solution to the equation

$$\gamma_i \phi_i(\mathbf{r}_1) = \int_{A_\gamma} \Gamma_{f_s}(\mathbf{r}_1, \mathbf{r}_2) \phi_i(\mathbf{r}_2)\, d\mathbf{r}_2 \tag{5.4.6}$$

where γ_i are the eigenvalues. The normalization as defined in Chapter 2 ensures that

$$\sum_{i=0}^{\infty} \gamma_i = A_r = \frac{\pi}{4} D^2 \tag{5.4.7}$$

Inserting this into Equation (5.4.4) yields

$$\begin{aligned} A_c &= \sum_{i=0}^{\infty} \gamma_i \left| \int \phi_i(\mathbf{r}) f_L^*(\mathbf{r})\, d\mathbf{r} \right|^2 \\ &\leqslant \sum_{i=0}^{\infty} \gamma_i = A_r \end{aligned}$$

with the equality holding when the normalized diffraction pattern is a set of eigenfunctions.

Recall from Equation (1.4.3) that the noise power received from an isotropic source, using a diffraction-limited receiving antenna, is

$$P_n = \underbrace{N(f)B}\,\lambda^2$$

Comparing this to Equation (3.3.18), where we have the received power from an atmospherically distorted field, again using a diffraction-limited receiving antenna (heterodyne detection),

$$P_s = \underbrace{a\alpha a_r^2}\, A_c$$

we see that we can attribute a coherence area of λ^2 to the noise source. As a consequence, we would expect that as a signal becomes more and more distorted due to scattering, its coherence area continuously decreases, reaching an asymptotic value of λ^2. For this asymptotic value, there is no "gain" remaining in the receiving structure, which, of course, becomes omnidirectional. Since the coherence areas all become equal, we see that the source appears spatially noiselike, in that it becomes equally distributed over all solid angles. Physically this means that, in a severe scattering environment, the radiation literally permeates throughout the channel (as in a cloud). From Equation (5.4.7) we can also observe that all the eigenvalues are equal to λ^2.

For systems where scattering has been strong enough to cause the incoming radiation to have a large range of angles (the radiance function subtends a large solid angle) so that $A_c \to \lambda^2$, we would require $(a_r^2/[N(f)]B_n \gg 1$ to have a high enough signal-to-noise ratio for maintaining phase lock on the individual modes. If this is not the case, then the modes would have to be combined by means of a square-law device (incoherently) with a greatly diminished performance. As we will see in Chapter 7, incoherent optical detection is appropriate for this case, whenever wide-field-of-view, narrowband filters can be obtained.

References

1. A. Ishimaru, *Wave Propagation and Scattering in Random Media*, Vols. 1 and 2, Academic Press, New York (1978).
2. V. I. Tatarski, *Wave Propagation in a Turbulent Medium* (translated by R. A. Silverman), McGraw-Hill, New York (1961).
3. V. I. Tatarski, The Effects of the Turbulent Atmosphere on Wave Propagation (translated by Israel Program for Scientific Translations), U.S. Department of Commerce, National Technical Information Service, Springfield, Virginia (1971) [originally published in 1967].
4. R. Fante, Electromagnetic beam propagation in turbulent media, *Proc. IEEE* **63**, 1669–1692 (1975).
5. R. Fante, Electromagnetic beam propagation in turbulent media, an update, *Proc. IEEE* **68**, 1424–1443 (1980).
6. S. F. Clifford, The classical theory of wave propagation in a turbulent medium, in: *Laser Beam Propagation in the Atmosphere* (J. W. Strobehn, ed.), Springer-Verlag (1978).
7. D. Fried, Limiting resolution through the atmosphere, *J. Opt. Soc. Am.* **56**, 1380 (1966).
8. R. E. Hufnagel and N. R. Stanley, Modulation transfer function associated with image transmission through turbulent media, *J. Opt. Soc. Am.* **54**, 52–61 (1964).
9. R. E. Hufnagel and N. R. Stanley, Propagation through atmospheric turbulence, in: *The Infrared Handbook* (W. L. Wolf and G. J. Zissis, eds.), The Environmental Institute of Michigan, Ann Arbor, Michigan (1978).
10. D. Fried, Optical heterodyne detection of an atmospherically distorted wavefront, *Proc. IEEE* **55**, 57–67 (1967).
11. R. Fante, Propagation of electromagnetic waves through a turbulent plasma using transport theory, *IEEE Trans. Antennas Propagat.* **AP-2**, 750–755 (1973).
12. R. Lutomirski and H. Yura, Propagation of a finite optical beam in an inhomogeneous medium, *Appl. Opt.* **10**, 1652–1658 (1971).
13. Philip M. Morse and Herman Feshbach, *Methods of Theoretical Physics*, McGraw-Hill, New York (1953), pp. 804–806.
14. R. Fante, Mutual coherence function and frequency spectrum of a laser beam propagating through atmospheric turbulence, *J. Opt. Soc. Am.* **64**, 592–598 (1974).
15. H. Yura, Mutual coherence function of a finite cross section optical beam propagating in a turbulent medium, *Appl. Opt.* **13**, 1399–1406 (1972).
16. G. Parry, Measurement of atmospheric turbulence induced intensity fluctuations in a laser beam, *Optical Acta* **28**, 715–728 (1981).
17. R. L. Phillips and L. C. Andrews, Measured statistics of laser-light scattering in atmospheric turbulence, *J. Opt. Soc. Am.* **71**, 864–870 (1981).
18. E. Jakeman and P. N. Pusey, Significance of K-Distributions in Scattering Experiments, *Phys. Rev. Lett.*, **40**(9), 546–550 (1978).
19. G. Parry and P. N. Pusey, K-distributions in atmospheric propagation of laser light, *J. Opt. Soc. Am.* **69**(5), 796–798 (1979).
20. E. Jakeman, On the statistics of K-distributed noise, *J. Phys. A: Math. Gen.* **13**, 31–48 (1980).
21. R. Barakat, Weak-scatter generalization of the K-density function with application to laser scattering in atmospheric turbulence, *J. Opt. Soc. Am. A* **3**, 401–409 (1986).

22. R. Barakat, Weak-scatter generalization of the K-density function. II. Probability density of total phase, *J. Opt. Soc. Am. A* **4**(7), 1213–1219 (1987).
23. E. Jakeman and R. J. A. Tough, Generalized K-distribution: a statistical model for weak scattering, *J. Opt. Soc. Am. A* **4**(9), 1764–1772 (1987).
24. D. Fried, Aperture averaging of scintillation, *J. Opt. Soc. Am.* **57** (1967).
25. Paul H. Deitz and Neal J. Wright, Saturation of scintillation magnitude in near-earth optical propagation, *J. Opt. Soc. Am.* **59**(5), 527–535 (1969).
26. H. Yura, Short term average optical-beam spread in a turbulent medium, *J. Opt. Soc. Am.* **63**, 567–572 (1973).
27. D. Fried, Statistics of geometric representation of wavefront distortion, *J. Opt. Soc. Am.* **55**, 1427 (1965).
28. H. Hodara, Refractive Index Fluctuations in Seawater, AGARD Lecture Series 61 on Optics of the Sea, Neuilly Sur Seine, France (1973), pp. 2.2-1–2.2.-12.
29. L. Lieberman, The effect of temperature inhomogeneities in the ocean on the propagation of sound, *J. Opt. Soc. Am.* **23**, 563 (1951).
30. R. Fante, Intensity, Coherence and Frequency Spectrum of a Focused Beam in a Random Media, AFCRL Technical Report, AFCRL-TR-7 4-0335, Physical Sciences Research Papers No. 598 (1974).
31. R. D. Anderson and L. Stotts, Underwater measurements of off-axis radiance compared with various analytical treatments of the radiative transfer equation, *J. Opt. Soc. Am.* **72**, 738–746 (1982).
32. M. Abramowitz and I. A. Stegun, *Handbook of Mathematical Functions*, Dover Publications, New York (1965), p. 505.
33. R. M. Gagliardi and S. Karp, *Optical Communications*, Wiley-Interscience, New York (1976).

6

The Optical Scatter Channel and Its Properties

With the introduction of ultraviolet, visible, and infrared technologies in the areas of communications and surveillance, it has become necessary to account for the atmospheric and oceanic influences on such systems—in specific terms, the effects of particulate multiple scattering on optical radiation transfer. This is because the haze, fog, rain, and clouds comprising most of the atmospheric channel, and the yellow substance and small "sea animals" in the marine channel, cause the majority of the degradation suffered by the information-containing signal traversing that channel. Recall from the previous chapter that atmospheric and marine turbulence can generate significant wavefront distortions of a signal when optical paths of 50 meters or more are involved. However, the total angular and spatial spreading which can be experienced by an initially collimated pencil beam rarely exceeds 100 microradians and a meter or two, respectively.(1) Also, coherence distances are still on the order of centimeters (requiring only modest diversity of a coherent receiver), and no experimental evidence of pulse broadening exists, even down to picosecond lengths.(2,3) In contrast, particulate multiple scatter can induce dispersion in angle of arrival the order of tens of degrees, beam spreading in the hundreds of meters, degradation of spatial coherence down to lengths of microns or less, and multipath time spreading in the tens of microseconds.(4) Because of the magnitude of these larger effects, the mutual coherence function approach to channel characterization cannot be universally applied (we will shortly find it to be valid only over a finite range of scattering thicknesses), and additional mathematical techniques must be used to model optical propagation through these individual channels. In this chapter, we shall review the inherent and characterizing properties of the three basic components of the optical scatter channel: the atmosphere, the ocean, and the air/sea interface. A clear understanding of these building blocks must be possessed if one is to model satisfactorily the transfer of optical radiation through their individual and/or combined parts. Chapter 7 will describe how this information has been, and is currently, used to quantify energy propagation through the invididual and combined structures of this channel. Whenever possible, we have compared measured characteristics of the channel and its components with predictions from pertinent analytical or empirically devised models.

6.1. Mie Scattering Theory

To appreciate how the atmosphere and the ocean can affect optimum radiation transfer, we will need to have a working knowledge of Mie scattering theory. For the sake of completeness, we shall digress momentarily and outline its main aspects in this section. Many excellent texts can be found detailing the complete Mie theory; see Refs. 5-8 as examples.

Mie scattering theory is the application of the Maxwell equation[9,10] to the problem of a homogeneous sphere irradiated by a plane wave from a single direction.† In particular, given the diameter D of the spherical particle, its complex refractive index $m = n_r - in_i$ relative to the surrounding medium, and the wavelength λ of the incident plane wave, Mie theory yields the complex scattered field amplitudes for any direction θ from the direction of incidence. These expressions reduce to their simplest form when the electric vectors of both the incident and scattered waves are resolved into components perpendicular and parallel to the scattering plane defined by the directions of incidence and scattering. In this case, the scattered field amplitudes are written as

$$S_1(m, x, \theta) = \sum_{k=1}^{\infty} \frac{2k+1}{k(k+1)} a_k(m, x)\pi_k(\theta) + b_k(m, x)\xi_k(\theta) \quad (6.1.1)$$

and

$$S_2(m, x, \theta) = \sum_{k=1}^{\infty} \frac{2k+1}{k(k+1)} a_k(m, x)\xi_k(\theta) + b_k(m, x)\pi_k(\theta) \quad (6.1.2)$$

where the subscript 1 denotes the perpendicular component and the subscript 2, the parallel component. The quantity x is the Mie size parameter and is given by

$$x = \frac{\pi D}{\lambda}$$

† This theory also applies to continuous media whose refractive index varies with radius.

which is the ratio of the particle diameter to the wavelength of the incident radiation. It will be seen shortly that this parameter defines the basic scattering characteristics of the particulate medium at the most fundamental level. The coefficient sets $\{a_k(m, x)\}$ and $\{b_k(m, x)\}$ are called the Mie coefficients and are certain Ricatti–Bessel functions of real and complex arguments.[7] The functions $\pi_k(\theta)$ and $\xi_k(\theta)$ are spherical harmonics which are usually expressable as Legendre polynomials. In considering the quantitative description of a stream of electromagnetic energy of arbitrary characteristics, one can choose from several existing schemes depending on the nature of the problem of interest.[6] For the case of particulate scattering, the Stokes vector and scattering matrix representations are commonly used. This is because they are well suited for analyzing electromagnetic propagation through regions containing idealized Rayleigh and larger-sized particles, idealized meaning those composed of optically homogeneous, isotropic material with perfect spherical symmetry.[7] In the above situation, the scattered radiance is written as

$$\mathbf{N}(\theta)\Delta\Omega = \boldsymbol{\sigma}(\theta) \cdot \mathbf{N}_0 \Delta\Omega_0 \Delta\Omega \tag{6.1.3}$$

where $\mathbf{N}$ and $\mathbf{N}_0$ are column vectors of the form

$$\mathbf{N} = \begin{bmatrix} N_1 \\ N_2 \\ U \\ V \end{bmatrix}$$

Here $\mathbf{N}_0$ is the incident radiance distribution, $\Delta\Omega_0$ is the elemental solid angle occupied by the illuminating source (assumed to be at a point where the incident energy is in a "plane wave" state), and $\Delta\Omega$ is the elemental solid angle in which the scattering is considered. $\mathbf{N}$ and $\mathbf{N}_0$ are known as Stokes vectors and possess elements, called the Stokes parameters, given by

$$\begin{aligned} N_1 &= \overline{f_1 f_1^*} \\ N_2 &= \overline{f_2 f_2^*} \\ U &= 2\,\mathrm{Re}\,\{\overline{f_1 f_2^*}\} \\ V &= 2\,\mathrm{Im}\,\{\overline{f_1 f_2^*}\} \end{aligned} \tag{6.1.4}$$

with f_1 and f_2 being the perpendicular and parallel components, respectively, of the electric field. For example, an unpolarized incident radiance distribution would be written as

$$\mathbf{N}_0 = I_0 \begin{bmatrix} 1 \\ 1 \\ 0 \\ 0 \end{bmatrix}$$

with

$$I_0 = \overline{f_{10} f_{10}^*} = \overline{f_{20} f_{20}^*}$$

On the other hand, a left circularly polarized incident distribution would be given by

$$\mathbf{N}_0 = I_0 \begin{bmatrix} 1 \\ 1 \\ 0 \\ 2 \end{bmatrix}$$

The matrix $\boldsymbol{\sigma}(\theta)$ is called the Stokes scattering matrix and transforms the initial distribution $\mathbf{N}_0$ into its scattered form based on the scattering characteristics of the particulate medium involved. Mathematically it is defined as

$$\boldsymbol{\sigma}(\theta) = \begin{bmatrix} \sigma_1(\theta) & 0 & 0 & 0 \\ 0 & \sigma_2(\theta) & 0 & 0 \\ 0 & 0 & \sigma_3(\theta) & \sigma_4(\theta) \\ 0 & 0 & -\sigma_4(\theta) & \sigma_3(\theta) \end{bmatrix} \tag{6.1.5}$$

where

$$\sigma_1(\theta) = |S_1(\theta)|^2/k_0^2 \tag{6.1.6a}$$

$$\sigma_2(\theta) = |S_2(\theta)|^2/k_0^2 \tag{6.1.6b}$$

$$\sigma_3(\theta) = \text{Re}\,\{S_1(\theta)S_2^*(\theta)\}/k_0^2 \tag{6.1.6c}$$

$$\sigma_4(\theta) = -\text{Im}\,\{S_1(\theta)S_2^*(\theta)\}/k_0^2 \tag{6.1.6d}$$

$$k_0 = 2\pi/\lambda$$

Let us now relate the elements of $\boldsymbol{\sigma}(\theta)$ to the inherent optical properties of the particulates.[†] Assume initially a monodispersion. In this case, the total scattering cross section of an individual particle is given by

$$\sigma_s(m, x) = \frac{1}{2}\iint_{\Omega} \frac{|S_1(\theta)|^2 + |S_2(\theta)|^2}{k_0^2}\, d\Omega \tag{6.1.7}$$

$$= \frac{2\pi}{k_0}\sum_{n=1}^{\infty}(2n+1)[|a_n|^2 + |b_n|^2] \tag{6.1.8}$$

$$= \frac{\pi D^2}{4} Q_s(m, x) \tag{6.1.9}$$

with Q_s being the scattering efficiency defined as the scattering cross section divided by the geometrical cross section of the spherical particle. However, scattering is not the only perturbation that can occur when the incident field interacts with the particle. Because the refractive index of the particle may have an imaginary component, some energy may be lost through absorption and transferred to heat. In other words, the incident radiance

[†] Recently, many researchers have begun using an alternate form of Equations (6.1.4) and (6.1.5). Specifically, they defined the incident and scattered Stokes vectors to be given by

$$\mathbf{N}' = \begin{bmatrix} I \\ Q \\ U \\ V \end{bmatrix} \quad \text{and} \quad \mathbf{N}_0 = \begin{bmatrix} I_0 \\ Q_0 \\ U_0 \\ V_0 \end{bmatrix}$$

respectively. In these equations, I represents the total power in the incident field and Q the relative power difference between the perpendicular and parallel components of the field. Equations (6.14) and (6.15) can be transformed into new equations through the relations

$$\mathbf{N}' = \mathbf{TN}$$

and

$$\boldsymbol{\sigma}(\theta) = \mathbf{T}\boldsymbol{\sigma}(\theta)\mathbf{T}^{-1}$$

respectively, where

$$\mathbf{T} = \begin{bmatrix} 1 & 1 & 0 & 0 \\ 1 & -1 & 0 & 0 \\ 0 & 0 & 1 & 0 \\ 0 & 0 & 0 & 1 \end{bmatrix} \quad \text{and} \quad \mathbf{T}^{-1} = \begin{bmatrix} \frac{1}{2} & \frac{1}{2} & 0 & 0 \\ \frac{1}{2} & -\frac{1}{2} & 0 & 0 \\ 0 & 0 & 1 & 1 \\ 0 & 0 & 0 & 1 \end{bmatrix}$$

can be modified through either the absorption or scattering process. This combined modification is known as the extinction process. That is, the incident radiance distribution is becoming extinct by the absorption and scattering of energy. The particle's extinction cross section is given by

$$\sigma_e(m, x) = \frac{2\pi}{k_0^2} \sum_{n=1}^{\infty} (2n+1) \operatorname{Re}\{a_n + b_n\} \tag{6.1.10}$$

$$= \frac{\pi D^2}{4} Q_e(m, x) \tag{6.1.11}$$

with Q_e being the extinction efficiency. The ratio of the scattering efficiency to the extinction efficiency is called the single-scatter albedo, denoted by the symbol ω_0. We will find this parameter a key aspect in the developments to come in the next chapter. It is clear from Equation (6.1.9) and (6.1.11) that the total absorption efficiency is given by

$$Q_a = Q_e - Q_s \tag{6.1.12}$$

and likewise the absorption cross section by

$$\sigma_a = \sigma_e - \sigma_s \tag{6.1.13}$$

Figures 6.1 and 6.2 illustrate the absorption, scattering, and extinction efficiencies as a function of particle radius for the wavelengths 0.53 μm and 3.8 μm, respectively.[11] As is evident in Figure 6.1, the scattering and extinction efficiencies for nonabsorbing spheres are equal. This is because a real refractive index implies that the only aspect of the extinction process is scattering. Notice also in these curves that extinction efficiency (scattering plus absorption, or scattering alone) increases with particle radius, or more specifically Mie parameter, to the fourth power. This suggests that the larger the scattering target, the more degrading the interaction. However, these curves show that there is a limit to this degradation. Specifically, we find that the scattering cross section can be no larger than twice the geometrical cross section $D^2/4$, and this extends directly to the extinction cross section, for large x. This is indicated by $Q_e \sim 2$ for $x \to \infty$. This implies that any particle comparatively large with respect to the wavelength of the incident radiation will scatter that radiation in a predominantly forward direction. Let us discuss this point in more detail.

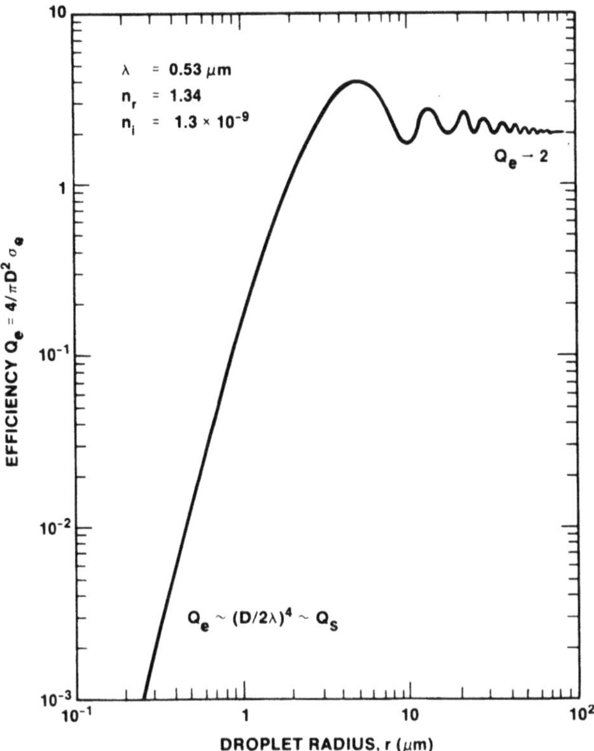

Figure 6.1. Absorption, scattering, and extinction efficiencies versus particle size for $\lambda = 0.53\ \mu$m.

Equating Equations (6.1.7) and (6.1.9), we have

$$\frac{\pi D^2}{4} Q(m, s) = \frac{1}{2} \iint\limits_{\Omega} \frac{[|S_1(\theta)|^2 + |S_2(\theta)|^2]}{k_0^2} \, d\Omega$$

or

$$\iint\limits_{\Omega} \left[\frac{P_1(\theta)}{4\pi} + \frac{P_2(\theta)}{4\pi} \right] d\Omega = 1 \tag{6.1.14}$$

where

$$P_j(\theta) = \frac{2|S_j(\theta)|^2}{x^2 Q_s} \pi \tag{6.1.15}$$

for $j = 1, 2$. The sum

$$\frac{P(\theta)}{4\pi} = \frac{P_1(\theta)}{4\pi} + \frac{P_2(\theta)}{4\pi}$$

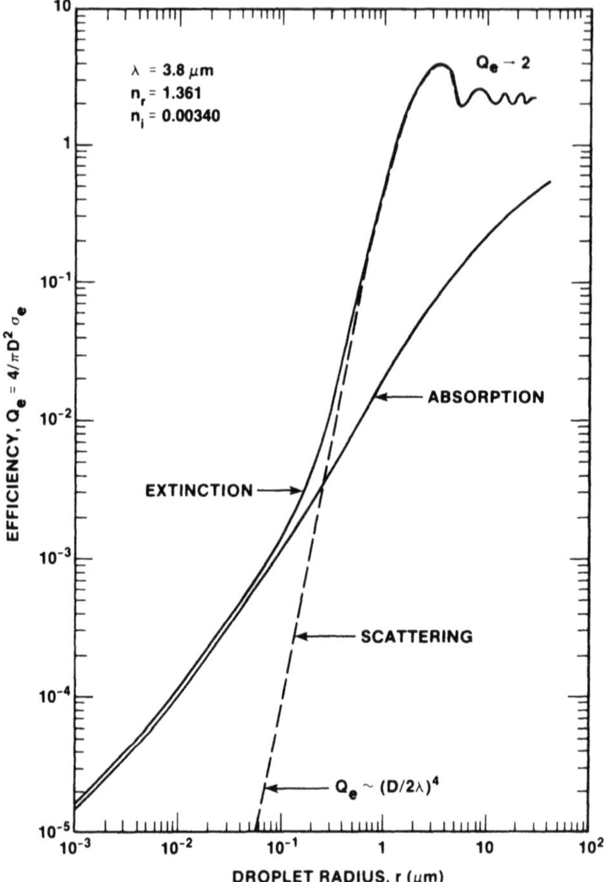

Figure 6.2. Absorption, scattering, and extinction efficiencies versus particle size for λ = 3.8 μm.

given in Equation (6.1.15) is called the normalized scalar phase function of the particle. The scalar phase function gives the energy scattered per unit angle at the angle θ from the direction of incidence divided by the total scattered energy. Loosely, it might be interpreted as the normalized power pattern of the particulate "antenna." Figure 6.3 depicts the behavior of the normalized scalar phase function for three different Mie parameters. It is apparent from these graphs (done for a fixed index of refraction) that the larger a Mie parameter is, the more peaked its scalar phase function is in the forward direction, i.e., $P(\theta)/4\pi$ increases. This implies that one can capture more energy from this "antenna" with a smaller field-of-view receiver as its geometrical size increases. This point will become more important in our discussions as the chapter develops.

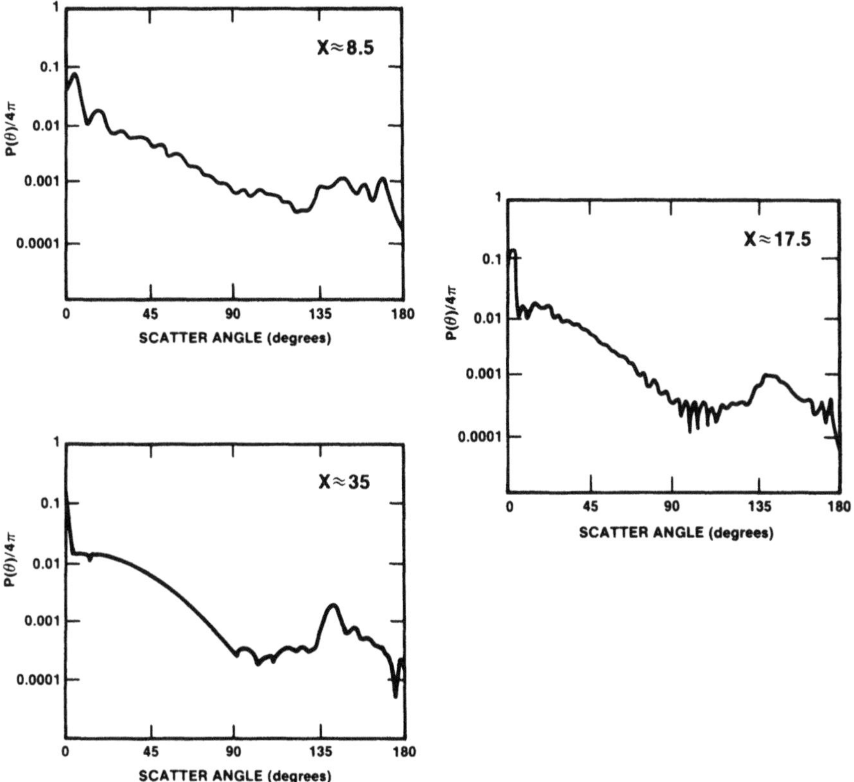

Figure 6.3. Mie scattering functions for various values of the Mie parameter $x = \pi D/\lambda$.[12]

If we now expand the definition given in Equation (6.1.15) to include the products $\text{Re}\,\{S_1 S_2^*/k_0^2\}$ and $-\text{Im}\,\{S_1 S_2^*/k_0^2\}$, namely, by defining

$$P_3(\theta) = \frac{2\,\text{Re}\,\{S_1(\theta),\, S_2^*(\theta)\}}{x^2 Q_s} \tag{6.1.16}$$

and

$$P_4(\theta) = \frac{-2\,\text{Im}\,\{S_1(\theta) S_2^*(\theta)\}}{x^2 Q_s} \tag{6.1.17}$$

then in the case of a single-particle medium, Equation (6.1.5) can be rewritten as

$$\boldsymbol{\sigma}(\theta) = \frac{D^4 Q_s}{16}\begin{bmatrix} P_1(\theta) & 0 & 0 & 0 \\ 0 & P_2(\theta) & 0 & 0 \\ 0 & 0 & P_3(\theta) & P_4(\theta) \\ 0 & 0 & -P_4(\theta) & P_3(\theta) \end{bmatrix} \tag{6.1.18}$$

For a polydisperse environment, which both the atmosphere and the ocean are, Equation (6.1.18) still holds if we redefine the $P_j(\theta)$'s to be given by

$$P_1(\theta) = \frac{4\pi^2}{k_0^2 b} \int_0^\infty \rho(m, x)|S_1(m, x, \theta)|^2 \, dx \tag{6.1.19a}$$

$$P_2(\theta) = \frac{4\pi^2}{k_0^2 b} \int_0^\infty \rho(m, x)|S_2(m, x, \theta)|^2 \, dx \tag{6.1.19b}$$

$$P_3(\theta) = \frac{4\pi^2}{k_0^2 b} \int_0^\infty \rho(m, x) \operatorname{Re}\{S_1(m, x, \theta) S_2^*(m, x, \theta)\} \, dx \tag{6.1.19c}$$

and

$$P_4(\theta) = \frac{-4\pi^2}{k_0^2 b} \int_0^\infty \rho(m, x) \operatorname{Im}\{S_1(m, x, \theta) S_2^*(m, x, \theta)\} \, dx \tag{6.1.19d}$$

where

$$b = \frac{\pi}{k_0^2} \int_0^\infty x^2 \rho(m, x) Q_s(m, x) \, dx \tag{6.1.20}$$

In these relations, $\rho(m, x)$ represents the particle distribution function for the medium, that is, the partial concentration per unit volume per unit increment of x. The total concentration of particles is given by

$$N_p = k_0^{-1} \int_0^\infty \rho(x) \, dx \tag{6.1.21}$$

in units of particles per unit volume. Figure 6.4 depicts the normalized scalar phase function of a Deirmendjian C.1 cloud at the wavelengths 0.45 μm, 10.6 μm, and 1.3 mm.[11] Figure 6.5 gives the particle distribution for the model C.1 cloud.[7] The mode (or average) radius of the distribution, denoted by $\langle r \rangle$, is 4 μm. We see in Figure 6.4 that as the wavelength increases from below the mode radius to well above it, the scalar phase function moves from a highly peaked, or forward-biased, function to an isotropic, or uniformlike, one. This is not surprising. Previously we saw that as the Mie parameter $x = \pi D/\lambda$ gets larger, the associated phase function $P(\theta)/4\pi$ becomes more peaked in the straight-through direction. Figure 6.4 shows that this type of behavior holds for polydispersion media as well. Thus, for discussion purposes, one can replace the diameter D in the Mie parameter relation with the mode diameter $\langle D \rangle$ (equal to twice the mode radius). One finds generally in a polydispersion that the extinction rate of an optical beam within any elemental volume is controlled by the effective area of the droplets within that volume, while the extent of the large-diameter-particle distribution dictates the tightness of its scalar phase function. That is, a large number density of small particles scatters considerably

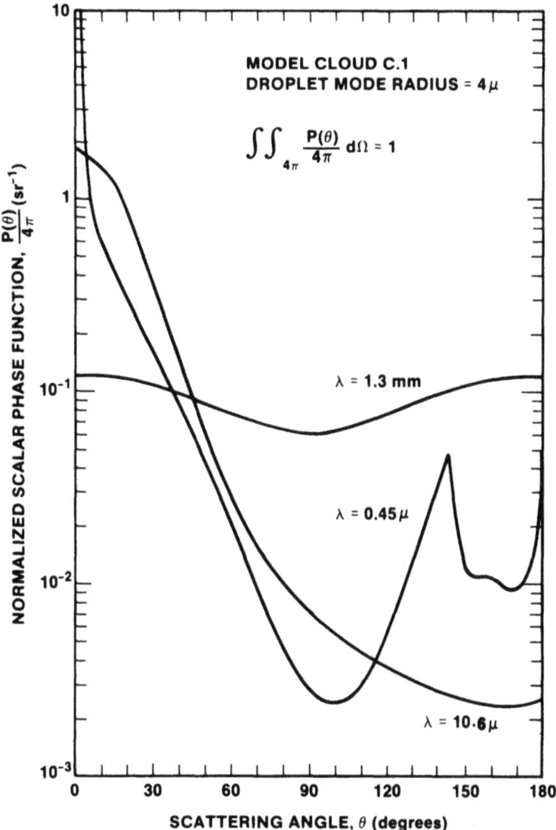

Figure 6.4. Typical scalar phase functions for short, medium, and long wavelengths.

more than an equal water content of large drops, and a large-particle cutoff in any size distribution implies a strong forward bias in the resultant phase function profile. It should also be noted that the imaginary part of the refractive index for either a single particle or a collection of particles affects the above inherent properties with the increase in absorption taking the place of the increase in scattering. These mechanisms help explain why the ocean and the atmosphere behave as they do relative to energy transfer.[13] However, before investigating this, we need to know more about each. Let us begin with the atmosphere.

6.2. General Characteristics of the Atmosphere

The principal degrading constituents of the natural atmosphere are haze, fog, clouds, and rain, with the first three occurring the most frequently.

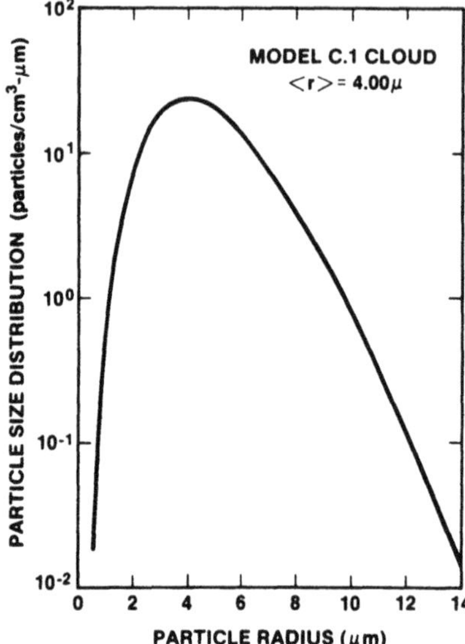

Figure 6.5. Particle distribution for a Deirmendjian C.1 cloud.

Although some data exist on the extinction rate for a few of these entities,[14] measurements of their associated single-scatter albedo and scalar phase function are quite hard to obtain *in situ*; hence, they must be obtained by other means. Fortunately, the particles involved in all four are primarily composed of water and are nearly spherical in shape. Thus, by knowing their size distributions, complex refractive index, and particle number density, one can use the Mie scattering theory to calculate the desired inherent properties of haze, fog, clouds, and rain. Ancillary definitions to those given in the above section are:

(a) the volume extinction coefficient, denoted by the letter c, given by

$$c = \frac{\pi}{k_0^2} \int_0^\infty x^2 \rho(m, x) Q_e(m, x)\, dx \tag{6.2.1}$$

which describes the intensity loss per unit length for any elemental cylindrical volume through the extinction process.

(b) the volume absorption coefficient, denoted by the letter a, given by

$$a = \frac{\pi}{k_0^2} \int_0^\infty x^2 \rho(m, x) Q_a(m, x)\, dx \tag{6.2.2}$$

which describes the intensity loss per unit length for any elemental cylindrical volume through the extinction process.

It follows from the above two equations that the parameter b defined in Equation (6.1.20) is the volume scattering coefficient and quantifies the intensity loss per unit length in any elemental cylindrical volume through the scattering process. Thus from the previous definitions, we have

$$a = c - b \tag{6.2.3}$$

$$\omega_0 = b/c \tag{6.2.4}$$

As we shall soon see, a represents the irreversible energy lost to the medium, while b does not in some applications, e.g., communications. However, our current subject is the atmosphere and its various constituents; let us return to that discussion. Figure 6.6 illustrates representative particle distributions

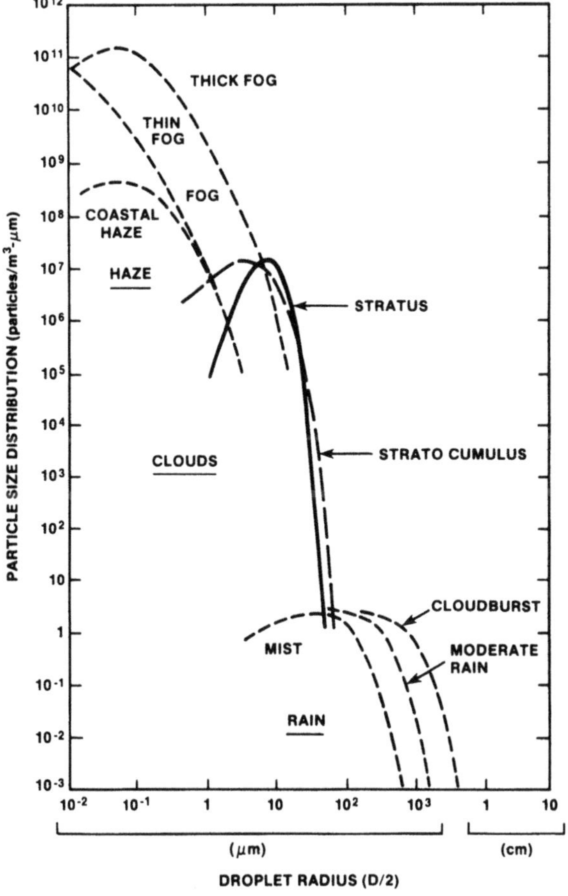

Figure 6.6. Several cloud and precipitation drop size distributions.

for fog, haze, clouds, and rain. Table 6.1 gives the typical range of mode radius for these entities. It is apparent from Figure 6.6 that fog and haze are more dominated by small particles than either clouds or rain, while fog is more densely populated and broader spectrumed. Thus we expect haze to be the least forward-biased in scattering, followed closely by fog, and clouds and rain the most forward-biased. This can be seen by comparing the relative magnitudes of the effective Mie parameter (for fixed wavelength) using the mode radius values given in Table 6.1. On the other hand, Figure 6.6 also suggests that fog, clouds, and rain will exhibit the largest extinction rates; atmosphere haze, the smallest. Figure 6.7 illustrates this point for clouds, rain, and haze using the classic Deirmendjian particle size distributions[7] and the complex refractive indices shown in Table 6.2. Both of these aspects were introduced in the previous section and will be developed in more detail.

6.2.1. Atmospheric Haze

Atmospheric haze can be broken down into two distinct types, continental and marine. Although haze is generally characterized by small particles with mode radii on the order of 0.5 to 1.0 μm and particle distribution cutoffs between 1 and 5 μm, both can exhibit some forward scatter-type behavior in the visible and infrared light regimes.[15] Deirmendjian proposed a family of particle size distributions for the natural atmosphere of the form

$$n(r) = A_0 r^{\alpha} \exp\{-A_1 r^{\gamma}\} \tag{6.2.5}$$

where $n(r)$ is a modified gamma distribution defined by $0 \leqslant r < \infty$. Table 6.3 gives the numerical values for the parameters of Equation (6.2.5) for various atmospheric constituents. Estimates of the particle density vary from 100 to 2300 particles per cubic centimeter in the lower atmosphere, e.g., hazes C, M, and L, and are in the tens of particles per cubic centimeter in the upper atmosphere, e.g., haze H. Thus the extinction coefficient in the visible may range from 0.106 km^{-1} for a haze M with particle density

Table 6.1. Size Ranges of Precipitation

Precipitation	Range of the mode, r_m (μm)
Haze	0.05–0.07
Fog (various)	0.01–15
Clouds (various)	4–40
Rain (above 500 ft)	10
Drizzle (below 500 ft)	20
Rain (3–15 mm/h, below 500 ft)	200–1000

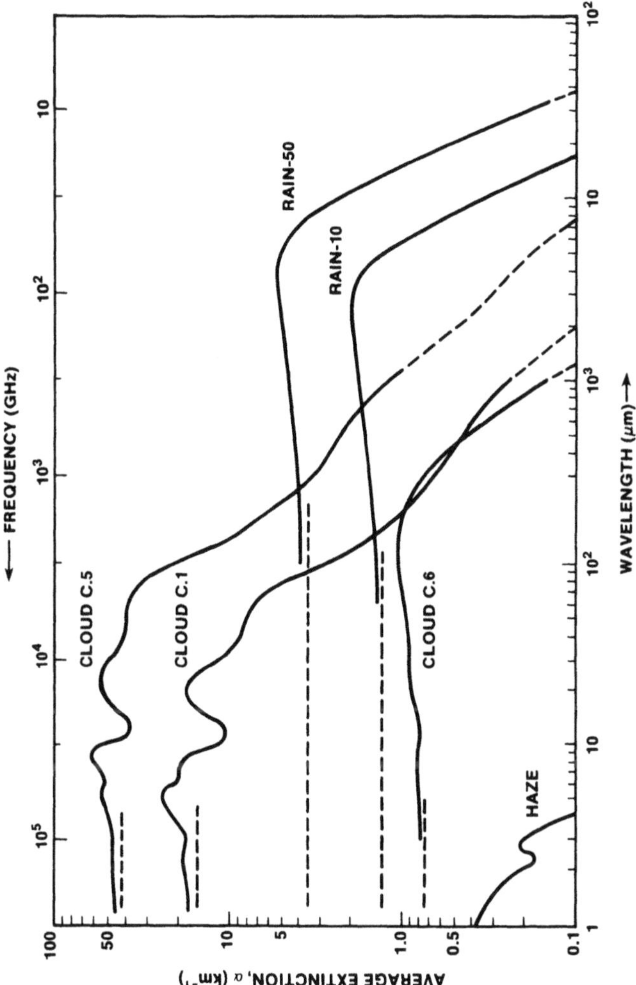

Figure 6.7. Theoretical extinction coefficients according to three cloud models, two precipitation models, and a haze model.

Table 6.2. Optical Constants of Water

λ (μm)	A_i	h_r	λ (μm)	n_i	n_r	λ (μm)	h_i	n_r
0.200	1.1×10^{-7}	1.396	3.40	0.0195	1.420	9.8	0.0479	1.229
0.225	4.9×10^{-8}	1.373	3.45	0.0132	1.410	10.0	0.0508	1.218
0.250	3.35×10^{-8}	1.362	3.50	0.0094	1.400	10.5	0.0662	1.185
0.275	2.35×10^{-8}	1.354	3.6	0.00515	1.385	11.0	0.0968	1.153
0.300	1.6×10^{-8}	1.349	3.7	0.00360	1.371	11.5	0.142	1.126
0.325	1.08×10^{-8}	1.346	3.8	0.00340	1.361	12.0	0.199	1.111
0.350	6.5×10^{-9}	1.343	3.9	0.00380	1.357	12.5	0.259	1.123
0.375	3.5×10^{-9}	1.341	4.0	0.00460	1.315	13.0	0.305	1.146
0.400	1.86×10^{-9}	1.339	4.1	0.00562	1.316	13.5	0.343	1.177
0.425	1.3×10^{-9}	1.338	4.2	0.00688	1.342	14.0	0.370	1.210
0.450	1.02×10^{-9}	1.337	4.3	0.00845	1.338	14.5	0.388	1.241
0.475	9.35×10^{-10}	1.336	4.4	0.0103	1.334	15.0	0.402	1.270
0.500	1.00×10^{-9}	1.335	4.5	0.0134	1.332	15.5	0.414	1.297
0.525	1.32×10^{-9}	1.334	4.6	0.0147	1.330	16.0	0.422	1.325
0.550	1.96×10^{-9}	1.333	4.7	0.0157	1.330	16.5	0.428	1.351
0.575	3.60×10^{-9}	1.333	4.8	0.0150	1.330	17.0	0.429	1.376
0.600	1.09×10^{-8}	1.332	4.9	0.0137	1.328	17.5	0.429	1.401
0.625	1.39×10^{-8}	1.332	5.0	0.0124	1.325	18.0	0.426	1.423
0.650	1.64×10^{-8}	1.331	5.1	0.0111	1.322	18.5	0.421	1.443
0.675	2.23×10^{-8}	1.331	5.2	0.0101	1.317	19.0	0.414	1.461
0.700	3.35×10^{-8}	1.331	5.3	0.0098	1.312	19.5	0.404	1.476
0.725	9.15×10^{-8}	1.330	5.4	0.0103	1.305	20.0	0.393	1.480
0.750	1.56×10^{-7}	1.330	5.5	0.0116	1.298	21.0	0.382	1.487
0.775	1.48×10^{-7}	1.330	5.6	0.0142	1.289	22	0.373	1.500
0.800	1.25×10^{-7}	1.329	5.7	0.0203	1.277	23	0.367	1.511
0.825	1.82×10^{-7}	1.329	5.8	0.0330	1.262	24	0.361	1.521
0.850	2.93×10^{-7}	1.329	5.9	0.0622	1.248	25	0.356	1.531
0.875	3.91×10^{-7}	1.328	6.0	0.107	1.265	26	0.350	1.539
0.900	4.86×10^{-7}	1.328	6.1	0.131	1.319	27	0.344	1.545
0.925	1.06×10^{-6}	1.328	6.2	0.0880	1.363	28	0.338	1.549
0.950	2.93×10^{-6}	1.327	6.3	0.0570	1.357	29	0.333	1.551
0.975	3.48×10^{-6}	1.327	6.4	0.0449	1.347	30	0.328	1.551
1.0	2.89×10^{-6}	1.327	6.5	0.0392	1.339	32	0.324	1.546
1.2	9.98×10^{-6}	1.324	6.6	0.0356	1.344	34	0.329	1.536
1.4	1.38×10^{-4}	1.321	6.7	0.0337	1.329	36	0.343	1.527
1.6	8.55×10^{-5}	1.317	6.8	0.0327	1.324	38	0.361	1.522
1.8	1.15×10^{-4}	1.312	6.9	0.0322	1.321	40	0.385	1.519
2.0	1.1×10^{-3}	1.306	7.0	0.0320	1.317	42	0.409	1.522
2.2	2.89×10^{-4}	1.296	7.1	0.0320	1.314	44	0.436	1.530
2.4	9.56×10^{-4}	1.279	7.2	0.0321	1.312	46	0.462	1.541
2.6	3.17×10^{-3}	1.242	7.3	0.0322	1.309	48	0.488	1.555
2.65	6.7×10^{-3}	1.219	7.4	0.0324	1.307	50	0.514	1.587
2.70	0.019	1.188	7.5	0.0326	1.304	60	0.587	1.703
2.75	0.059	1.157	7.6	0.0328	1.302	70	0.576	1.821
2.80	0.115	1.142	7.7	0.0331	1.299	80	0.547	1.886
2.85	0.185	1.149	7.8	0.0335	1.297	90	0.536	1.924
2.90	0.268	1.201	7.9	0.0339	1.294	100	0.532	1.957

continued

Table 6.2. (*continued*)

λ (μm)	A_i	h_r	λ (μm)	n_i	n_r	λ (μm)	h_i	n_r
2.95	0.298	1.292	8.0	0.0343	1.291	110	0.531	1.966
3.00	0.272	1.371	8.2	0.0351	1.286	120	0.526	2.004
3.05	0.240	1.426	8.4	0.0361	1.281	130	0.514	2.036
3.10	0.192	1.467	8.6	0.0372	1.275	140	0.500	2.056
3.15	0.315	1.483	8.8	0.0385	1.269	150	0.495	2.069
3.20	0.0924	1.478	9.0	0.0399	1.262	160	0.496	2.081
3.25	0.0610	1.467	9.2	0.0415	1.255	170	0.497	2.094
3.30	0.0368	1.450	9.4	0.0433	1.247	180	0.499	2.107
3.35	0.0261	1.432	9.6	0.0454	1.239	190	0.501	2.119
						200	0.504	2.130
						300	0.55	2.14
						400	0.64	2.18
						500	0.74	2.22
						600	0.81	2.27
						700	0.89	2.32
						800	0.95	2.38
						900	1.02	2.44
						1000	1.09	2.50
						2000	0.90	2.56
						3000	1.18	2.77

100 cm^{-3} to 0.121 km^{-1} for a haze C with particle density 2300 cm^{-3}. Figures 6.8 and 6.9 illustrate the normalized scalar phase function at $\lambda = 0.45\ \mu m$ and $\lambda = 1.19\ \mu m$ for Deirmendjian hazes M and L, respectively. Figure 6.10 gives the particle distributions for these hazes (as well as for haze H). Haze M represents a marine-type and haze L, a continental haze. It follows from Figures 6.8 and 6.9 that as the wavelength increases (or, alternatively, the effective Mie parameter decreases), the scalar phase function decreases in forward-scattering efficiency, even when absorption is present. Comparing these two figures in light of Figure 6.10, we see that the greater the percentage of large particles in a distribution (as in marine-type hazes), the more forward-biased the distribution. In addition, the larger the particle density (especially in the small-particle regime), the greater the extinction coefficient.

The distribution models described are intended for use in evaluating radiation transfer in uniformly distributed particulate media. Unfortunately, the clear atmosphere is a gradient particulate medium, beginning with its largest density near the earth's surface and monotonically decreasing in value as one moves radially into space.(16,17) One of the first complete models to emerge for the gradient atmosphere was developed by Wells *et al.*(18) In this work, the continental and maritime components were represented by Junge haze and Deirmendjian modified gamma function models, respectively.(17) The aerosol radius growth rate with relative humidity was

Table 6.3. Deirmendjian Size-Distribution Models[a]

Distribution type	N	a	$\langle r \rangle$	α	γ	b	$n(\langle r \rangle)$
Haze M	100 cm^{-3}	5.3333×10^4	0.05 μm	1	$\frac{1}{2}$	8.9443	360.9 cm^{-3} μm^{-1}
Rain M	1000 m^{-3}	5.3333×10^5	0.05 mm	1	$\frac{1}{2}$	8.9443	3609 m^{-3} mm^{-1}
Haze L	100 cm^{-3}	4.9757×10^6	0.07 μm	2	$\frac{1}{2}$	15.1186	446.6 cm^{-3} μm^{-1}
Rain L	1000 m^{-3}	4.9757×10^7	0.07 mm	2	$\frac{1}{2}$	15.1186	446 m^{-3} mm^{-1}
Haze H	100 cm^{-3}	4.0000×10^5	0.10 μm	2	1	20.0000	541.4 cm^{-3} μm^{-1}
Hail H	10 m^{-3}	4.0000×10^4	0.10 cm	2	1	20.0000	54.14 m^{-3} cm^{-1}
Cumulus cloud, C.1	100 cm^{-3}	2.37330	4.00 μm	6	1	$\frac{1}{2}$	24.09 cm^{-3} μm^{-1}
Corona cloud, C.2	100 cm^{-3}	1.0851×10^{-2}	4.00 μm	8	3	$\frac{1}{24}$	49.41 cm^{-3} μm^{-1}
MOP cloud, C.3	100 cm^{-3}	5.5556	2.00 μm	8	3	$\frac{1}{3}$	98.82 cm^{-3} μm^{-1}
Double corona cloud, C.4	100 cm^{-3}	5.5556	4.00 μm	8	3	$\frac{1}{3}$	98.82 cm^{-3} μm^{-1}

[a] From Ref. 7.

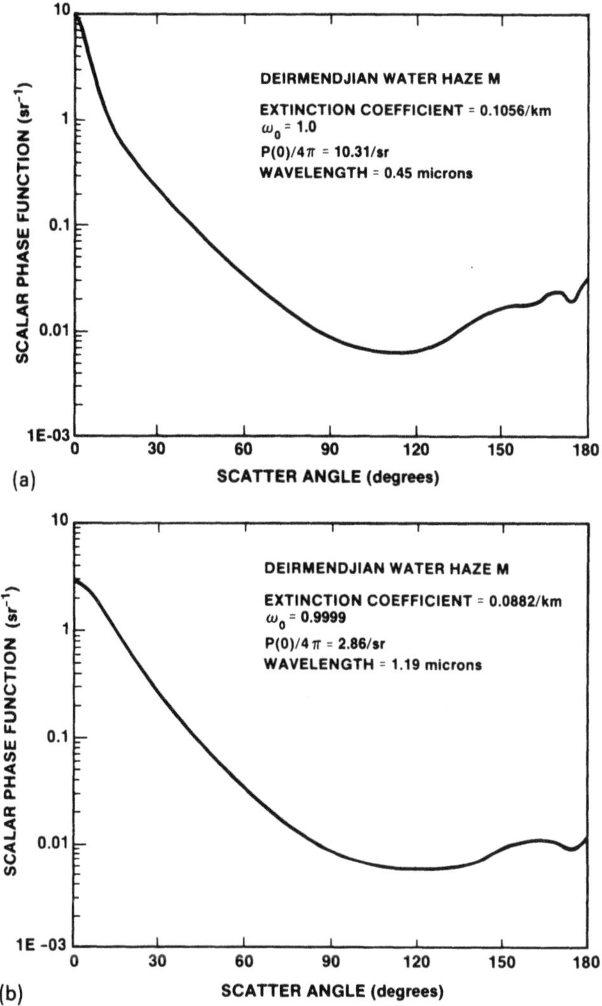

Figure 6.8. Normalized scalar phase function for a Deirmendjian water haze M at (a) $\lambda = 0.45\ \mu m$ and (b) $\lambda = 1.9\ \mu m$.

assumed to be that of Barnhardt and Streete[19] under a NaCl dry-particle assumption. In the continental component, the constants of the Junge distribution were adjusted to agree with the work of Barnhardt and Streete for a relative humidity of 80% and a continental-to-maritime mixture of 1:2.5. Mathematically, the model was given by

$$n(r) = \frac{\beta}{F}\left[0.47\left(\frac{r}{F}\right)^{-4} \exp\left\{\frac{h}{h_c}\right\} + 5.75(c_1 + c_2 V^{\delta})\left(\frac{r}{F'}\right) \exp\{-W\}\right] \tag{6.2.6}$$

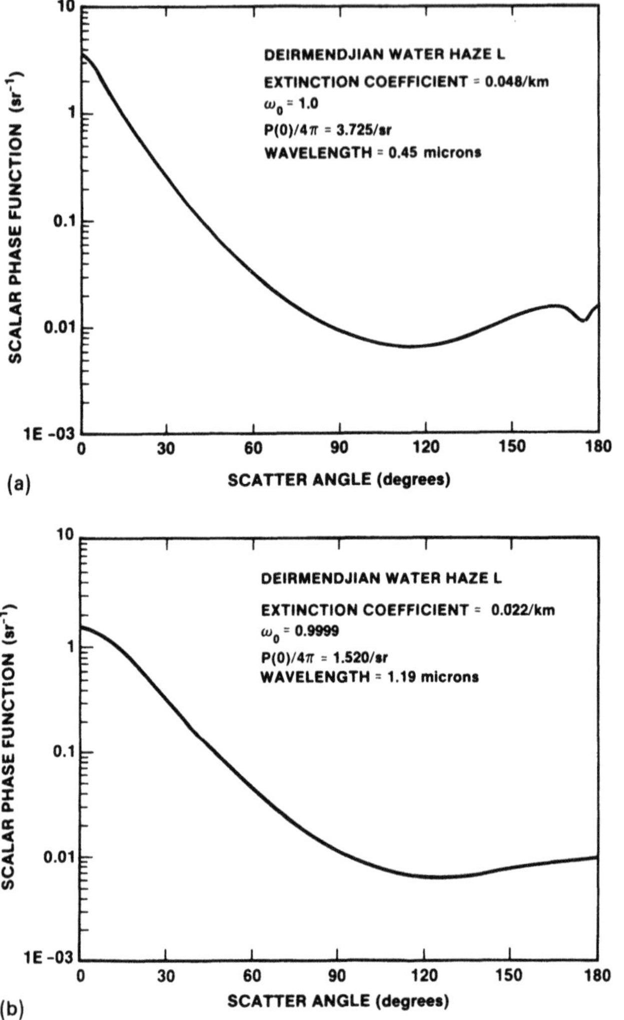

Figure 6.9. Normalized scalar phase function for a Deirmendjian water haze L at (a) $\lambda = 0.45\ \mu$m and (b) at $\lambda = 1.19$ m.

where

$$W = \frac{8.5 r^{\gamma(v)}}{F'} + \frac{h}{h_m} \tag{6.2.7}$$

$$F = 1 - 0.9 \ln\left\{1 - \frac{\mathrm{RH}}{100}\right\} \tag{6.2.8}$$

$$F' = \frac{F(\mathrm{RH})}{F(80)} \tag{6.2.9}$$

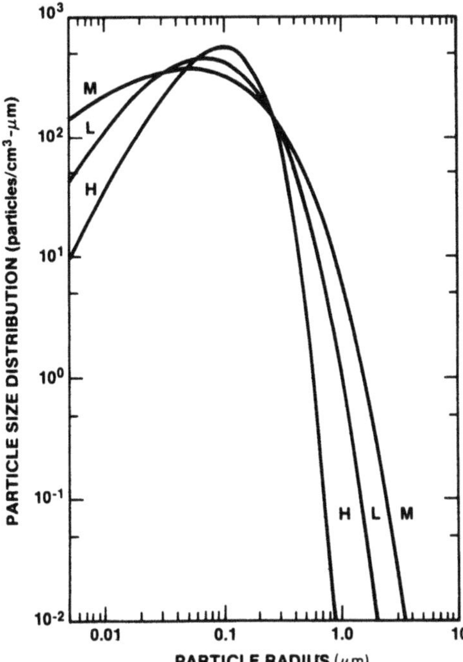

Figure 6.10. Particle size distribution for Deirmendjian H, L, and M hazes.

with

$$\text{RH} = \text{relative humidity in percent, and}$$

$$\gamma(v) = 0.384 - 2.93 \times 10^3 v^{1.25} \tag{6.2.10}$$

In Equations (6.2.6) and (6.2.7), v is wind speed, h is the altitude, and h_c and h_m are the size-dependent scale heights for continental and maritime hazes, respectively. At sea level for $r > 1\ \mu$m, $h_c = -0.63$ km (increase in relative proportion of large particles with increasing height) and $h_m = 0.8$ km. The factor β is a normalization factor which is inversely proportional to visibility.[20,21] For $v < 7$ m/s, the constants in Equation (6.2.6) have the values $c_1 = 2.5 \times 10^2$, $c_2 = 7.5 \times 10^2$, $\delta = 1.16$; for $v \geqslant 7$ m/s, $c_1 = 0$, $c_2 = 6.9 \times 10^3$, $\delta = 0.29$.

It should be noted that Equation (6.2.6) had to be applied with some care at large particle sizes. Specifically, the Junge model for the continental haze component, i.e., the first term in Equation (6.2.6), was not valid beyond $r \sim 20\ \mu$m. Physically, particles of larger sizes than this tend to settle out under the influence of gravity. Similar arguments applied to the water-coated salt particles of the maritime haze component. Thus, Equation (6.2.6) was valid only for radii of $\leqslant 20\ \mu$m. This is an especially important consideration in the computation of the scalar phase function as its value at $\theta = 0^0$ depends on $r^4 p(r)\, dr$ and thus is particularly sensitive to the shape of the tail of the size distribution. For example, Figure 6.11 gives the scalar phase function for the Wells, Gal, and Munn maritime haze at $\lambda = 1.06\ \mu$m. By increasing the particle cutoff from 20 μm to 30 μm, one finds that the extinction coefficient only increases by $< 15\%$, but the value of $P(\theta)/4\pi$ doubles. The r^{-4} dependence of the continental haze component in Equation (6.2.6) is also particularly troublesome in this regard if applied without a particle size cutoff. Comparing Figures 6.8 and 6.11, it is clear that the former's phase function is less peaked than the latter's. This is because Wells *et al.* assume a stronger large-particle density distribution than Deirmendjian does.

Katz modified the constants in Equation (6.2.6) using a larger data set of meteorological parameters obtained from surface weather ships.[17] The visibility scaling factor was removed because of the unreliable visibilities reported in the data set. The scale height term in the continental component was also eliminated, leaving the altitude variations as functions of the humidity and exponential scale height in only the maritime component. In addition, the aerosol growth model was changed from that of Barnhardt and Streete to that of Fitzgerald.[21] The resultant model, known as the Munn–Katz model, is described by[17]

$$\rho(r) = 1.7\left(\frac{r}{\alpha}\right)^{-4} + 1.62(c_1 + c_2 v_0^8)\exp\left\{\frac{-h}{h_0 F} - 8.5\left(\frac{r}{\alpha}\right)^{\Gamma}\right\} F^{-1}\left(\frac{r}{\alpha}\right) \tag{6.2.11}$$

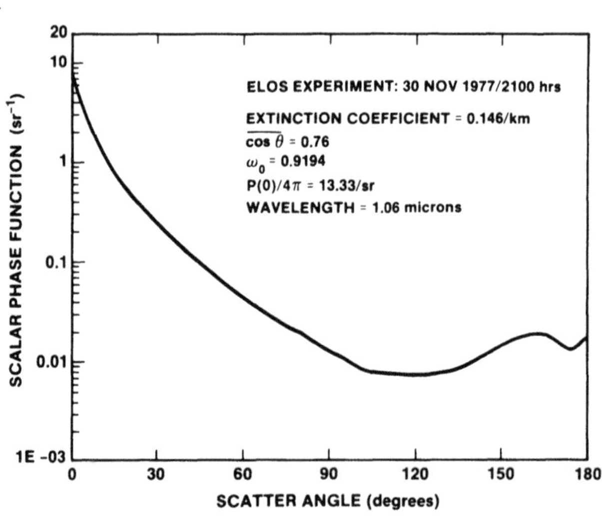

Figure 6.11. Normalized scalar phase function for a Wells, Gal, and Munn maritime haze.

where h_0 is the scale height, set at 800 m for $h < 1$ km, $\alpha = 0.81 \exp\{0.0665/1.058S\}$, S is the saturation ratio, given by

$$S = \frac{\text{RH}}{100}$$

v_0 is the wind factor scaled with surface wind v:

$$v_0 = \begin{cases} 0.5 \text{ m/s} & \text{for } 0 \leqslant v \leqslant 4 \text{ m/s} \\ (v - 3.5) \text{ m/s} & \text{for } v > 4 \text{ m/s} \end{cases}$$

and

$$\begin{aligned} F &= 1 + \left(\frac{v_0}{60}\right)^3 \\ &= 0.384 - 0.00293 v_0^{1.25} \end{aligned}$$

For $v_0 < 7$ m/s, $c_1 = 350$, $c_2 = 10^3$, $\delta = 1.15$; for $v_0 > 7$ m/s, $c_1 = 0$, $c_2 = 6900$, $\delta = 0.29$. Reference 17 compares Equation (6.2.11) with measured particle distributions and shows reasonable agreement between them. Let us look at haze's close cousin, fog.

6.2.2. Atmospheric Fogs

Atmospheric fogs are created by a thickening of the atmosphere due to high-humidity conditions appearing. The atmospheric hazes described above exist at relative humidities less than 80%. Light fogs begin to occur after that point and progress to thick fogs at saturation or 100% relative humidity. Fogs have a similar composition to haze, with mode radii between 0.01 and 15 μm, but are more dense in particles and possess larger distribution cutoffs. This can be clearly seen in Figure 6.6. Thus we expect fog to attenuate more than haze and to be more forward-biased in scattering. Figure 6.12 shows a typical particle size distribution for a marine fog measured off the coast of San Diego, California. Figure 6.13 gives the inherent properties for this distribution. It follows from this latter figure that both of the above points are borne out. In general, atmospheric fogs are highly peaked in their scattering about $\theta = 0^0$ and possess extinction coefficients between 10 and 100 km^{-1}. However, unlike atmospheric haze, fogs are usually uniformly distributed near the earth's surface and rarely extend very high in the atmosphere. Particulate media which are located above the surface of the earth are the clouds, and they are the topic we will address next.

6.2.3. Atmospheric Clouds

Atmospheric clouds are characterized by size distributions much larger than that of either haze or fog. They are finite in physical extent, many

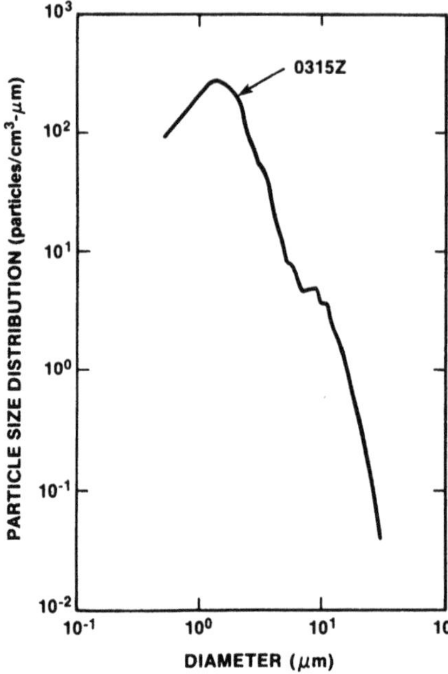

Figure 6.12. Particle distribution for a marine fog measured in Point Loma, California.

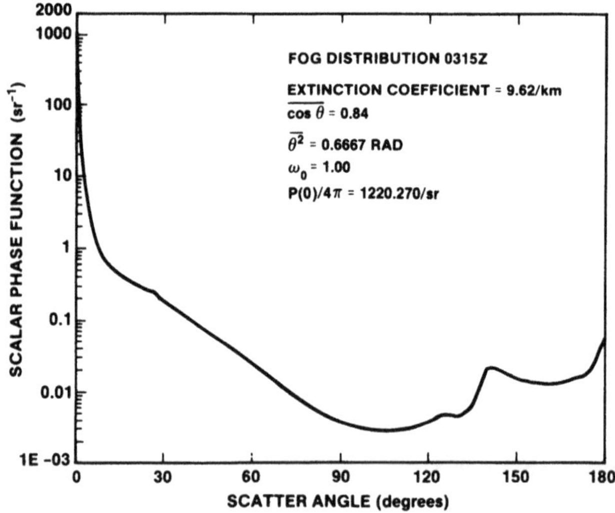

Figure 6.13. Inherent properties for the particle size distribution given in Figure 6.12.

times in all three dimensions, and are located usually above 1 to 3 km from the earth's surface. Table 6.4 gives rought estimates of cloud thickness and mean free paths for the various cloud types found in nature.[11] Experimental measurements of solar transmission for each cloud type from various sources[11] are also given. We see that clouds possess the same range of extinction coefficients as fog. Deirmendjian developed a number of particle size distributions for the various cloud types given in Table 6.4. Although derived from the literature up to 1967, these models, denoted by the symbols C.1, C.2, C.3, and C.4, are still widely used in the radiative transfer community today. The particular form of these models was previously presented in Equation (6.2.5) using the numerical values given in Table 6.4. Figure 6.14 illustrates these distributions graphically. We see that the C.1, C.2, C.3, and C.4 clouds are dominated, although varyingly, by large particles. This suggests that the associated scalar phase functions are forward scattering in the visible and near-infrared wavelength regimes. This is indeed the case,[7] as illustrated for the model C.1 cloud in Figure 6.4.

Besides clear atmospheric conditions, clouds are the most common constituent of the atmospheric channel. Unfortunately, they are also the most stressing to optimum communication system performance and need to be considered from a statistical occurrence point of view. This is because their large dimensional extent (relative to that of fogs) and multiple layering potential within any chosen column of atmosphere can manifest themselves into a significant integrated entity which most system designs have trouble accommodating. Ciany *et al.* used Table 6.4 and the Air Force Global Weather Central (AFGWC) 3DNEPH Data Base to develop a first-order cloud climatology map for cloud optical and physical thickness statistics.[22] Figure 6.15 shows the general layout of the 3DNEPH data base configuration. Specifically, 3DNEPH reports the types of low, middle, and high clouds; present weather; maximum cloud top and minimum cloud base altitudes; total cloud cover; and percentage cloud coverage in each of 15 layers. It is clear from this figure that certain cloud types can be found in one or more layers simultaneously and may be terrain following. The statistical input which generates this data base can come from surface reports, upper air sounders, and aircraft and satellite sources, and is manipulated in such a way as to minimize the inconsistencies inherent to these various raw data sources. Figure 6.16 illustrates the basic program flow. The output of each individual data processor provides one portion of the 22 cloud parameters in the 3DNEPH data base. By integrating these various outputs in a weighted manner, the final output data record is obtained for inclusion in 3DNEPH. Figures 6.17 and 6.18 compare optical thickness distribution functions produced by Ciany *et al.* with inferred distribution functions derived from experimental data for St. Louis, Missouri, and Malta, respectively. Both figures show good agreement between the computed

Table 6.4. Optical and Physical Thicknesses for Various Cloud Types

Cloud name	Extinction length (m)	Average thickness at Netherlands (m)	Average thickness at Malta (m)	Optical thickness at Netherlands	Optical thickness at Malta	Optical thickness derived from Haurwitz	Optical derived from experimental global radiation data for Malta
Cumulonimbus	25	2970	3200	119	128		114
Nimbostratus	55	2800	3000	51	55	79	51
Stratus	15	1120	700	75	47	43	43
Altostratus	45	1620	1270	36	28	22	30
Stratocumulus	20	740	390	37	20	27	20
Altocumulus	65	1080	1320	17	20	15	16
Cumulus	50	940	450	19	9.1		10
Cirrostratus	350	1890	1750	5.4	5.0	4.7	
Cirrocumulus	350	1670	1900	4.8	5.4		
Cirrus	350	930	875	2.7	2.5	4.5	

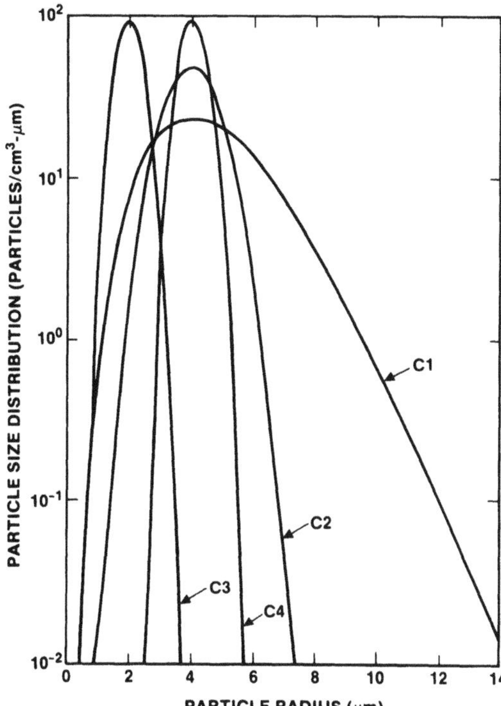

Figure 6.14. Particle distributions for Deirmendjian model C.1, C.2, C.3, and C.4 clouds.

statistics and experiment. Appendix E contains midseason cumulative probability distributions for optical and physical cloud thickness at various selected locations in the world. These figures should prove quite useful to those readers developing first-order system availability assessments of potential communication systems designed for the atmospheric channel. Figure 6.19 shows two comparisons of midseason and yearly optical thickness cumulative distributions and illustrates the relative accuracy of Appendix E for general availability calculations.

6.2.4. Atmospheric Rain

Atmospheric rain is the last constituent of the atmosphere that we will consider. From the contents of our previous discussions, it is clear what the primary aspects of the inherent properties of rain are. The extinction coefficients of rain are similar in magnitude to those of clouds (as seen in Figure 6.7), and its characteristic phase functions are highly peaked in the forward direction because of the large size of droplets. Should the reader require representative size distributions of this entity for computational

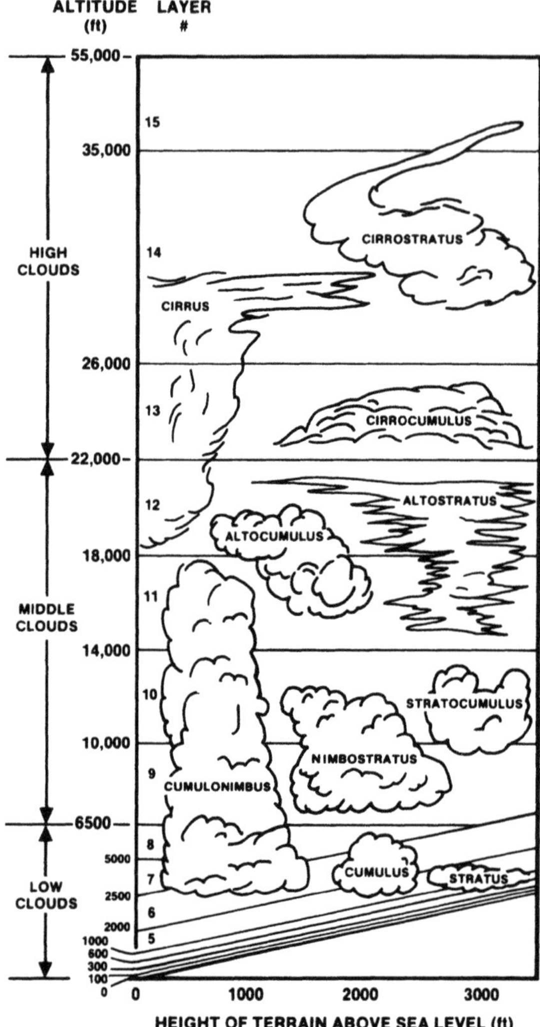

Figure 6.15. 3DNEPH designated layers and cloud types.

purposes, he is referred again to the Deirmendjian models given in Table 6.1 (and Ref. 7). Let us now turn to the basic characteristics of seawater and its inherent properties.

6.3. General Characteristics of the Ocean

The marine environment, unlike the atmosphere, has two significant mechanisms affecting the propagation of light through it. One is scattering,

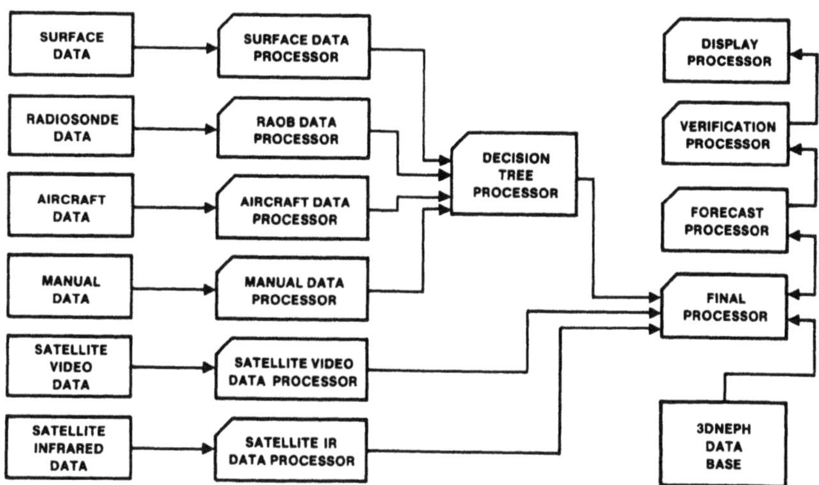

Figure 6.16. 3DNEPH program flow.

which we were exposed to in the previous section, and the other is absorption. The agents creating these effects are the water itself and the dissolved and suspended material in it. In this section, we shall describe those agents and their effect on the inherent properties of seawater. References 23 and 24 contain excellent discussions on this subject and provided much of the material drawn upon in the following review. Before proceeding, we need

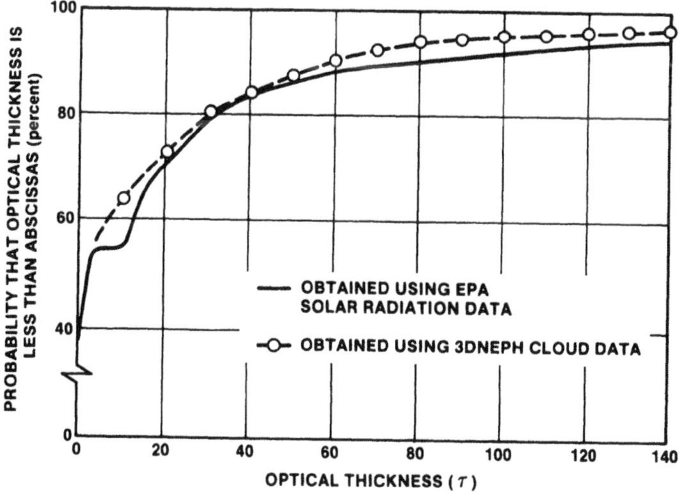

Figure 6.17. Optical thickness distribution function for St. Louis, Missouri, 1976; comparison of EPA solar radiation and 3DNEPH data.

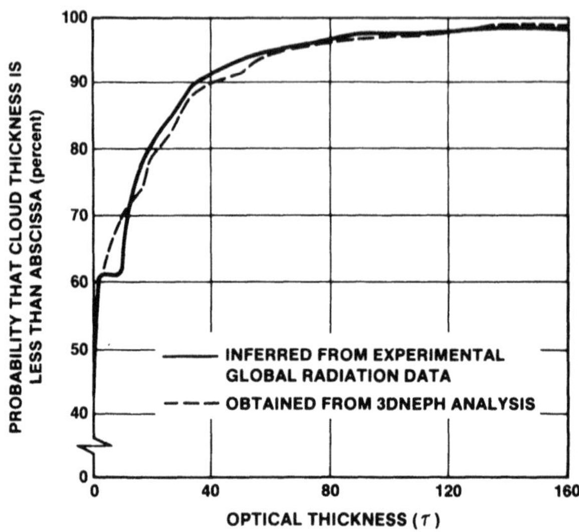

Figure 6.18. Optical thickness distribution function for Malta, 1972; comparison of global radiation and 3DNEPH data.

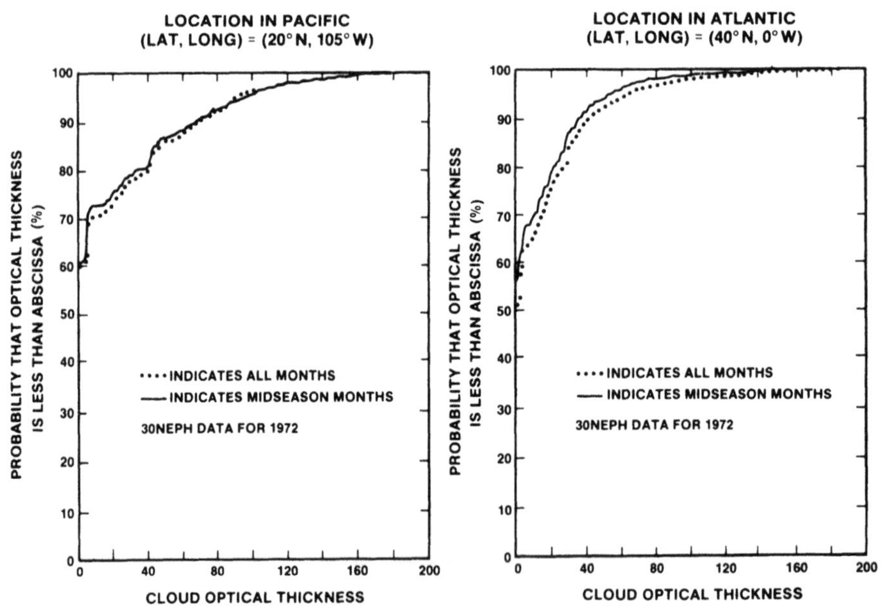

Figure 6.19. Comparisons of midseason and yearly optical thickness distributions.

to define one more parameter, the volume scattering function. Its mathematical form is

$$\beta(\theta) = \frac{4\pi^2}{k_0^2} \int_0^\infty n(m, x)|S(m, x, \theta)|^2 \, dx \tag{6.3.1}$$

where

$$|S(m, x, \theta)|^2 = \frac{1}{4\pi}|S_1(m, x, \theta)|^2 + \frac{1}{4\pi}|S_2(m, x, \theta)|^2 \tag{6.3.2}$$

and is normally used by the optical oceanography community to describe the scattering profile of an elemental volume rather than the normalized scaler phase function. Referring back to Equations (6.1.19a) and (6.1.19b), we see that

$$b = 2\pi \int_0^{2\pi} \beta(\theta) \sin\theta \, d\theta \tag{6.3.3}$$

where b is the volume scattering coefficient. Hence, we can write

$$\beta(\theta) = bP(\theta)/4\pi \tag{6.3.4}$$

Optical propagation through pure water is essentially energy transfer through a Rayleigh scattering medium, that is, a medium composed of "particles" (here molecular water) much smaller than the wavelength of the incident light. Table 6.5 shows some of the inherent properties of pure water for various wavelengths.[21] It is apparent from this table that the scattering coefficients roughly follow the expected λ^{-4} dependence. In

Table 6.5. Inherent Properties of Pure Water

	Coefficient (10^{-3} m^{-1})			
Wavelength (nm)	c	b	a	β_w (90°)
400	43	5.81	37.19	363
425	33	4.47	28.53	280
450	19	3.49	15.51	218
475	18	2.76	15.24	173
500	36	2.22	33.78	138
525	41	1.79	39.21	112
550	69	1.49	67.51	93
575	91–109	1.25	89.75–147.75	78
600	186–272	1.09	184.9–270.9	68

addition, it is also clear that the extinction process is dominated by absorption, with its minimum near 475 nm. This last point helps explain the deep blue color of the ocean. The volume scattering function for pure water can be shown to be closely approximated by

$$\beta_w(\theta) = \beta(\theta)|_{\theta=90^0}\left[1 + \frac{1-\delta}{1+\delta}\cos^2\theta\right] \tag{6.3.5a}$$

$$\approx \beta(90^0)[1 + 0.8349\cos^2\theta] \tag{6.3.5b}$$

using the results of Morel,[25] which can be further quantified for several wavelengths utilizing Table 6.5. Table 6.6 provides a comparison of the volume scattering coefficients and the volume scattering functions at $\theta = 90^0$ for pure water and pure seawater at various wavelengths.[25] The parameter S cited indicates the salinity of the seawater. It follows from this table that pure seawater scatters about 30% more than pure water does, but follows the same λ^{-4} law. Thus one can use Equation (6.3.4) (with, for example, Table 6.5) to compute the volume scattering function of pure seawater. From Morel, it appears that sea salt exhibits strong absorption in the ultraviolet and negligible absorption in the visible. Hence, all the inherent properties of pure water are basically known. Unfortunately, the water of the world are not always pure, especially near the continental land masses. In these situations, one finds the greatest degradation of optical radiation transfer. The agents responsible for this degradation are dissolved organic matter and, more importantly, suspended materials such as phytoplankton. Let us look at the former first.

Table 6.6. Volume Scattering Function at 90° and Total Scattering Coefficient for Pure Water and Seawater as a Function of the Wavelength

	Pure water		Pure seawater (S = 35–39‰)	
Wavelength (nm)	$\beta_0(90°)$ (10^{-4} m^{-1})	b_0 (10^{-4} m^{-1})	β_0 (90°) (10^{-4} m^{-1})	b_0 (10^{-4} m^{-1})
350	6.47	103.5	8.41	134.5
375	4.80	76.8	6.24	99.8
400	3.63	58.1	4.72	75.5
425	2.80	44.7	3.63	58.1
450	2.18	34.9	2.84	45.4
475	1.73	27.6	2.25	35.9
500	1.38	22.2	1.80	28.8
525	1.12	17.9	1.46	23.3
550	0.93	14.9	1.21	19.3
575	0.78	12.5	1.01	16.2
600	0.68	10.9	0.88	14.1

The effects of numerous *organic materials* dissolved in seawater have been investigated by several researchers.[24] Dissolved humic acids, by-products of decayed terrestrial and oceanic entities, appear to be the most abundant and the most degrading of this class of material.[23] These complex mixtures are classified under the general title of "yellow substance" and are formed from carbohydrates by a "Maillard" reaction. Figure 6.20 gives the absorption coefficient for yellow substance as a function of wavelength.[23] We see that these materials have their strongest effect in the ultraviolet and that their absorption exponentially decreases with increasing wavelength to negligible levels above 625 nm. On the other hand, the scattering of light by yellow substance is minimal across the spectrum, compared to what we found above. Thus we see that this type of material augments the inherent properties of pure seawater only in absorption and not in scattering. Particulate matter suspended in the ocean affects both.

Suspended materials such as phytoplankton organic detritus and zooplankton are the most detrimental to optical energy transfer in the marine

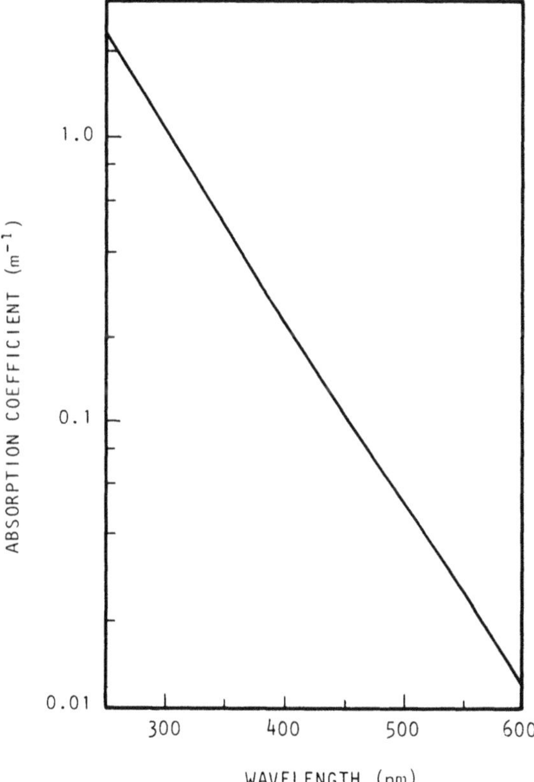

Figure 6.20. Absorption curve for yellow substance.

environment. Unfortunately, the shapes of these materials are not usually spherical, a mandatory requirement for the application of Mie theory. However, it has been shown by Penndorf that Mie theory can be applied to irregular particles of refractive index less than 2.0 if the orientations of the particles are random.[24,26] In addition, Hodkinson[27] and Holland and Gagne[28] have showed that the forward scattering of randomly oriented irregular particles can be modeled as a suspension of spherical particles with equivalent radii. On the other hand, the large effect of particle shape on the backscattering position of the volume scattering function has only been demonstrated,[28] and very little success in fitting the observed profile using Mie theory has emerged.

The cumulative particle size distribution of the ocean has been suggested by Bader to be approximately

$$\rho(D) = \rho_0 D^{-\gamma} \tag{6.3.6}$$

in units of particles per unit volume larger than diameter D.[24] The parameter ρ_0 is the number of particles per unit volume greater than 1 μm and γ is the slope of the size distribution, a constant. Figure 6.21 compared Equation (6.3.6) for various parameters with measured cumulative distributions. It is apparent from this figure that this analytical expression is only a crude approximation to reality. In this figure, the constant γ ranges between 0.7 and 6; the average value is around 2.5. Kullenberg suggests that the two limiting factors of Equation (6.3.6) are its ability to accurately predict the number of large particles and its estimation of the small particles which are removed by the processes of dissolution, absorption, and flocculation.[23] Thus caution must be employed in its use.

Besides particle shape, the indices of refraction for suspended material can play havoc with any calculation of the inherent properties of the ocean using Mie theory. Several researchers have attempted to determine an effective complex index for a collection of suspended particles from observed inherent properties.[24] Zanevold *et al.* have developed a method for computing a refractive index distribution. Results from this work indicate indices (relative to that of water) above 1.10 and below 1.05, possibly corresponding to skeletal and inorganic material for the former and organic matter for the latter.[24] Let us now develop the inherent properties of suspended material in seawater.

Figure 6.22 illustrates a number of volume scattering function measurements in different oceanic waters.[23] It is quite interesting to note in this figure the similarities of the forward scattering profiles for the diverse marine conditions. This type of behavior is identical to that found for the scalar phase functions of the atmosphere, that is, a forward-scattering bias. However, the imaginary portion of the refractive index plays a stronger role

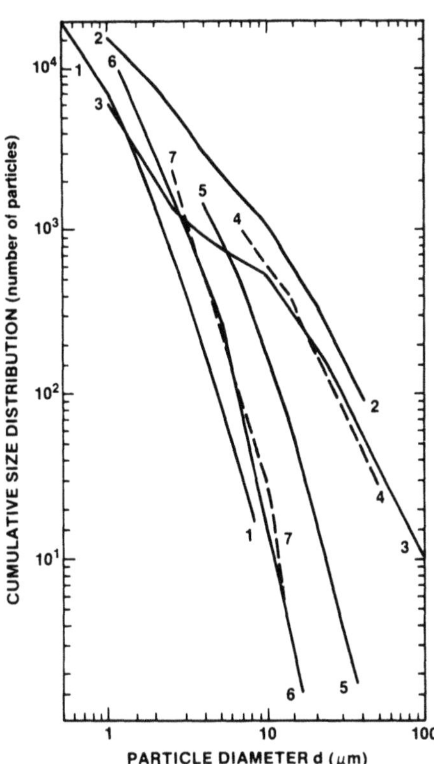

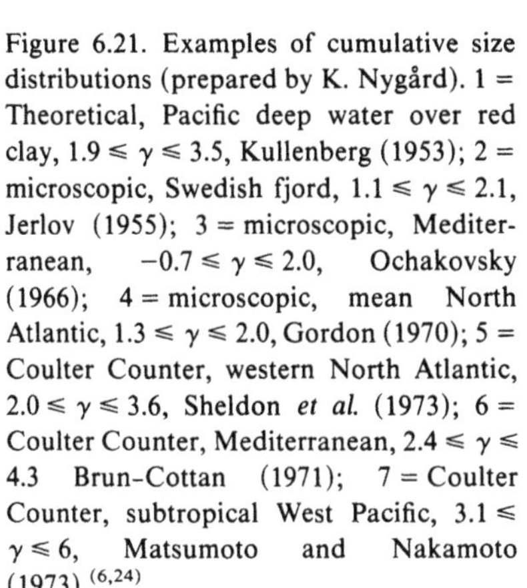
Figure 6.21. Examples of cumulative size distributions (prepared by K. Nygård). 1 = Theoretical, Pacific deep water over red clay, $1.9 \leqslant \gamma \leqslant 3.5$, Kullenberg (1953); 2 = microscopic, Swedish fjord, $1.1 \leqslant \gamma \leqslant 2.1$, Jerlov (1955); 3 = microscopic, Mediterranean, $-0.7 \leqslant \gamma \leqslant 2.0$, Ochakovsky (1966); 4 = microscopic, mean North Atlantic, $1.3 \leqslant \gamma \leqslant 2.0$, Gordon (1970); 5 = Coulter Counter, western North Atlantic, $2.0 \leqslant \gamma \leqslant 3.6$, Sheldon *et al.* (1973); 6 = Coulter Counter, Mediterranean, $2.4 \leqslant \gamma \leqslant 4.3$ Brun-Cottan (1971); 7 = Coulter Counter, subtropical West Pacific, $3.1 \leqslant \gamma \leqslant 6$, Matsumoto and Nakamoto (1973).[6,24]

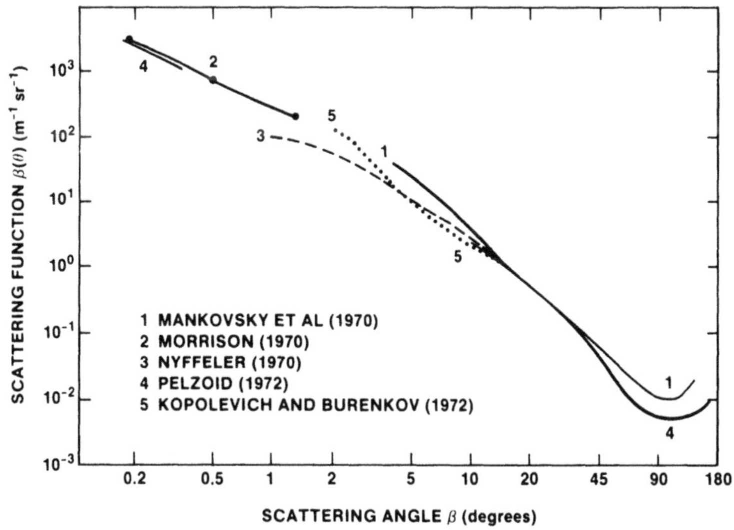

Figure 6.22. Comparison between different scattering functions (normalized at 15°).

in defining the shape of the scattering function than it does in the case of the atmosphere because of the increased absorption. For example, Mueller has shown theoretically that a three-layer absorbing sphere (modeled to resemble phytoplankton) exhibits strongly wavelength-dependent absorption and non-zero-angle scattering, but in a complementary fashion.[23] That is, as the absorption increases, the profile of the scattering function becomes more peaked (less diffuse). Augmenting this effect are, of course, the influences of the real portions of the refractive index distribution and the large particle density, just as in the case of the atmosphere. This is clearly shown in Figures 6.23 and 6.24, respectively. These figures show that the lower-sloped size distributions and the lower indices of refraction create strongly forward-biased volume scatttering functions. Because of the close relationship between $\beta(\theta)$ and b, the volume scattering coefficient exhibits the same parametric dependencies. The overall conclusion is that the cleanest waters possess the least forward-scattering and attenuating effects

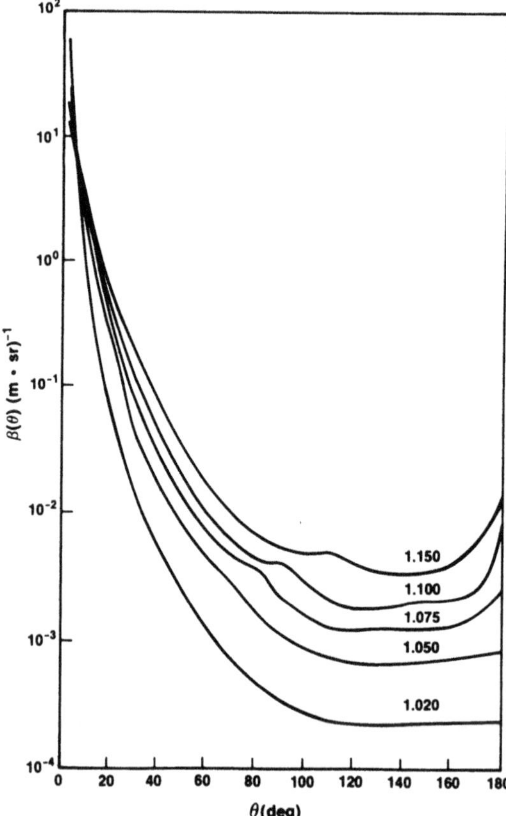

Figure 6.23. Normalized volume scattering functions for various refractive indices with Junge exponent $C = 2.7$.

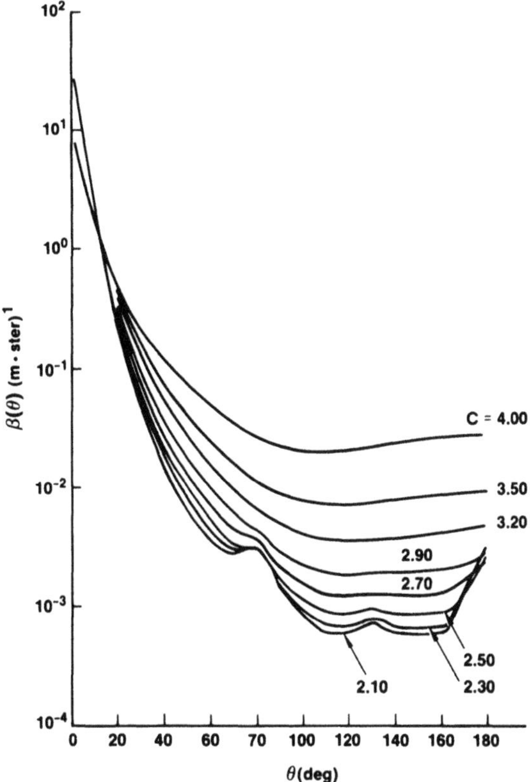

Figure 6.24. Normalized volume scattering functions for various particle size distributions characterized by the Junge exponent C with refractive index 1.075.

on optical propagation; the dirtiest, the most. This is supported by scattering function measurements shown in Figures 6.25 and 6.26. In the first case, we see this to be true for different experimental sites.(29) In the second case, we see it is true as one goes deeper to cleaner waters.(30) This latter facet is typical in nature. That is, in those marine environments containing a large number of particulates, the majority of the suspended material will usually be concentrated in the first 30 to 50 meters, with the water becoming progressively cleaner after that as one goes deeper. In any event, what we are trying to suggest is that the marine channel is a forward-scattering medium just like the atmospheric channel, with the strongest effects occurring when the particulate population is high.

Table 6.7 gives the inherent properties of several diverse oceanic conditions.(24) In this table, the parameter c_w denotes the extinction coefficient for pure seawater, b_p the scattering coefficient of the particulate

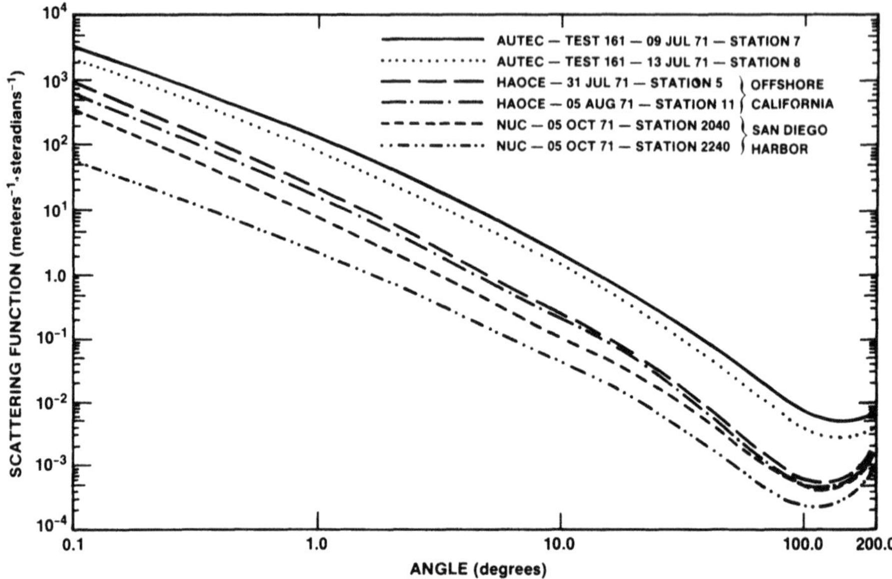

Figure 6.25. Volume scattering function of various water types.

material, a_p the absorption coefficient for those materials, and a_y the absorption coefficient for yellow substance. Clearly, the overall extinction coefficient is given by

$$c = c_2 + a_y + a_p + b_p \tag{6.3.7}$$

We also see that $(c - c_w)$ ranges from $0.05\ \text{m}^{-1}$ for clean/deep waters to $1.72\ \text{m}^{-1}$ for coastal waters, in accordance with our previous discussions. Ocean water is a more attenuating medium than we found the atmosphere to be. The wavelength dependence of $(c - c_w)$ for the data shown in Table 6.7 has been suggested to be given by[(23)]

$$(c - c_w)_{380\ \text{nm}} = 1.8(c - c_w)_{655\ \text{nm}} \tag{6.3.8}$$

This expression supports Morel's hypothesis of an inverse wavelength relationship for all but the cleanest of waters.[(25)] Thus it appears that the extinction process is dominated by the absorption of particulates and yellow substance, and only affected slightly by the particulate scattering.

The inherent properties are not the only aspects of the marine channel important to its characterization. So are the ocean's apparent properties.

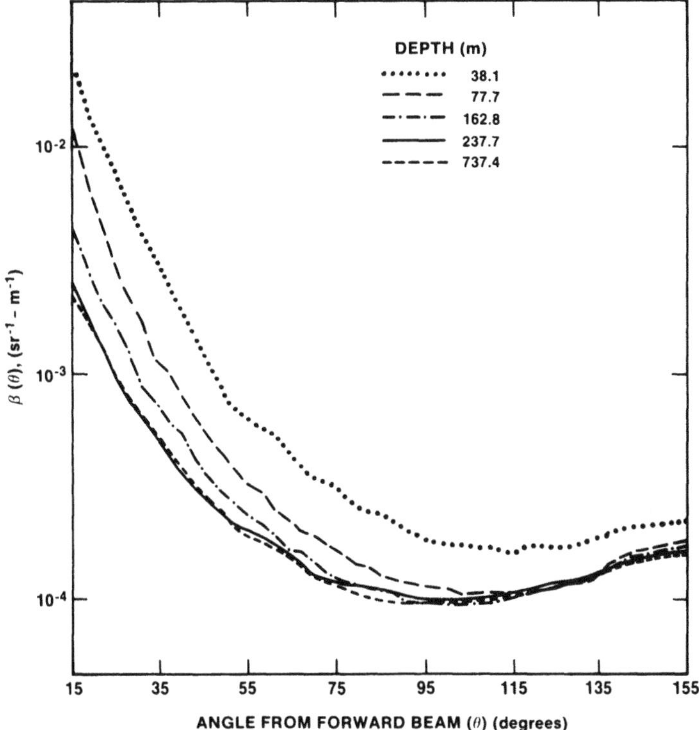

Figure 6.26. Volume scattering coefficient as a function of scattering angle and depth. *USS Rexburg*, 22–23 August 1967.

6.4. The Apparent Properties of the Sea

Apparent properties are those perceived as general features of the light distribution found within a particulate multiple-scattering medium.(31,32) They can be developed either through solutions of the radiative transfer equation (which will be derived in the next chapter) or through direct measurements of specific quantities in the field. However, these properties are most useful in describing situations where absorption is clearly present and the scattering is approaching, or is in, the diffusion regime.(13) In these cases, they allow simple, practical answers for a wide range of multiple-scattering problems to be developed and exact mathematical relationships between themselves and the more fundamental inherent properties to be formulated. In this section, we will define the apparent properties normally associated with optical oceanography and focus in on the most important one used in underwater communications analysis, the diffuse attenuation coefficient.

Table 6.7. A Comparison of Inherent Optical Properties[a]

Location	Wavelength (nm)	Values[b] (m^{-1})				
		$c - c_w$	a_p	b_p	c_p	a_y
Caribbean Sea	655	0.06	0	0.06	0.06	
	440	0.09		0.06		
Equator, central Pacific	440	0.09		0.05		
Romanche Deep	440	0.12		0.07		
Galapagos	655	0.11	0.04	0.04	0.11	
	440	0.24		0.08		
Off Peru (64 miles)	700	0.39			0.39	0
	400	0.73			0.64	0.09
Galapagos	700	0.16			0.16	0
	400	0.25			0.21	0.04
Northwestern Galapagos	700	0.07			0.07	0
	400	0.11			0.08	0.03
Continental slope	668	0.05				0
	365	0.10			0.08	0.02
Bermuda	655	0.10			0.10	0
	380	0.20			0.17	0.03
Kattegat	655	0.23	0.08	0.15	0.23	0
	380	0.54	0.27	0.16	0.44	0.11
South Baltic Sea	655	0.27	0.07	0.20	0.27	0
	380	1.15	0.28	0.21	0.49	0.68
Bothnian Gulf	655	0.38	0.10	0.28	0.38	0
	380	1.72	0.33	0.31	0.64	1.08
North Atlantic	655					
	420					0.03
North Sea	655					0.01
	420					0.10
Baltic Sea	665					0.02
	420					0.33
Sargasso Sea	633	0.03	0.01	0.02	0.03	
	484	0.03	0.01			
	440	0.05	0.01	0.04		
	375			0.02		
Western Mediterranean area	655	0.08	0.04	0.04	0.08	0
	375	0.06		0.04		

[a] Ref. 24.
[b] $c_{observed} = c_w + a_p + b_p + a_y$.

The temporal and spatial variations in the inherent properties of natural waters are directly traceable to the fluctuations in concentration of suspended particles due to currents, as long as the time scale of interest is short compared to those of the setting, biological, and chemical processes of the medium.[25] This is quite fortunate since these conditions generally yield coherence lengths of the horizontal variations (typically in the hundreds of meters, for noncoastal waters) much larger than that of the vertical variation. Hence, these properties can be assumed functionally dependent on depth alone. Figures 6.27–6.31 illustrate a sequence of typical inherent properties profiles for nontropical waters. Tropical waters are generally more uniform with depth. These particular measurements were taken by personnel from the Visibility Laboratory, Scripps Institution of Oceanography, off Santa Catalina Island on 29 June 1975.[33] It is apparent in these figures that the volume absorption coefficient is fairly uniform with depth. On the other

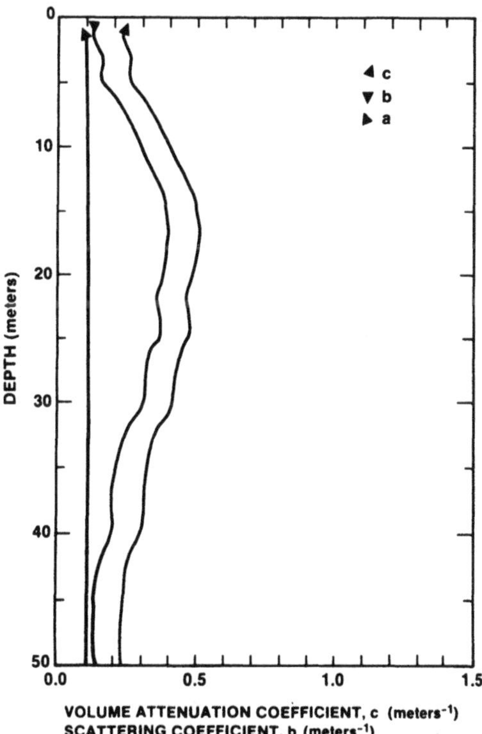

Figure 6.27. Ocean optical properties, Santa Catalina Island, 20 June 1975, 0950 PDT.

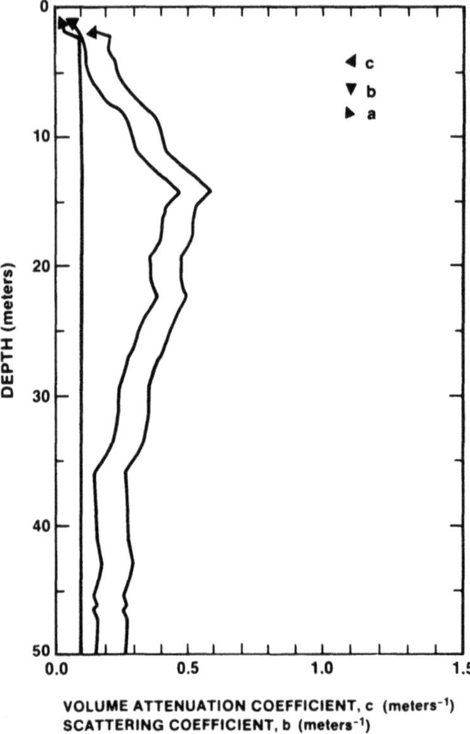

Figure 6.28. Ocean optical properties, Santa Catalina Island, 20 June 1975, 1050 PDT.

hand, the volume scattering coefficient (hence the attenuation coefficient c) varies with depth and has its greatest magnitude in the 10 to 30-m depth range. It is also clear that the scattering layer becomes more concentrated and shallower as the day warms. This type of behavior is generally tied to the diurnal nature of photosynthesis by phytoplankton and the temperature variation within the column. Specifically, the biology migrates toward the surface during the day and becomes more uniformly distributed at night. In addition, when the water column possesses a temperature gradient with a strong transition layer, dead "animals" and other organic matter will be inhibited in their diffusion downwards. The result is a confined scattering region growth in the upper portions of the water column—the type of situation exhibited above. This region of strong temperature change or gradient is known as the thermocline. It is usually present in nontropical waters and absent in the warmer oceans. One also finds in many situations that more particulates exist above the thermocline than below. In other

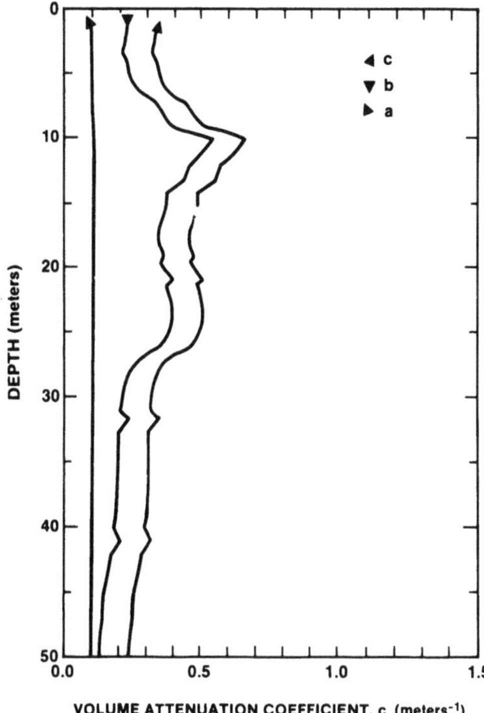

Figure 6.29. Ocean optical properties, Santa Catalina Island, 20 June 1975, 1200 PDT.

words, the water becomes "cleaner," and hence clearer, with depth. This is an important point when dealing with underwater communications. Figure 6.32 is a typical plot of the volume scattering function for the same location. However, unlike the above properties, the shape of the scattering function was essentially independent of depth and time, as expected.

Since the apparent optical properties are jointly dependent on the inherent properties and the radiance structure within a specific water column, a consistent set of initial illumination conditions must be established in order that we keep the above-cited depth dependence, and that different groups of data can be compared and analyzed together. Most researchers use clear sky illumination, when the sun is near-overhead and the sea state is low (≤ 2), whenever it is convenient. Given this, the spatial radiance will be of the form $N(z, \theta, \varphi)$ and possessing variability solely on the particulate structure of the water. Let us now turn to the definitions of the apparent properties of natural waters.

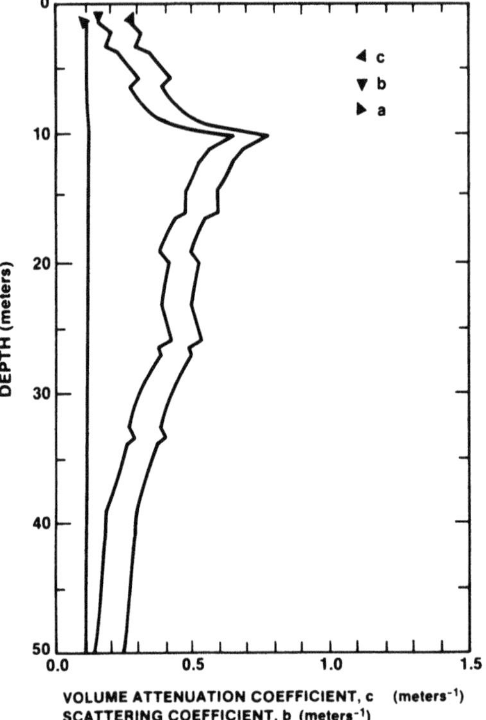

Figure 6.30. Ocean optical properties, Santa Catalina Island, 20 June 1975, 1250 PDT.

6.4.1. Reflectance

The reflectance of water is defined as the ratio of the upwelling to downwelling irradiance at depth z. Mathematically, we have

$$R = \frac{E_u(z)}{E_d(z)} \tag{6.4.1}$$

where

$$E_u(z) = -\int_0^{2\pi} \int_0^{\pi/2} N(z, \theta, \varphi) \cos\theta \, d\Omega \tag{6.4.2}$$

$$E_d(z) = -\int_0^{2\pi} \int_{\pi/2}^{\pi} N(z, \theta, \varphi) \cos\theta \, d\Omega \tag{6.4.3}$$

$$d\Omega = \sin\theta \, d\theta \, d\varphi \tag{6.4.4}$$

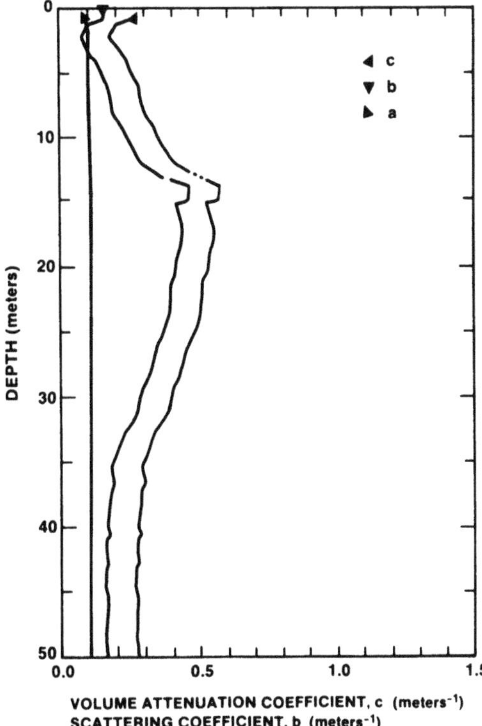

Figure 6.31. Ocean optical properties, Santa Catalina Island, 20 June 1975, 1345 PDT.

will $N(z, \theta, \varphi)$ the radiance distribution at depth. In the above, E_u represents the flux per unit area per unit wavelength interval measured by a horizontally oriented cosine collector facing downward to accept upwelling radiance, and E_d the reverse.[24,25] Equation (6.4.1) is also known as the irradiance ratio. A more useful definition of reflectance, at least as far as remote sensing applications are concerned, is the ratio of upwelling radiance to downwelling irradiance. This definition is called the directional reflectance and is written as

$$\rho = \frac{\pi N_u(z)}{E_d(z)} \tag{6.4.5}$$

Equation (6.4.5) gives the ratio of the radiant flux returned (reflected) by the water column to that which would be reflected into the same solid angle

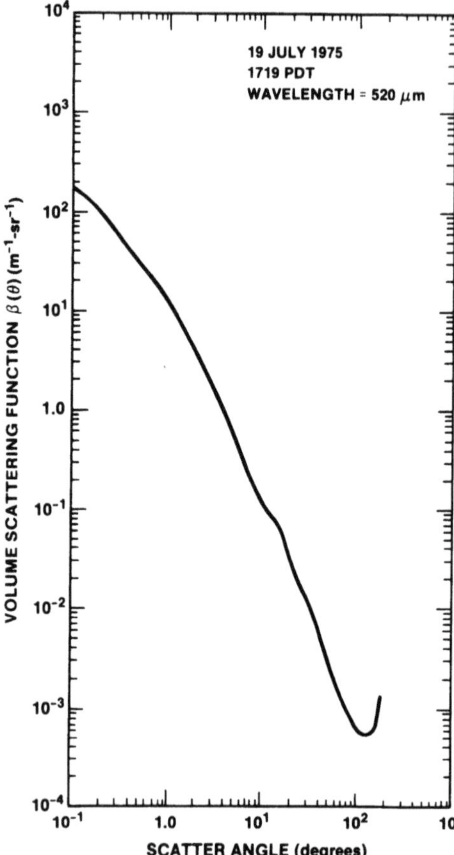

Figure 6.32. Typical volume scattering function for Santa Catalina Island area, 19 July 1975.

under the same conditions of incidence by a perfect diffuser.[34] Austin *et al.* have shown that the above reflectances can be related through the equation

$$R = \frac{E_u}{E_d} = \frac{N_u Q}{E_d}$$

$$= \rho Q \tag{6.4.6}$$

where Q is the compensation factor for the fact that the upwelling radiance distribution may not be uniform at the depth of interest.[35] It is typically between 0.95 and 3.82 and is wavelength dependent.

6.4.2. Irradiance Attenuation Coefficients

The irradiance attenuation coefficients for natural waters are defined as the logarithmic depth derivatives of the upwelling and downwelling irradiance. Mathematically, we write

$$k_u(z) = \frac{-1}{E_u(z)}\left[\frac{dE_u(z)}{dz}\right] \tag{6.4.7}$$

and

$$k_d(z) = \frac{-1}{E_d(z)}\left[\frac{dE_d(z)}{dz}\right] \tag{6.4.8}$$

for the upwelling and downwelling irradiance attenuation coefficients, respectively. In practice, these coefficients are usually determined from log-irradiance versus depth plots, $k_{u,d}$ being the slope of the resulting curves.

The last two paragraphs constitute the basic definitions of the most commonly measured and used apparent optical properties: the upwelling and downwelling irradiances, their associated attenuation coefficients, the irradiance ratio, and the directional reflectance. However, in underwater communication analysis, only the downwelling irradiance attenuation coefficient is important. This can be explained as follows:

Radiation transfer in natural waters to depths greater than a few scattering length ($\propto 1/b$) provides one with probably the finest example of diffusion propagation found in nature. This contrasts with the type of multiple scattering that is usually seen in the atmospheric channel, which is more a mixing phenomenon than a diffusive one. The reason for this is the relative amount of absorption present in the two different media. To illustrate this, we will describe how these two phenomena manifest themselves physically and show mathematically why absorption plays the defining role.

On a relatively overcast day, the "sky" looks uniformly dark and is somewhat time-of-day independent as long as the cloud conditions remain unchanged. This is because the incident solar illumination on the cloud top has been scattered so many times within the cloud layer that it has lost track of where it entered and a high percentage of it is reflected in the backward direction. However, both the top and bottom surfaces are isotropic sources emitting equal intensity in all hemispherical viewing directions, differing only in their total integrated power levels. This type of optical radiator is called a Lambertian surface. The mixing phenomenon described above stems from the negligible absorption experienced with all visible light while interacting with all the particulates within the cloud. For most clouds, the single-scatter albedo is >0.999.[36,37] This implies that the light essentially "rattles" around in these clouds until it emerges either through the top or bottom of the layer (this is why thin and/or finite extent clouds appear "white"). The average number of particle interactions is generally given by the cloud's optical thickness, defined to be the volume extinction coefficient ($\sim b$) times the cloud physical thickness z. Mathematically, it is given by $\tau = cz$ and represents the number of extinction lengths comprising the cloud's thickness.

When light first enters natural waters, it begins to experience a mixing process. However, this light redistribution takes on a more diffusive profile as the number of particulate interactions increases. The extra path length of the scattered radiation increases the possibility (or probability) of that radiation being absorbed rather than being rescattered. Hence, the light making it to depth takes the path of least resistance (Fermat's principle). The result is a blurred solar disk always directly overhead. That is, the sun is always overhead independent of time of day. This asymptotic radiance pattern observed at these "deep" depths is the diffusion pattern. Figure 6.33 illustrates the above transition process for lake data taken by Tyler.[38] Thus, we see a uniform response for clouds and a diffusion pattern for water. Let us look at this mathematically.

In an unbounded medium with uniform scattering and absorption properties, the asymptotic radiance distribution can be shown to equal[13]

$$N_{\infty}(z, \theta) \approx S_1 N(u) \exp\{-kz\} + S_2 N(-u) \exp\{kz\} \qquad (6.4.9)$$

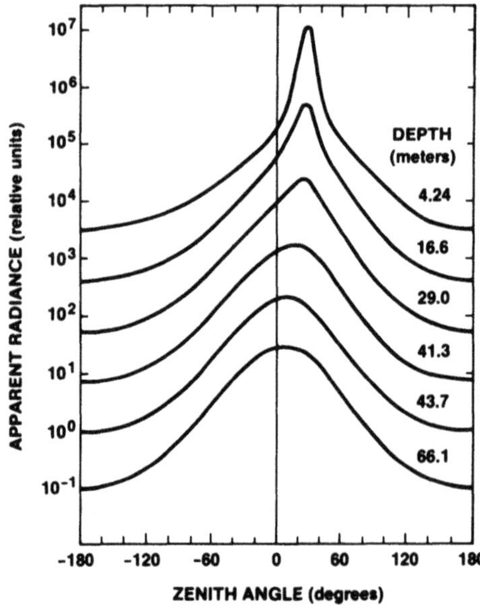

Figure 6.33. In-water radiance distribution in the plane of the sun obtained from Ref. 38.

where $u = \cos\theta$, S_1 = "diffusion stream" in positive z-direction, S_2 = "diffusion stream" in negative z-direction, k = asymptotic radiance attenuation coefficient, and $N(u)$ = diffusion pattern, characteristic for medium's scalar phase function and albedo. Here k is the smallest eigenvalue and $N(u)$ the corresponding eigenfunction of the integral equation

$$(1 - ku)N(u) = \tfrac{1}{2}\int_{-1}^{+1}\int_{-1}^{+1} h(u, v)N(v)\,dv \qquad (6.4.10)$$

with $h(u, v)$ being the redistribution functions defined as the average of the medium's phase function overazimuth. From the form of Equation (6.4.8), it is clear that

$$k_d(z) = k = -k_u(z)$$

in the asymptotic limit. That is, the radiance and irradiance attenuation coefficients become equal in magnitude at the asymptotic depth limit. In optical oceanography, k is known as the diffuse attenuation coefficient. It is most useful in communication analysis since Equation (6.4.8) gives the exact irradiance/radiance loss rate at deep depths and a conservative estimate of that rate in the diffusive transition region. Let us now see how this parameter relates to the described phenomena.

Figure 6.34 depicts the ratio of the diffuse attenuation coefficient to the volume extinction coefficient as a function of the asymmetry factor g

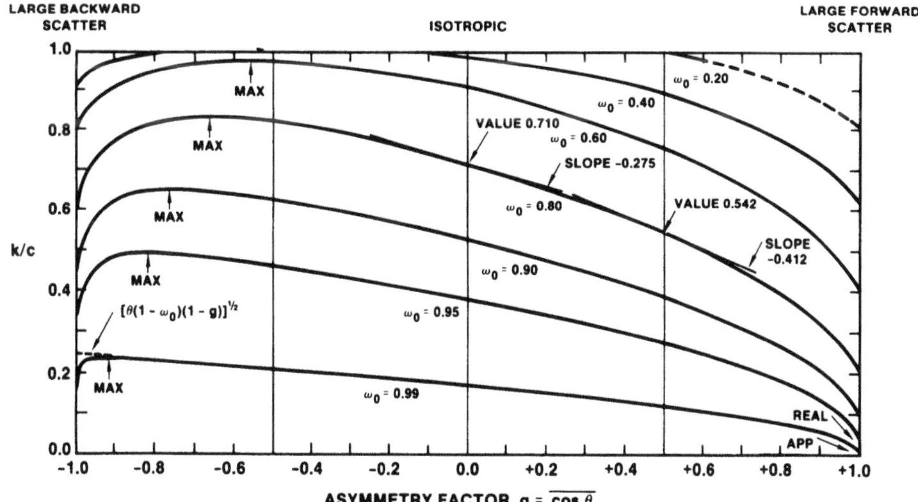

Figure 6.34. Typical values of k/c for various asymmetry factors and single-scatter albedos.

for selected values of the single-scatter albedo. The asymmetry factor is the average cosine of the scalar phase function of the medium, i.e.,

$$g = \overline{\cos\theta} = \int_0^{2\pi} \int_0^{\pi} \cos\theta \frac{P(\theta)}{4\pi} \, d\Omega \tag{6.4.11}$$

and is one of the more often used ways of describing the scattering nature of the medium. As the value of g approaches 1, the more forward scattering the medium; conversely, the closer g is to -1 the more backward scattering the medium. Isotropic scattering media possess asymmetry factors near zero. This plot illustrates the relative magnitude of diffusive scattering for various media. For example, a C.3 cloud (high-altitude "Mother of Pearl") can be approximated by

$$k \sim c(1 - \omega_0)^{0.65}$$

Thus for $\omega_0 = 0.989$, $k \sim 0.11c$ and does not vary greatly for other cloud types. If the cloud of interest has a 61-m extinction length and a physical thickness of 2500 m, the diffusive loss is equal to

$$\exp\{-0.011(2500/61)\} = 0.64 \text{ or } 10\log_{10}(0.64) = -1.96 \text{ dB}$$

compared to the -5.88-dB loss normally associated with that type of cloud. This indicates why clouds have a uniform emission nature. The ratio of k to c is typically between 0.2 and 1.0, depending on the water type. This is shown in Figure 6.35 for some representative scattering media. The diffusive loss is in the tens of decibels for most reasonable geometries and exceeds the normal mixing loss by several orders of magnitude. Hence, the diffusive profile for the radiance emerges at "deep" depths.

Detailed spectral irradiance measurements from which k_d and k_u values may be obtained have been made by many researchers,[24,25] most notably Tyler and Smith, e.g., Refs. 39 and 40, and Morel and his coworkers, e.g., Refs. 41 and 42. Unfortunately, as complete as they are, they have been performed in a very limited number of locations with very few measurements in open ocean waters.

Jerlov has devised a classification scheme for oceanic waters which divides them into five specific classes: Types I (cleanest), IA, IB, II, and III (dirtiest). This class division was based on spectral irradiance transmittance measurements made in the upper portions (0–10 m) of the ocean. Figure 6.36 illustrates the attenuation rate of the above Jerlov water types as a function of wavelength, while Figure 6.37 gives the worldwide distribution.[24] It is clear from Figure 6.36 that the cleaner waters (I, IA, IB) transmit better in the blue than at other wavelengths, while the dirtier

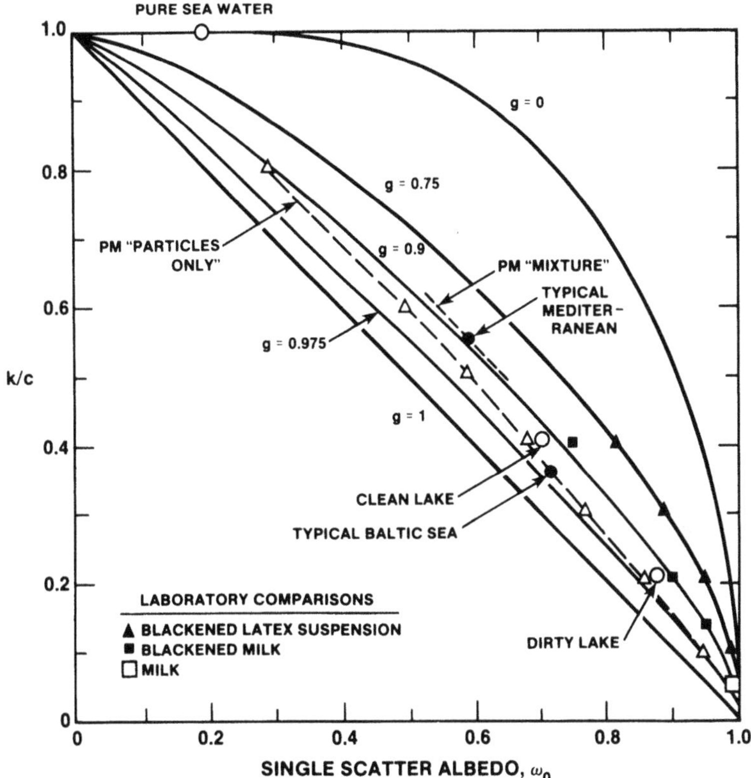

Figure 6.35. Plot of k/c as a function of single scatter albedo.

waters (II, III) have less attenuation in the blue-green wavelength regime. This type of behavior was pointed out previously in the inherent properties section. The most interesting aspect of these curves is that 475 nm appears to be a minimum attenuation wavelength in all the Jerlov water types. Figure 6.38 illustrates that water reflectivity ρ is a function of wavelength for the Jerlov I, IA, IB, II, III, 3, and 5 water types and pure water, as derived by Petzold.[43] It is apparent from this graph that water reflectivity is highest for clearest waters and lowest for the dirtiest waters. Not surprisingly, it can be shown to first order that $\rho \propto 1/k$.

Although the above Jerlov water model is generally quite useful in a first-order link budget assessment, it suffers from several shortcomings as a data base for statistical optical communications analysis. First, it is not clear that the oceans of the world possess spectral transmittance characteristics which can be easily separated into just five effective water-type groups. Second, the amount of data used to create this data base was very limited and the spatial variability of $k_d(z)$ cannot be quantified with any great

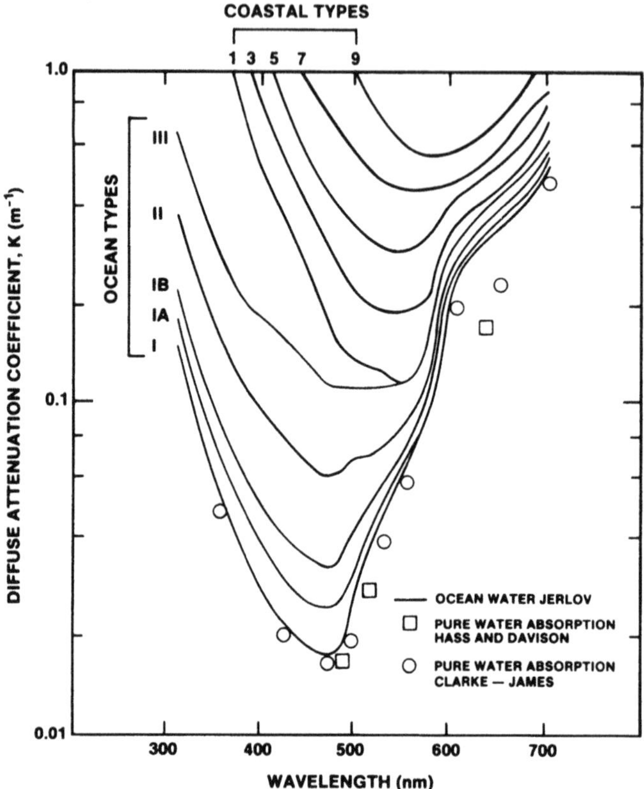

Figure 6.36. Diffuse attenuation coefficient versus wavelength.

accuracy. Third, there is no information in this model about the values of the diffuse attenuation coefficient below 10 meters. Finally, no diurnal or seasonal knowledge about k_d exists in this data base. To alleviate these model inadequacies, R. W. Austin and his coworkers at the Visibility Laboratory (VISLAB), Scripps Institution of Oceanography, have undertaken an effort to develop a more complete statistical data base for the diffuse attenuation coefficient.(11-16) Specifically, they are using remotely sensed radiance measurements from the Coastal Zone Color Scanner (CZCS) onboard the NIMBUS-7 satellite to infer spectral k_d. Orbit parameters and general characteristics for the NIMBUS-7 are shown in Figures 6.39 and 6.40 and Table 6.8.(11) Figures 6.41 and 6.42 illustrate two comparisons between measured k_d values and inferred values derived from coincident ground-truth measurements and NIMBUS-7 overflights. Figure 6.41 is a comparison with the initial "k-algorithm" (circa 1978), and Figure 6.42 is a comparison using the current improved version. It is apparent from these graphs that both VISLAB k-algorithms produce values of k_d in good

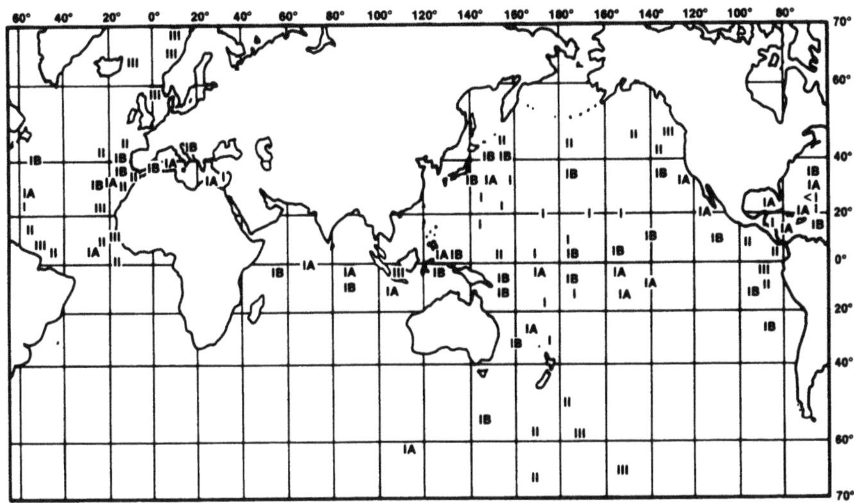

Figure 6.37. Worldwide distribution of Jerlov water types.

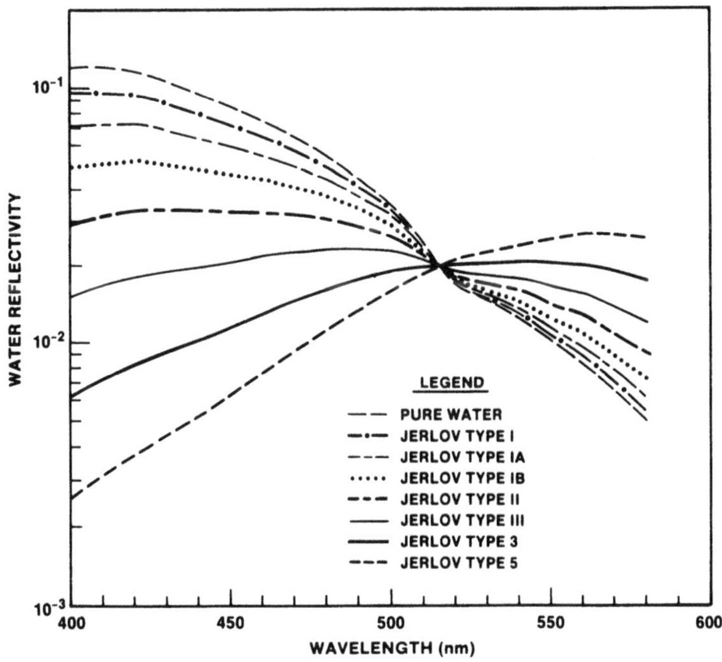

Figure 6.38. Water reflectivity ρ as a function of wavelength for Jerlov water types I, II, III, 3, and 5.

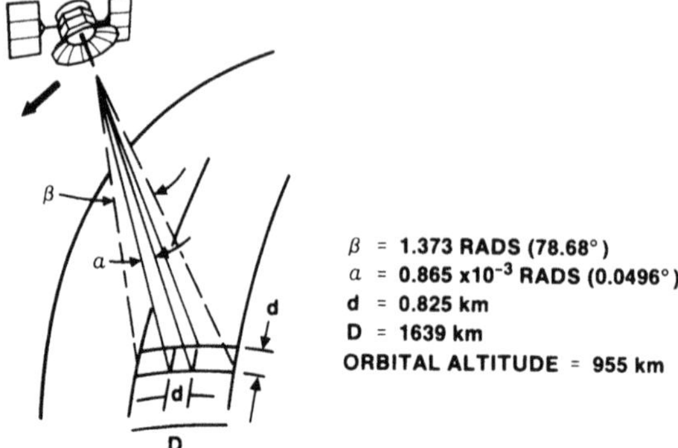

Figure 6.39. CZCS scanning arrangement.

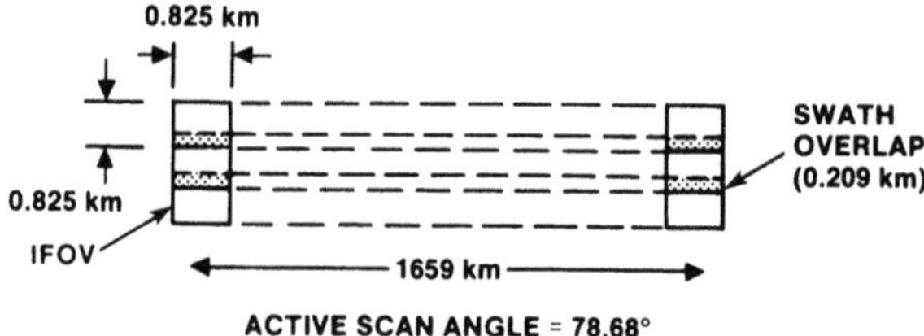

Figure 6.40. CZCS earth scan pattern.

agreement with measured values, the values from the latter algorithm being the more accurate. From the data processed so far, the VISLAB has been able to ascertain five basic "Jerlov" water types. These are shown as a function of wavelength in Figure 6.43 and in Table 6.9. Figure 6.44 illustrates a first-order k map for the Pacific Ocean at $\lambda = 490$ nm. Table 6.10 gives a

Table 6.8. Orbit Parameters Used for Coverage Plots

Nominal orbit parameters	
Altitude	955 km
Inclination	99.2°
Period	104.15 min
Orbits per day	13.82 (results in daily ascending node separation of 4.55°)
Ascending node separation for adjacent orbits	26.04° (westward motion)
Launch parameters	
Ascending mode time	1152
Date	25 Nov 1978
Resulting sun angle	8°

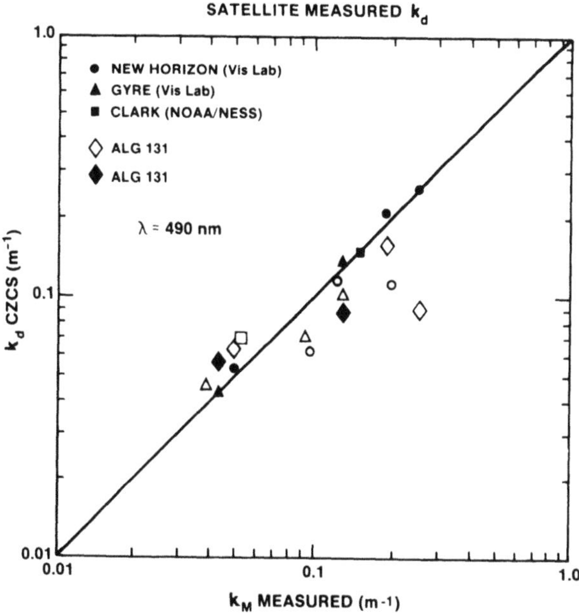

Figure 6.41. Comparison of k values derived form CZCS observations with k values determined from surface station measurements. Solid points (●, ▲, ■) overpass coincident with surface measurements. Open points (○, △, □) overpass on different date than surface measurement.[43]

first-order water column model for this data base which was derived from a Scripps/VISLAB CZCS cruise which occurred in the summer of 1980.

6.4.3. Secchi Disk

Another source of optical properties for oceanic waters is the body of Secchi depth determinations acquired during the past hundred or so years around the world. This technique involves dropping a 30-cm disk over the sunny side of a boat and noting the depth in which it "disappears" with one's head in the water.[47] Although subjective in nature, these have been the most common measurements of water clarity and do provide a relative measure of the volume extinction and irradiance attenuation coefficients.[48] The most frequently used relations are

$$cD_s \approx 6.0 \tag{6.4.12}$$

and

$$k_d D_s \approx N^* \tag{6.4.13}$$

where N^* is an empirically derived constant having a value between 1.4 and 1.7, and D_s is the Secchi depth.[49] Reference 50 suggests the value of

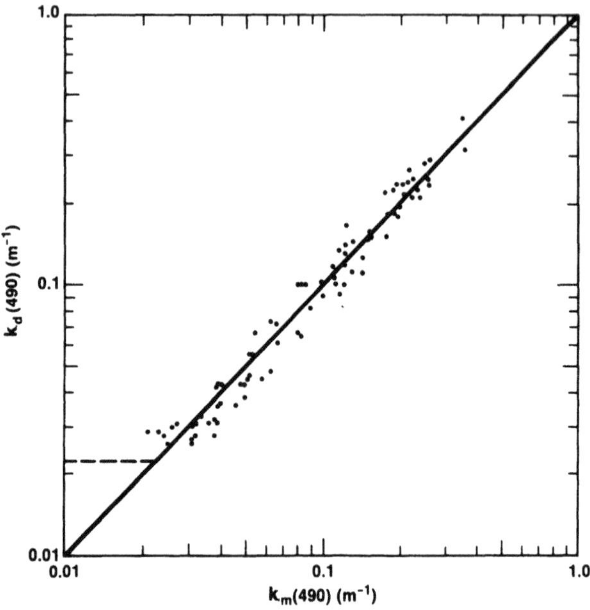

Figure 6.42. Values of k_d computed from "k" algorithm.

N^* may be around 3.35. Table 6.11 is a compilation of averaged volume extinction coefficients, Secchi depth, and their product taken in the inhomogeneous waters off Santa Catalina Island in 1975.(33) The averaged value of 6.96 is in reasonable agreement with Equation (6.4.12), given the amount of data and water variability.(33) However, note its standard deviation. It is apparent that both equations have technical uncertainty associated with their accuracy. However, given that cautionary note, one can get a good relative feel for the water clarity of the world's oceans through the atlas contained in Reference 48.

To conclude this section, we will review some of the semiempirical relationships developed between the apparent and inherent optical properties. The most comprehensive set of relationships created to data are those given by Gordon *et al.*, in a paper documenting the results of their analysis of many Monte Carlo simulations of the oceanic environment.(51) For a homogeneous ocean illuminated by collimated irradiance, they showed that

$$\frac{k_d(z)}{c}\cos\theta' = 1 - \omega_0 F = a + b_b \tag{6.4.14}$$

and

$$R = \frac{1}{3}\frac{\omega_0\beta}{1-\omega_0 F} = \frac{1}{3}\frac{b_b}{a+b_b} \tag{6.4.15}$$

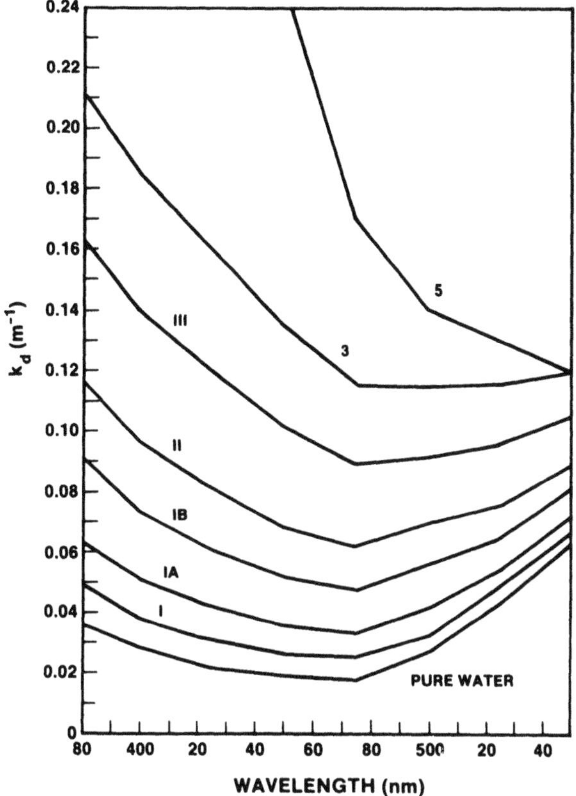

Figure 6.43. Diffuse attenuation coefficients for new Jerlov water types as function of wavelength.

where θ_0' is the illumination angle in water and b_b is the backward-scattering coefficient. Danielson has developed an approximate expression for k/c for multiple-scattering situations where $\omega \geqslant 0.99$; specifically,

$$\frac{k}{c} \sim [3(1-\omega_0)(1-g)]^{1/2} \tag{6.4.16}$$

where ω_0 and g were defined previously.[13] This relationship is illustrated with computed data in Figure 6.26.[†]

[†] Notice that $(1-g) = \int_0^{2\pi}\int_0^{\pi}\left(\frac{\theta^2}{2}\right)\left[\frac{P(\theta)}{4\pi}\right] d\Omega \approx \frac{\overline{\theta^2}}{2}$ for "peaked" scalar phase functions, with $(\overline{\theta^2})^{1/2}$ being the rms scattering angle for the medium. The accuracy of this approximation is very good for g close to 1.

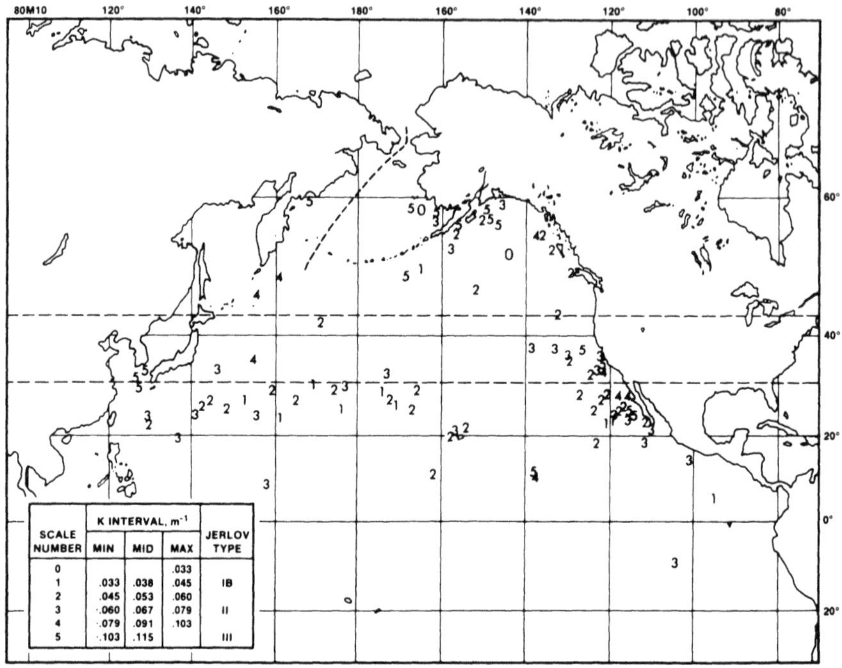

Figure 6.44. Pacific Ocean distribution of new Jerlov water types (all seasons), 15 July 1982.[42]

Table 6.9. Spectral K_d Values (m^{-2}) for Jerlov Water Types

Wavelength (nm)	Jerlov type: IB		IB-II		II		II-III		III
425	0.042	0.051	0.061	0.071	0.081	0.100	0.120	0.140	0.160
430	0.040	0.049	0.059	0.069	0.079	0.097	0.115	0.135	0.155
440	0.038	0.047	0.056	0.064	0.073	0.091	0.109	0.127	0.145
450	0.036	0.044	0.052	0.060	0.068	0.085	0.102	0.118	0.135
460	0.035	0.042	0.050	0.057	0.065	0.081	0.097	0.112	0.127
470	0.033	0.040	0.048	0.055	0.063	0.077	0.092	0.106	0.120
475	0.033	0.040	0.048	0.055	0.062	0.072	0.082	0.099	0.116
480	0.035	0.041	0.048	0.056	0.064	0.077	0.090	0.103	0.116
490	0.038	0.045	0.053	0.060	0.067	0.079	0.091	0.103	0.115
500	0.042	0.049	0.056	0.063	0.070	0.081	0.092	0.103	0.115
510	0.047	0.053	0.060	0.066	0.072	0.082	0.093	0.104	0.115
520	0.052	0.057	0.063	0.069	0.075	0.085	0.095	0.105	0.116
525	0.054	0.059	0.065	0.070	0.076	0.086	0.096	0.106	0.116

Table 6.10. Relationship of Surface k_d to $k_d(z)$ at Depth[a]

Value at surface ($\leqslant$ 50 ft)	Value at depth (> 50 ft)
$k_d(z) = k_0 \leqslant 0.0555\ \mathrm{m}^{-1}$	$k = k_0$
$k_d(z) = k_0 > 0.0555\ \mathrm{m}^{-1}$	$k = 0.0323\ \mathrm{m}^{-1} + 0.417k_0$

[a] Ref. 44.

Wilson has reviewed some of the relationships created from experimental data.(52) Preisendorfer(31) has showed from one set of data that

$$\frac{k}{c} \approx \frac{4}{3}\left(1 - \frac{b}{c}\right) \tag{6.4.17}$$

Sorenson *et al.*(53) used a larger data base to generate the relationship $k/c \approx 1 - 0.83(b/c)$. Using laboratory data rather than field measurements, Timofeeva and Gorobetz(54) came up with

$$\frac{k}{c} \approx \left[\xi\left(1 - \frac{b}{c}\right)\right]^{b/2c}\Bigg|_{\xi=0.25} \tag{6.4.18}$$

Table 6.11. Product of Beam Attenuation Coefficient, $\bar{c}$, and Secchi Depth, D_s

Date	Time (PDT)	$\bar{c}$ (m^{-1})	D_m[a] (m)	D_s (m)	$\bar{c}D_s$
					6.4
18 June	1230	0.3556	18	18	6.86
	1330	0.327	22	21	
19 June	1030	0.4464	16	15	6.69
	1130	0.4294	15.5	15	6.44
	1230	0.4803	15.1	14	6.724
	1430	0.4294	15	18	7.729
20 June	0950	0.3991	19	19	7.58
	1054	0.3880	19	19	7.37
	1205	0.4631	15	15	6.94
	1250	0.472	16	15	7.076
	1345	0.342	15	15	5.131
24 June	1211	0.3187	22.4	22	7.0115
26 June	1345	0.569	16.1	15	8.535
					6.960 Avg
					0.8011 SD

[a] D_m = Maximum depth of averaging.

Comparing these last two equations, Wilson[52] showed that

$$\frac{k}{c} \approx 1 - \frac{b}{c} \cdot A_0 \tag{6.4.19}$$

with $A_0 = 0.85$ for a minimum error of 5% for $b/c < 0.85$. Thus, Equation (6.4.19) gives a working definition of k/c for most reasonable oceanic conditions from an experimental point of view (see Figure 6.35).

References

1. A. L. Buck, Effect of the atmosphere on laserbeam propagation, *Appl. Opt.* **6**, 703-708 (1967).
2. J. R. Kerr, P. J. Titterton, and C. M. Brown, Atmospheric distortion of short lasers, *Appl. Opt.* **8**(11), 2233-2239 (1967).
3. Brookner, Atmospheric propagation and communication channel model for laser wavelengths, *IEEE Trans. Commun. Technol.* **COM-18**, 396-416 (1970).
4. E. A. Bucher, Computer simulation of light pulse propagation for communication through thick clouds, *Appl. Opt.* **12**, 2391-2400 (1973).
5. H. C. van de Hulst, *Light Scattering by Small Particles*, John Wiley and Sons, New York (1957).
6. M. Kerker, *The Scattering of Light and Other Electromagnetic Radiation*, Academic Press, New York (1969).
7. D. Deirmendjian, *Electromagnetic Scattering on Spherical Polydispersion*, Elsevier, New York (1969).
8. J. A. Stratton, *Electromagnetic Theory*, McGraw-Hill, New York (1941).
9. M. Born and E. Wolf, *Principles of Optics*, Fourth Edition, Pergamon Press, New York (1970).
10. J. D. Jackson, *Classical Electrodynamics*, John Wiley and Sons, New York (1975), Chap. 6, 209-254.
11. S. Karp, A Test Plan for Determining the Feasibility of Optical Satellite Communications through Clouds at Visible Frequencies, Naval Ocean Systems Center Technical Note, TN279 (July 1, 1978).
12. G. W. Kattawar, Monte Carlo methods in radiative transfer, in: Multiple Light Scattering in Atmospheres, Oceans, Clouds and Snow, Institute for Atmospheric Optics and Remote Sensing, Short Course No. 420, Williamsburg, Virginia, December 4-8, 1978.
13. H. C. van de Hulst, Multiple Light Scattering in Atmospheres, Oceans, Clouds and Snow, Institute for Atmospheric Optics and Remote Sensing, Short Course No. 420, Williamsburg, Virginia, December 4-8, 1978.
14. T. S. Chu and D. C. Hogg, Effects of precipitation on propagation at 0.63, 3.5 and 10.6 microns, *Bell System Tech. J.* **47**, 723-759 (1968).
15. W. S. Ross, W. P. Jaeger, J. Nakai, T. T. Nguyen, and J. H. Shapiro, Atmospheric optical propagation—an integrated approach, *Opt. Eng.* **21**, 775-785 (1982).
16. C. W. Fairall, K. L. Davidson, and G. E. Schacher, Meteorological models for optical properties in the marine atmospheric boundary layer, *Opt. Eng.* **21**, 847-857 (1982).
17. H. G. Hughes and J. H. Richter, Extinction coefficients calculated from aerosol size distributions measured in a marine environment, *Proc. Soc. Photo-Opt. Instrum. Eng.* **195**, 39-45 (1979).

18. W. C. Wells, G. Gal, and M. W. Munn, Aerosol distribution in maritime air and predicted scattering coefficients in the infrared, *Appl. Opt.* **16**, 654-659 (1977).
19. E. A. Barnhardt and J. L. Streete, A method of predicting atmospheric aerosol scattering coefficients in the infrared, *Appl. Opt.* **9**, 1337-1344 (1969).
20. R. M. Lerner and A. E. Holland, The optical scatter channel, *Proc. IEEE* **58**, 1547-1563 (1970).
21. J. W. Fitzgerald, Approximate formulas for the equilibrium size of an aerosol particle as a function of its dry size and composition and the ambient relative humidity, *J. Appl. Meteorol.* **14**, 1044-1049 (1975).
22. C. Ciany, R. Farrell, and G. Saum, Laser Communications Climatology Study, McDonnell-Douglas Astronautics, Final Report, Contract No. N00014-78-C-0539 (August 31, 1979).
23. H. R. Gordon, R. C. Smith, and J. R. V. Zaneveld, Introduction to ocean optics, *Proc. Soc. Photo-Opt. Instrum. Eng.* **208**, 14-55 (1980).
24. N. G. Jerlov, *Marine Optics*, Elsevier, Amsterdam (1976), p. 251.
25. A Morel, Optical properties of pure water and pure sea water, in: *Optical Aspects of Oceanography* (N. G. Jerlov and E. S. Nielsen, eds.), pp. 1-24, Academic Press, New York (1974).
26. P. Penndorf, On the Phenomenon of the Colored Sun, Air Force Cambridge Research Center, Technical Report No. 53-7 (1953).
27. J. R. Hodkinson, Light scattering and extinction by irregular particles larger than the wavelength, in: *I.C.E.S. Electromagnetic Scattering* (M. Kerker, ed.), Vol. 5, pp. 87-100 Pergamon Press, London (1963).
28. A. C. Holland and G. Gagne, The scattering of polarized light by polydisperse systems of irregular particles, *Appl. Opt.* **9**, 1113-1121 (1970).
29. T. J. Petzold, Volume Scattering Functions for Selected Ocean Waters, Scripps Institute of Oceanography, SIO Ref. 72-28 (1972).
30. S. P. Tucker, Measurements of the absolute volume scattering function for green light in southern California coastal waters, Ph.D. Dissertation, Oregon State University, Corvallis (1973).
31. R. W. Preisendorfer, *Hydrologic Optics*, Vol. V, Properties, U.S. Government Printing Office, Boulder, Colorado (1976).
32. J. G. Tyler and R. W. Preisendorfer, Transmission of energy within the sea, in: *The Sea*, (M. N. Hill ed.), Vol. 2, Interscience Publishers, New York (1962).
33. R. G. Driscoll, J. N. Martin, and S. Karp, OPSATCOM Field Measurements, Technical Document 490, Naval Electronics Laboratory Center (June 1, 1976).
34. D. B. Judd, Terms, definitions and symbols in reflectometry, *J. Opt. Soc. Am.* **57**, 445-452 (1967).
35. R. W. Austin, S. Q. Duntley, W. H. Wilson, C. F. Edgerton, and S. E. Moran, in: *Ocean Color Analysis*, Scripps Institution of Oceanography, SIO Ref. 7.4-10 (1974).
36. R. E. Davidson, D. R. Moore, and H. C. van de Hulst, The transfer of visible radiation through clouds, *J. Atmos. Sci.* **26**, 1078-1087 (1968).
37. S. Twomey, Private communication.
38. J. E. Tyler, Radiance distribution as a function of depth in an underwater environment, *Bull. Scripps Inst. Oceanog.* **7**, 363-412 (1960).
39. R. C. Smith and J. E. Tyler, Optical properties of clear natural waters, *J. Opt. Soc. Amer.* **57**, 595-602 (1966).
40. J. G. Tyler and R. C. Smith, *Measurement of Spectral Irradiance Underwater*, Ocean Series, 1, Gordon and Breach Science Publishers, New York (1970).
41. A. Morel and L. Caloumanos, Mesures d'Éclairements Sous-marins Flux du Photons et Analyse Spectrale, Campagnes Hartmattan et Cancen II, 3e Parte, Présentation des Résultats, Rapport no. 11, Laboratoire d'Océanographie Physique, Villefranche-sur-Mer, France (1970).

42. A. Morel and L. Prieur, Analyse Spectrale des Coefficients d'Absorption et de Rétrodiffusion pour Diverses Régions Marines, Rapport No. 17, Laboratoire d'Océanographie Physique, Villefranche-sur-Mer, France (1975).
43. T. J. Petzold, Prediction of Ocean Water Reflectance Factor from "Water Type" or Diffuse Attenuation Coefficient, Science Applications, Inc., Interim Report No. 33 (Sept. 18, 1981).
44. R. W. Austin, Private communication.
45. R. W. Austin, Coastal zone color scanner radiometry, *Proc. Soc. Photo-Opt. Instrum. Eng.: Ocean Optics VI* **208**, 170–177 (1980).
46. R. W. Austin, Remote sensing of the diffuse attenuation coefficient of ocean water, in: *Special Topics in Optical Propagation, AGARD Conference Proceedings No.* 300, pp. 18-1–18-9, Technical Editing and Reproduction Ltd., London (1981).
47. J. E. Tyler, The Secchi disc, *Limnology and Oceanography* **XIII**, 1-6 (1960).
48. M. A. Frederick, An atlas of Secchi disc transparency measurements and Forel-Vle color codes for the oceans of the world, U.S. Naval Postgraduate masters thesis (Sept., 1970).
49. H. R. Gordon and A. W. Wouters, Some relationships between Secchi depth and inherent optical properties of natural waters, *Appl. Opt.* **17**, 3341–3343 (1978).
50. V. I. Man'Kovskty, Empirical formula for estimating the light attenuation coefficient in sea water, *Oceanology: Methods and Instruments* **18** (4), pp. 493–494 (1978).
51. H. R. Gordon, R. Brown, and M. M. Jacobs, Computed relationships between the inherent and apparent optical properties of a flat homogeneous ocean, *Appl. Opt.* **19**, 417–427 (1975).
52. W. H. Wilson, Spreading of light in ocean water, *Proc. Soc. Photo-Opt. Instrum. Eng.: Ocean Optics VI* **208**, 64–72 (1980).
53. G. P. Sorenson, R. C. Honey, and J. R. Payne, Analysis of the Use of Airborne Laser Radar for Submarine Detection and Ranging, Stanford Research Institute, Report 55E3 (1966).
54. V. A. Timofeeva and F. I. Gorobetz, The relationship between the coefficient of extinction of collimated and diffuse fluxes of light, *Izv. Akad. Nauk SSSR, Ser. Geofiz.* **3**, 291 (1967).

7

Mathematical Models for Energy Propagation in the Optical Scatter Channel

With the basic composition of the optical scatter channel defined in Chapter 6, we can now turn to how this information is used to quantify radiance and irradiance propagation in that medium. Unfortunately, the extensive amount of material currently available on the subject prohibits our being complete and all-inclusive in one chapter of a book. Therefore, we shall limit our discussions to those mathematical approaches and results which have found, and still find, great utility in optical communication systems analysis. We will begin the chapter with a formulation of the mutual coherence function for multiple-forward-scatter media, as derived by Lutomirski.(1) This development will be discussed in terms of its physical implications and also its validity in predicting real-life phenomena. The discussion will then move into a radiative transfer analysis of energy transport in particulate media, and the basic limitations of the closed-formed solutions derived by the small-angle scattering/Huygens–Fresnel approximations will be considered. The conclusion one draws at this point is that the aforementioned techniques can provide insight and answers to optical propagation problems if used properly, but can give misleading results if not. Other mathematical techniques can then be employed if one expects channel characterizations outside the validity range of these closed-form solution sets. Some of the more useful analytical methods of this type will be highlighted and discussed. The result of this discussion will be an in-depth look at two Monte Carlo-based analyses which provide function sets of engineering equations for general atmospheric and marine communication system performance assessments. The next section of this chapter will describe three mathematical techniques which can be applied to energy transfer through the air/sea interface. The final section of this chapter will illustrate how these propagation models can be integrated to yield a total picture of radiation transport in the optical scatter channel. Throughout the chapter, comparisons between model predictions and experimental data will be made whenever possible.

For a more complete look at energy transfer through the various individual and combined components of the optical scatter channel, the interested reader is directed to the journal articles authored by Fante,[2,6] Karp,[3] Ishimaru,[4] Stotts and Titterton,[5] and Lutomirski and Snead,[7] and to the books and reports written by Chandrasekhar,[8] Preisendorfer,[9,10] Van de Hulst,[11] Ishimaru,[12] Jerlov and Nielsen,[13] Jerlov,[14] the Naval Ocean Systems Center,[15] and Fante.[16]

7.1. The Mutual Coherence Function for Multiple-Forward-Scatter Media

In Chapter 2 we found that the general formula for the irradiance distribution emerging from an inhomogeneous random medium can be written in terms of the mutual coherence function (MCF) as

$$I(\mathbf{p}, z) = \left(\frac{k_0}{2\pi z}\right) \int\int d^2\rho\, \Gamma_s(\boldsymbol{\rho}, z) \exp\left\{-\left(\frac{ik_0}{z}\right)\boldsymbol{\rho}\cdot\mathbf{p}\right\} \times \int\int d^2R\, f_a(\mathbf{R}+\tfrac{1}{2}\boldsymbol{\rho}) f_a^*(\mathbf{R}-\tfrac{1}{2}\boldsymbol{\rho}) \exp\left\{\left(\frac{ik_0}{z}\right)\boldsymbol{\rho}\cdot\mathbf{R}\right\} \tag{7.1.1}$$

where $f_a(\mathbf{r})$ is the field in the transmitting aperture and $\Gamma_s(\boldsymbol{\rho}, z)$ is the mutual coherence function created in the transmitting aperture from a hypothetical spherical wave source at receiver point $(\boldsymbol{\rho}, z)$. Lutomirski and Yura were the first to successfully use this formulation to characterize finite beam propagation through a turbulent medium.[17] Lutomirski,[1] among others,[18,19] extended this work to the characterization of finite beam propagation in a particulate medium. Although not the only means of formulating the MCF in this case, it is by far the most amenable to a general engineering analysis and gives some clear insight into irradiance propagation effects in a particulate medium.[20] In the following subsections, we shall outline and discuss the Lutomirski results in detail.

7.1.1. The MCF Model

One of the major conclusions which can be drawn from Chapter 6 is that both the atmospheric and marine channels possess polydispersions which effectively scatter light predominately in the forward-scatter direction. This implies that the initial light distribution emerging from any source plane will become only slightly distorted after several scattering events have occurred. The implication of this fact is that the paraxial (Huygens–Fresnel) approximation used to characterize irradiance propagation in the turbulence channel may also apply to the marine and atmospheric channels as well. However, this fact was not perceived immediately and was not used in the first formulation of the MCF for polydispersion media.

To the best of our knowledge, the first development of the mutual coherence function created by independent particulate multiple scatterings was by Wells[(21)] in his 1969 paper considering the loss of visual resolution in water by the light scattering of microorganisms. In the early 1970s, Arnush[(22)] and Fante[(23)] independently reconsidered this problem using radiative transport theory (to be discussed in Section 7.2) to derive the mutual coherence function under the small-angle scattering approximation. Lutomirski reformulated the turbulence problem in terms of the scattering properties of a polydispersion shortly thereafter and derived a general set of approximate relations for the particulate scattering MCF using the Huygens–Fresnel approximation.[(1,24)] All four developments give consistent results and are all apparently correct, if interpreted correctly. However, the latter results give some simple insights into the effect of particulate scattering on coherency and will therefore be the main thrust of what follows.

In his derivation, Lutomirski showed that the MCF for spherical wave propagation through a spatially uniform distribution of scatterers can be written as

$$\Gamma_s(\rho, z) = \exp\left\{-az - z\int_0^\infty dx\, n(x)\pi x^2\left[Q_s(0, x) - \int_0^1 Q_s(u\rho, x)\, du\right]\right\} \tag{7.1.2a}$$

$$= \exp\{-cz - bz\xi(\rho)\} \tag{7.1.2b}$$

with

$$\xi(\rho) = b^{-1}\int_0^\infty dx\, n(x)\pi x^2 \int_0^1 Q_s(u\rho, x)\, du \tag{7.1.3}$$

In Equation (7.1.2), a and b are the volume absorption and scattering coefficients previously defined in Equations (6.2.2) and (6.1.20), respectively, and z is the separation between the source and receiving planes along the optical axis. The quantity Q_s in Equation (7.1.3) is given by

$$Q_s(\rho, x) = \frac{2}{k_0^2 x^2}\int_0^\pi d\theta \sin\theta |S(\theta, x)|^2 J_0(k\rho \sin\theta) \tag{7.1.4}$$

for spherically symmetric particles and reduces to the normal scattering efficiency Q_s defined in Equation 9 (6.1.9) for $\rho = 0$. After applying the small-angle scattering approximation to Equation (7.1.4) based on the anticipated highly peaked nature of $|S|^2$ in both the marine and atmospheric channels, we find that

$$Q_s(\rho, x) \approx \frac{2}{k_0^2 x^2} \int_0^\infty d\theta\, \theta |S(\theta, x)|^2 J_0(k_0 \rho \theta) \tag{7.1.5}$$

which is obviously the two-dimensional Fourier transform of $|S|^2$ for an azimuthally symmetric scattering function. Substituting Equation (7.1.4) into Equation (7.1.3) and inverting the order of integration, we write

$$\xi(\rho) = \frac{2\pi}{k_0 b} \int_0^\infty dx\, n(x) \int_0^\pi d\theta \sin\theta |S(\theta, x)|^2 g(k_0 \rho \sin\theta)$$
$$\times \frac{\int_0^\infty dx\, n(x) \int_0^\pi d\theta \sin\theta |S(\theta, x)|^2 g(k_0 \rho \sin\theta)}{\int_0^\infty dx\, n(x) \int_0^\infty d\theta \sin\theta |S'(\theta, x)|^2} \tag{7.1.6}$$

which reduces further to

$$\xi(\rho) \simeq \frac{2\pi}{k_0 b} \int_0^\infty dx\, n(x) \int_0^\infty d\theta\, \theta |\xi(\theta, x)|^2 g(k_0 \rho \theta)$$
$$\approx \frac{\int_0^\infty dx\, n(x) \int_0^\infty d\theta\, \theta |S(\theta, x)|^2 g(k_0 \rho \theta)}{\int_0^\infty dx\, n(x) \int_0^\infty d\theta\, \theta |S(\theta, x)|^2} \tag{7.1.7}$$

for multiply forward-scattered radiation. In Equations (7.1.6) and (7.1.7),

$$g(y) = \int_0^1 J_0(uy)\, du \tag{7.1.8}$$

Using the above form of the particulate MCF and its ancillary relations, Lutomirski developed some interesting approximate expressions for the MCF under some realizable operating conditions. These approximate expressions will be given in the following subsection.

7.1.2. Discussion of the Particulate MCF and Its Approximate Forms

Referring to the set of equations (7.1.2)-(7.1.8), it is apparent that when $\rho = 0$, $(0) = 1$ and Equation (7.1.2) reduces to

$$\Gamma_s(0, z) \simeq \exp\{-az\} \tag{7.1.9}$$

Equation (7.1.9) implies that the intensity at a point from a spherical wave source is reduced only by the absorption loss when collected from all directions. On the other hand, the opposite limit of $\rho = \infty$ gives

$$\Gamma_s(\infty, z) = \exp\{-cz\} \tag{7.1.10}$$

which indicates that points widely separated possess a limiting correlation value. The form of this residual coherence at these large separations implies that most of the radiation reaching each point is an incoherent sum of the scattered light produced by various particulates within the medium, and that the only coherent radiation present is the unabsorbed and unscattered light from the common initial source. This means that a detector with a narrow field of view looking at only the source will see the received point source field with its power diminished by this value.

Approximate forms of the function given by Equation (7.1.2) can also be derived under other limiting situations. The most interesting occur when $bz \gg 1$ and ρ is either very large or very small. Under these conditions, one can expand the Bessel function in either the low or high limit to yield approximate expressions which integrate easily for the mutual coherence functions. Specifically, we find that

$$\Gamma_s(\rho, z) \approx \begin{cases} \exp\left\{-az - \left(\dfrac{\rho}{\rho_0}\right)^2\right\} & \text{for } \rho \ll \dfrac{\rho_0}{(bz)^{1/2}} \quad (7.1.11a) \\ \exp\{-cz\} & \text{for } \rho \gg \dfrac{\rho_0}{(bz)^{1/2}} \quad (7.1.11b) \end{cases}$$

where

$$\rho_0 = \frac{\lambda}{\gamma_0 (bz)^{1/2}} \tag{7.1.12}$$

$$\gamma_0^2 = \iint_{4\pi} d\Omega\, \theta^2 \frac{P(\theta)}{4\pi} \tag{7.1.13}$$

Here γ_0 is the rms scatter angle of the medium's scalar phase function. From our discussions in the previous chapter, it is clear that the value of this parameter is highly dependent on the wavelength of light and the particle diameters involved. Typically γ_0 is on the order of degrees for marine environments and in the tens of degrees for the atmospheric. The beam spread of the Gaussian MCF shown in Equation (7.1.11a) is denoted by the parameter ρ_0 and is called the transverse coherence length of the medium. From Equation (7.1.12), ρ_0 is seen to be inversely related to the source-receiver separation and the rms scatter angle γ_0. In most practical situations, this parameter will be less than the wavelength of light. It is apparent from Equation (7.1.11a) that the MCF is essentially Gaussian in shape when ρ is small. The important point of this expression is that the MCF becomes independent of the particular form of the medium's scalar phase function after many scattering lengths have been traversed. However, one must be careful not to assume that this expression is true in all situations; the minimum value of the MCF is given by Equation (7.1.11b), the value obtained for large separations. Given these cautionary notes, let us proceed with our discussions.

Analyzing Equations (7.1.9) and (7.1.11), we come up with the following observations: For point separations (values of ρ) much less than the coherence length ρ_0, the resulting field after traversal of the scattering medium is coherent over this dimension. This implies that a receiving aperture of that general size would capture a coherent field whose intensity would be diminished only by the absorption losses incurred in transit. For those points whose separation are on the order of ρ_0, the emerging field will experience some degradation in field coherence as well as the aforementioned absorption loss. Finally, for those aperture locations displaced from the optical axis many times the medium's coherence length, the observed field will appear spatially incoherent, except for a small residual incident field. This latter field will be plane-wave-like by this time, and maximally degraded by the absorption and scattering processes within the medium.

In summary, the mutual coherence function for particulate medium can be written in approximation as

$$\Gamma_s(\rho, z) \approx \begin{cases} \exp\{-az - (\frac{\rho}{\rho_0})^2 & \rho \ll \frac{\rho_0}{(bz)^{1/2}} \\ \exp\{-cz\}, & \rho \gg \frac{\rho_0}{(bz)^{1/2}} \end{cases} bz \gg 1 \\ \exp\{-az\} \quad bz \ll 1 \tag{7.1.14}$$

comprising some very interesting scattering scenarios. For the intermediate scattering case where bz is say between 1 and 10, or when we are interested in the specific quantitative nature of the fields near the optical axis, the specific nature of the medium's scalar phase function must be employed in the computation of the MCF, i.e., we must calculate the complete MCF of the particulate medium using Equations (7.1.2) and (7.1.3).

7.2. Radiation Transport in Atmospheric and Marine Environments

By the mid-1970s, there were many published models for both radiance and irradiance propagation in the atmospheric and marine channels.[2,4,6]

All were either based on the Huygens–Fresnel or small-angle-scattering approximations. Researchers like Yura[25] showed the equivalency of the two formulations, and work began on comparing these models with experiments. To the best of our knowledge, the first comparison between these theories and experiment was reported in a paper authored by Bravo-Zhivotovsky *et al.* in 1969.[26] In this paper, solutions of the radiative transfer equation under the small-angle-scattering approximation were derived and quantified numerically, and these results were validated against several experimental data sets, e.g., that of Ganich and Levin.[27] The first published comparison between closed-form solutions of the radiative transfers equations under this approximation and data came in a Naval Electronics Laboratory Center (NELC) report describing optical communications between satellite and submerged terminals.[28] In this report, Karp showed a quite reasonable agreement between the Arnush small-angle scattering radiance model,[22] integrated on a spherical cap, and the Duntley tank data.[29] However, a subsequent NELC report describing a comparison between the Arnush model and a set of radiance measurements taken off Santa Catalina Island in 1975 failed to show this good agreement, unless a "shallow-water" correction was made to the model.[30] Anderson and Stotts performed a more detailed comparison of this model and selected NELC data using more complete ground-truthing information, and were not able to show any better predictive performance for the model.[31] Resolution of the apparent discrepancy between the former and later comparisons did not emerge until some three years later with the realization that the range of validity for closed-form small-angle scattering models had not been cited strong enough. Tam and Zardecki were the first to note that the closed-form scattering models proposed by Arnush and others produced numerical results inconsistent with complete solutions of the original equation of interest, even though the apparent range of validity criteria for the models were being met.[32] Stotts showed that this poor agreement resulted from the omission of key angular and spatial frequencies required to produce an accurate radiance distribution near the optical axis.[33] Thus a modification of the validity range criteria was required. Fante reconfirmed this modification for the Huygens–Fresnel-based theories in 1982.[34] In this section we shall review radiative transfer theory, the cousin of Huygens–Fresnel in analyzing light propagation through particulates, and establish why closed-form solutions must be used carefully in characterizing radiation transport in the optical scatter channel. In addition, we will highlight some of the theories which allow radiance and irradiance characterization beyond the bounds alluded to, and present the results of three Monte Carlo simulations which will give the reader some useful tools for link budget analysis of communication systems operating in the marine and/or atmospheric channels.

7.2.1. The Theory of Radiative Transfer in Mie Scattering Media

In a free-space environment, the radiance along any ray is invariant. This property no longer holds for radiance propagation in the optical scatter channel, and one finds the radiance continually altered by the refractive, reflective, and absorption processes inherent to the particles comprising the medium. Mathematically, we can write this alteration for a plane-parallel particulate medium as

$$N_z = N_0 T_z \tag{7.2.1}$$

where N_0 is the initial radiance distribution vertically incident on this slab and T_z is a positive number between 0 and 1. Clearly, $T_0 = 1$. The parameter T_z is known as the beam transmittance of the medium over the path z. For two paths placed end-to-end along any ray, given lengths z and z', respectively, the final residual radiance equals

$$N_{z+z'} = N_0 T_{z+z'} \tag{7.2.2a}$$

$$= (N_0 T_z) T_{z'} \tag{7.2.2b}$$

where

$$T_{z+z'} = T_z T_{z'} \tag{7.2.3}$$

If the beam transmittance varies smoothly with distance, then we have

$$\frac{dT_z}{dz} = \lim_{\Delta z \to 0} \left(\frac{T_{z+\Delta z} - T_2}{\Delta z} \right)$$

$$= T_z \lim_{\Delta z \to 0} \left(\frac{T_{\Delta z} - 1}{\Delta z} \right) = -cT_z \tag{7.2.4}$$

defining the volume extinction coefficient at that location. Integrating Equation (7.2.4), we obtain

$$T_z = \exp\left\{ -\int_0^z c(z')\, dz' \right\} \tag{7.2.5}$$

$$= \exp\{-cz\} \tag{7.2.6}$$

with the latter expression valid for constant c. For nonvertical illumination of the particulate medium, the beam transmittance is given by

$$T_z(\theta) = [T_z]^{\sec\theta}$$

$$= \exp\left\{ -\sec\theta \int_0^z c(z')\, dz' \right\} \tag{7.2.7}$$

for horizontally stratified media, where θ is the illumination angle off the vertical.† This relation is generally true for angles less than 85° (as the sun does not disappear as it approaches the horizon). For incidence angle between 85° and 90°, Equation (7.2.7) can be replaced by

$$T_z(\theta) = \exp\left\{-m(\theta)\int_0^z c(z')\,dz'\right\} \tag{7.2.8}$$

where

$$m(\theta) = \left[\pi\delta_0 + \left(\frac{\pi\delta_0}{z}\right)^2 \cos^2\theta\right]^2 - \left(\frac{\pi\delta_0}{z}\right)\cos\theta \tag{7.2.9}$$

with

$$\delta_0 = 1600$$

is an expression derived from experimental data.[35] The expression

$$\int_0^z c(z')\,dz'$$

signifies an effective integrated optical thickness over the path. Table 7.1 gives the integrated Rayleigh and Mie optical thickness at $\lambda = 500$ nm for an optical path originating from space as a function of altitude above the earth's surface.[36] The reader is directed to Chapter 6 and Refs. 36–38 for more information on the volume extinction coefficient and its calculation.

It is apparent from the development in the previous paragraph that $cN(z)\Delta z \sec\theta$ gives the loss of radiance each elemental increment of path due to absorption and scattering. However, this path radiance can also be increased by other radiance distributions being partially deflected by particulate scattering along the same ray direction of $N(z)$. Let us be more specific.

Let ΔV be an elemental volume of particulates at the point $\boldsymbol{r} = (x, y, z)$. If this volume is illuminated by an initial radiance distribution $N(\boldsymbol{r}; \theta', \varphi')$ along the ray direction (θ', φ'), the fractional increase in radiance at $\boldsymbol{r}$ per unit length along the ray direction (θ, φ) can be written as

$$dN'(\boldsymbol{r}; \theta, \varphi) = N(\boldsymbol{r}; \theta; \varphi')\beta(\boldsymbol{r}; \theta', \varphi', \theta, \varphi)\,d\Omega'$$

†See Appendix F for a more realistic optical loss model relevant to atmospheric channel communication analyses.

Table 7.1. Parameters for the Model of a Clear Standard Atmosphere of λ = 500 nm

Altitude (km)	Rayleigh optical thickness (h-00)	Ext optical thickness (h-00)
0	0.1452	0.370
1	0.1288	0.233
2	0.1140	0.166
3	0.1006	0.130
4	0.0885	0.108
5	0.0776	0.092
6	0.0678	0.081
7	0.0590	0.071
8	0.0512	0.063
9	0.0443	0.056
10	0.0381	0.050
11	0.0327	0.044
12	0.0279	0.039
13	0.0239	0.035
14	0.0204	0.031
15	0.0174	0.028
16	0.0149	0.025
17	0.0128	0.022
18	0.0109	0.020
19	0.0093	0.018
20	0.0080	0.016
21	0.0068	0.014
22	0.0058	0.012
23	0.0050	0.011
24	0.0043	0.009
25	0.0037	0.008
26	0.0032	0.007
27	0.0027	0.006
28	0.0023	0.005
39	0.0020	0.004
30	0.0017	0.004
31	0.0015	0.003
32	0.0013	0.003
33	0.0011	0.002
34	0.0010	0.002
35	0.0008	0.002
36	0.0007	0.001
37	0.0006	0.001
38	0.0005	0.001
39	0.0005	0.001
40	0.0004	0.001
41	0.0004	0.001
42	0.0003	0.000
43	0.0003	0.000
44	0.0002	0.000
45	0.0002	0.000
46	0.0002	0.000
47	0.0002	0.000
48	0.0001	0.000
49	0.0001	0.000
50	0.0001	0.000

where $d\Omega'$ is the elemental solid angle in the (θ', φ') direction. This implies that the radiance generated at $\mathbf{r}$ per unit length in the (θ, φ)-direction by the surrounding light field is given by

$$N'(\mathbf{r}; \theta, \varphi) = \iint_{4\pi} N(\mathbf{r}; \theta', \theta')\beta(\mathbf{r}; \theta', \varphi'; \theta, \varphi)\, d\Omega' \qquad (7.2.10)$$

with $\beta(\mathbf{r}; \theta', \varphi'; \theta, \varphi)$ being the volume scattering function. The local rate of change for the radiance along any arbitrary path in the particulate medium can therefore be written as

$$\cos\theta \frac{dN(\mathbf{r}; \theta, \varphi)}{dz} = -cN(\mathbf{r}; \theta, \varphi) + \iint_{4\pi} N(\mathbf{r}; \theta', \varphi')\beta(\mathbf{r}; \theta', \varphi', \theta, \varphi)\, d\Omega' \qquad (7.2.11)$$

or

$$\cos\theta \frac{dN(\mathbf{r}; \theta, \varphi)}{d\tau} = -N(\mathbf{r}; \theta, \varphi) + \omega_0 \iint_{4\pi} N(\mathbf{r}; \theta', \varphi') \frac{P(\mathbf{r}; \theta', \varphi'; \theta, \varphi)}{4\pi}\, d\Omega' \qquad (7.2.12)$$

using Equations (6.1.20), (6.2.4), (6.3.1), (6.3.2), and (6.3.3). Equations (7.2.11) and (7.2.12) are particular forms of the classical equation of radiative transfer.[18] These equations have been the interest of many researchers over the last 30 years, and several mathematical techniques for their solution have emerged.[8-11,39,40] Specific techniques developed include the method of discrete ordinates,[8,41,42] invariant bedding,[10,43] Monte Carlo,[44-48] the two-, four-, and n-flux models,[8,40,49] and finite differences.[50] Unfortunately, we cannot review all of these methods here and must therefore refer the reader to the excellent works cited for further information on these general solution techniques. In the next two subsections, we shall focus our attentions on two solution sets which have found great utility in communication systems analysis in the optical scatter channel: solutions of the radiative transfer equation in the small-angle-scattering limit and those derived from Monte Carlo simulation techniques.

7.2.2. Multiple-Forward-Scatter Characterizations Using the Small-Angle Scattering Approximation

In the late sixties, Arnush exploited the forward-scattering nature of Mie particle environments such as the atmospheric and marine channels to transform Equations (7.2.11) and (7.2.12) into the following form[51]:

$$\left[\frac{\partial}{\partial z} + \boldsymbol{\gamma} \cdot \frac{\partial}{\partial \mathbf{r}} + c\right] N(z; \mathbf{r}, \boldsymbol{\gamma}) \approx N_0(z; \mathbf{r}, \boldsymbol{\gamma}) + \omega_0 \iint_{-\infty}^{+\infty} p(|\boldsymbol{\gamma} - \boldsymbol{\gamma}'|) N(z; \mathbf{r}, \boldsymbol{\gamma}')\, d^2\boldsymbol{\gamma}' \tag{7.2.13}$$

where[51]

$$p = \frac{P}{4\pi} \tag{7.2.14}$$

Figure 7.1 depicts the spatial and angular coordinates of Equation (7.2.13) pictorially. Those readers not familiar with this equation transformation are referred to Appendix A for its derivation. Arnush and several other researchers then spent the next decade applying various solutions of this equation to the characterization of radiance and irradiance propagation in the optical scatter channel.[22-25,27,33-35,53-56] Let us consider these solutions in more detail.

7.2.2.1. *A First-Order Solution for Multiple Forward-Scattered Irradiance*

Assume we have an initial source radiance given by

$$N_0(\mathbf{z}; \mathbf{r}, \boldsymbol{\gamma}) \approx \frac{\omega_0 c p_T}{\pi r_0^2} \exp\left\{-\frac{r^2}{r_0^2} - \tau\right\} p(|\boldsymbol{\gamma}|) \tag{7.2.15}$$

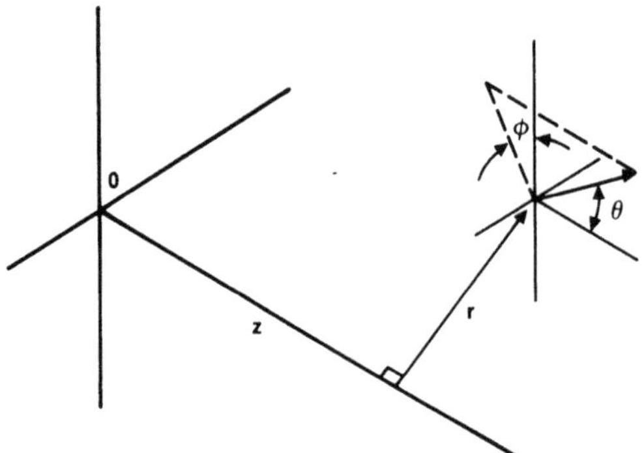

Figure 7.1. Spatial and angular coordinate system.

with p_T being the total transmitter power and r_0 the incident beam radius at $z = 0$. Equation (7.2.15) describes a single-scatter radiance source derived from an initially collimated laser source decaying exponentially within the medium of interest. Because of the anticipated peaked nature of the scalar phase function in the atmospheric and marine scatter channels, it is not unreasonable to expect the quasi-delta function-like nature of these scalar phase functions to negligibly alter the initial radiance distribution given in Equation (7.2.3) after several scattering lengths have been traversed. Given this scenario, what is the functional form of the received power captured by an optical receiver located in the $z = z_0$ plane?

Since most optical communication receivers have finite apertures and finite fields of view, one should expect these types of optical systems to detect optical radiation from multiple forward scatter in the angular range

$$0 \leqslant |\boldsymbol{\gamma}| \leqslant \theta' \approx \frac{r_1}{z_0} \tag{7.2.16}$$

beginning at the entrance plane of the medium, and decreasing asymptotically to the range

$$0 \leqslant |\boldsymbol{\gamma}| \leqslant \theta_{\text{fov}} \tag{7.2.17}$$

for those locations near the receiver. Mathematically, Equation (7.2.16) is assumed for $\theta' < \theta_{\text{fov}}$ and Equation (7.2.17) otherwise. Here r_1 is the radius of the receiver's entrance aperture and θ_{fov} its field-of-view half-angle. In other words, the scattered radiation captured by the receiver is that essentially single-scattered by the medium and confined to the angular region $0 < \theta < \theta_{\text{fov}}$. Rewriting Equation (7.2.13) to reflect this situation, we have

$$\begin{aligned} &\left[\frac{d}{d\tau} + 1\right] N(\tau, \mathbf{r}, \boldsymbol{\gamma}) \\ &\quad \approx N_0(\tau, \mathbf{r}, \boldsymbol{\gamma})/c + \omega_0 \iint_{|\boldsymbol{\gamma}-\boldsymbol{\gamma}'| \leqslant \theta_{\text{fov}}} d^2\boldsymbol{\gamma}^1 p(|\boldsymbol{\gamma} - \boldsymbol{\gamma}'|) N(\tau, \mathbf{r}, \boldsymbol{\gamma}') \end{aligned} \tag{7.2.18}$$

In addition, if we assume that $N(z; \mathbf{r}, \boldsymbol{\gamma})$ changes very slowly over the scalar phase function's range of significance, then we can expand $N(z, \mathbf{r}, \boldsymbol{\gamma})$ in a Taylor series about $\boldsymbol{\gamma}' \sim \boldsymbol{\gamma}$ and retain only the zeroth-order term. This reduces Equation (7.2.18) to

$$\left[\frac{d}{dz} + (1 - \omega_0 \Phi(\theta_{\text{fov}})\right] N(\tau, \mathbf{r}, \boldsymbol{\gamma}) \approx \frac{N_0(\tau, \mathbf{r}, \boldsymbol{\gamma})}{c} \tag{7.2.19}$$

where

$$\Phi(\theta_{\text{fov}}) = \int_0^{2\pi} \int_0^{\theta_{\text{fov}}} p(\theta) \sin\theta \, d\theta \, d\varphi$$

Restricting our attention to the total radiant flux, we find that Equation (7.2.19) becomes

$$\left[\frac{d}{d\tau} + (1 - \omega_0 \Phi)\right] F(z) \approx \omega_0 \Phi P_T \exp\{-\tau\} \tag{7.2.20}$$

in light of Equation (7.2.15). Solving this equation, we obtain

$$F(z) \approx P_T \exp\{-(1 - \omega_0\Phi)\tau\}[1 - \exp\{-\omega_0 \Phi \tau\}] \tag{7.2.21}$$

It is evident that Equation (7.2.21) defines a mean free path for the forward scatter radiation equal to

$$\frac{1}{\omega_0 c \Phi}$$

As we will see shortly, the rms transverse displacement produced by a multiple-scattering medium at a receiver location many scattering lengths away can be written as

$$r_{\text{rms}} = (\overline{r^2})^{1/2} \sim z_0^{3/2} (\omega_0 c \gamma_0^2)^{1/2} \tag{7.2.22}$$

where the mean free path in this case is given by

$$\bar{l} \sim \frac{1}{\omega_0 c}$$

This suggests that the rms transverse displacement for receiver locations in the few-scattering-length regime has the form

$$r_{\text{fs}} \sim z_0^{3/2} (\omega_0 c \Phi \theta_0^2)^{1/2} \tag{7.2.23}$$

with the variable θ_0 is the rms scatter angle for the scalar phase function of the medium under the small-angle scattering approximation. This scatter angle differs from the normal rms scatter angle γ_0 and is modeled as follows.

In general, one finds most real-world scalar phase functions to have large rms scatter angles γ_0 even though they are highly forward-biased in scattering. This is a direct consequence of the way γ_0 is defined. Recall from Equation (7.1.13) that the rms scatter angle is given by

$$\gamma_0^2 = \int_0^{2\pi} \int_0^{\pi} d\Omega \, p(\theta) \theta^2$$

It is apparent in this equation that the high-peaked nature of the scalar phase function is essentially lost by the θ^3 damping in the integrand when θ is small. This implies that the rms scatter angles calculated for peaked phase functions will be similar in value to those not as peaked. In other words, this definition does not reflect the forward-biased scattering nature of the medium. To satisfactorily model multiply forward-scattered light propagation in this approximation, a better measure of the rms scatter angle must be defined. Based on the assumptions needed to create Equation (7.2.18), we defined an effective single scatter angle θ_0 to be given by[56]

$$\theta_0 \equiv \left. \frac{1}{[\pi p(\theta)]^{1/2}} \right|_{\theta=0} \tag{7.2.24}$$

and write the power received by a finite aperture, finite field-of-view optical receiver adjacent to a particulate scattering medium of thickness z_0 as

$$P_R \approx \frac{P_T r_1^2 \, c^{-(1-\omega_0\Phi)\tau}[1 - e^{-\omega_0\Phi\tau}]}{\omega_0 \tau \Phi \theta_0^2 z_0^2} \tag{7.2.25}$$

where

$$\Phi(\theta_{\text{fov}}) = \int_0^{2\pi} \int_0^{\theta_{\text{fov}}} p(\theta)\, d\Omega \tag{7.2.26}$$

and θ_0 is goven by Equation (7.2.24). Equation (7.2.25) has the basic form and functional dependence suggested by Bucher in his 1975 paper describing a Monte Carlo computer simulation of light propagation in clouds[57] and by Gordon from his geometrical analysis of underwater propagation.[58] Figure 7.2 compares the normalized power (pathloss) predicted by Equation (7.2.25) with experimental data taken off San Diego, California,[59,60] and the pathloss result derived by Bucher[61] (to be described in a later subsection), as a function of optical thickness τ. The measurements were taken over a 0.96-km horizontal path through marine fog. The particle size distribution and scalar phase function for this fog are given in Figures 6.12 and 6.13, respectively. It is clear from these graphs that the multiple-forward-scatter solution is in better agreement with the experimental data than is the diffusion regime equation of Bucher. A comparison was also made between the measured pulse broadening incurred during the same experiments and predictions based on the multiple-forward-scatter and Bucher models.[59,60] The forward-scatter model used in the comparison was a modified version of the multipath time spread developed by Stotts.[62] This

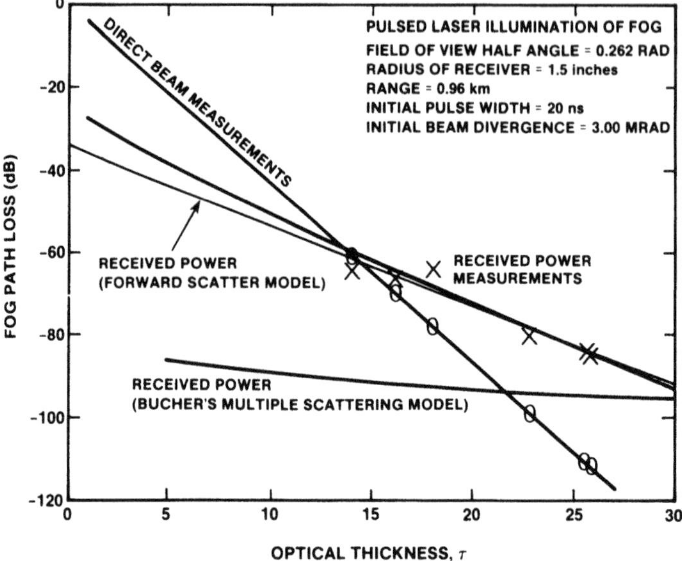

Figure 7.2. Normalized pathloss as a function of optical thickness.

equation has been shown[62,63] to be in good agreement with Monte Carlo simulations[61] and experimental results[61] and is given by

$$\Delta t = \frac{z_0}{c}\left\{\frac{0.30}{\omega_0\tau\gamma_0^2}[(1 + 2.25\ \omega_0 z\gamma_0^2)^{3/2} - 1] - 1\right\} \tag{7.2.27}$$

where c is the speed of light. Table 7.2 contains this comparison. Although the forward-scatter model fails to show as good agreement as it did in the pathloss comparison, it is clearly more in line with the experimental data than the Bucher result[61] by two orders of magnitude. Similar experimental

Table 7.2. Comparison of Experimental and Theoretical Multipath Time Spreading

Time	Optical thickness	Measured values (ns)	$\Delta T(\theta_0)^a$ (ns)	$\Delta T(\gamma_0)^b$ (μs)
2320	18	<20	47.1	6.53
2315	16.1	<20	46.5	6.07
2318	14	<20	45.9	5.53
2330	25.8	~55	49.4	8.23
2345	22.8	~30	48.5	7.61

^{a} $\Delta T(\theta_0)$ is the multipath time spread using the rms scatter angle θ_0 derived from the Gaussian approximation of the scalar phase function.

^{b} $\Delta T(\gamma_0)$ uses the rms scatter angle γ_0 derived from the complete scalar phase function.

results for pathloss and multipath time spreading have been reported by Elliot,[65] Matter and Bradley,[66] Paik *et al*,[67] Ryan and Carswell,[68] and Kennedy and Shapiro[69] and their comparisons with Monte Carlo simulations support the above findings. The importance of these results is that optical receivers operating in an atmospheric multiple-forward-scatter medium whose optical and physical thicknesses are less than 25 and 10 km, respectively, will require less entrance aperture, field of view, and slot integration time, for a given signal-to-noise ratio, then those receivers required to function in diffusion-type media of the same optical and physical thicknesses. One finds a somewhat analogous situation existing in the marine channel, where Mie particles and absorption tend to keep multipath and angular beam spreading in check to reasonable levels. Unfortunately, the pathloss is quite a bit higher although functionally the same. In any event, these results were good news to the optical designer who was usually plagued with communication receiver requirements for the optical scatter channel which demanded E/O technologies which were either stressing or beyond the state of the art. He was now able to work with a more reasonable level of propagation effects and thus provide a less risky and/or costly design.

Ross *et al.* reformulated the radiative equation given as Equation (7.2.13) to yield a less ad hoc solution for multiply forward-scattered light propagation. Unfortunately, a comparison between the predictions of their new model and those derived from Equation (7.2.25) with new experimental data did not show better predictive performance for the former over the latter.[70] The conclusion of their study was that a better understanding of scalar phase functions and their impact on multiple forward scatter was necessary if better comparisons were to be forthcoming. In truth, this quest for understanding had been going on for some time at that point, and specific results were just beginning to clarify which analytical methods were appropriate for characterizing multiply forward-scattered radiation near the optical axis at propagation distances less than 10 scattering lengths and which were valid elsewhere.

7.2.2.2. *Over-the-Horizon Propagation from Multiple Forward Scatter*

The consideration of the optical scatter channel as a possible medium for many optical communication and surveillance applications has generally been limited to those situations where line-of-sight operations occur and any interaction with particulate media is tolerable. Indeed, with few exceptions,[71-75] this viewpoint dominated the optical community for much of the last quarter-decade. However, recent research into over-the-horizon communications and surveillance suggests that multiple forward-scatter may provide a workable means for optical technology to play a role in these application areas.[59,60]

Over-the-horizon (OTH) optical propagation can occur by a number of different physical mechanisms. For example, anomalous refractivity gradients and structures in the marine boundary layer and above can result in refractive propagation beyond the geometric horizon (i.e., the horizon computed neglecting refraction), where such propagation manifests itself in the form of either ducting or simple downward curvature of the initial beam. The latter circumstance, experienced when the refractivity falls off monotonically and approximately linearly in height, can be accommodated analytically by the effective earth radius concept wherein rectilinear ray paths are used. The result is that the earth's radius is imagined to be somewhat larger than its initial value by an amount that increases as the magnitude of the surface refractivity gradient grows. Radiation transfer by atomospheric aerosols or by cloud bottoms are two other means by which one can scatter energy beyond the horizon. The radiative transfer equation in the small-angle scattering limit has been used to describe atmospheric aerosol-induced optical propagation for OTH applications, and we will focus on the analytical models which resulted for the remainder of this subsection.

As depicted in Figure 7.3, we distinguish between two physical scattering processes for OTH propagation: (1) particulate single scatter and (2) particulate multiple forward scatter. In general, one would expect this latter component to dominate at ranges in excess of a few scattering lengths due to the near-unity single-scatter albedo at most visible and near-IR wavelengths (see section 6.2). However, the proximity of the earth alters this situation. A horizon-directed beam will only be a small fraction of an extinction length above the surface; thus, any downward scattering beyond a very small angle off the forward direction will result in a surface encounter. This interaction will essentially remove light from the multiply forward-scattered beam via diffuse reflection and/or absorption. Hence the propagation phenomenon of dominance becomes a geometry-dependent, as well as meteorologically dependent, result. Based on this fact, two OTH optical

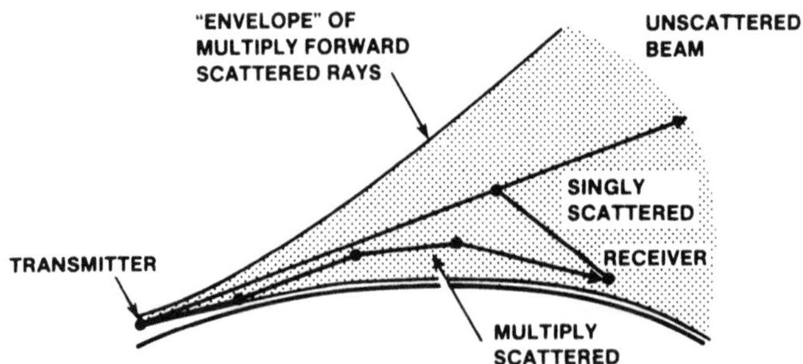

Figure 7.3. Beyond-the-horizon propagation mechanisms (cloud-free case).

propagation models were developed—one based on single-scattering events[76] and the other on multiple-scattering events. For convenience, we will cite only the basic models here. The ratio of received power to transmitted power (integrated pathloss) for the single-scatter situation is given by

$$\frac{P_R}{P_T}=\frac{8A\beta\exp\{-cD\}}{3\pi\lambda r}\sum_{i=1}^{3}\frac{c_i}{b_i^3}[J(x_i^{\max})-J(x_i^{\min})]\Big/\sum_{i=1}^{3}\frac{c_i}{b_i^2}U(b_i) \qquad (7.2.28)$$

where

$$x_i^{\max}=\frac{2\pi}{\lambda b_i}(\theta_{\min}+2\theta_{\text{fov}})x_i^{\min}=\frac{2\pi}{\lambda b_i}\sin(\theta_{\min})$$

$$J(x)\equiv L(x)-\frac{2}{(1+y)^3}L\left(\frac{x}{1+y}\right)+\frac{1}{(1+2y)^3}L\left(\frac{x}{1+2y}\right)$$

$$y_i=2\frac{k}{b_i}\left|\frac{m^2-1}{m^2+2}\right|$$

$$L(x)=[(1-\tfrac{5}{2}W+\tfrac{3}{2}W^2)K(W)-(1-2W)E(W)]\cdot W^{-3/2}$$

$$U(b)\equiv 1-\frac{4}{\left(1+\frac{g}{b}\right)^2}+\frac{6}{\left(1+\frac{2g}{b}\right)^2}-\frac{4}{\left(1+\frac{3g}{b}\right)^2}+\frac{1}{\left(1+\frac{4g}{b}\right)^2}$$

$$g\equiv k\left(\frac{4}{3}\right)^{1/4}\left|\frac{m^2-1}{m^2+2}\right|^{1/2}$$

$$b_i=\frac{d_i}{F'}$$

$$K(x)\equiv\int_0^{\pi/2}\frac{d\theta}{(1-x\sin^2\theta)^{1/2}}$$

$$E(x)\equiv\int_0^{\pi/2}d\theta(1-x\sin^2\theta)^{1/2}$$

$$W=\frac{4x^2}{(1+4x^2)}$$

and $k=2\pi/\lambda$, $\lambda\equiv$ wavelength of interest in microns $m\equiv$ refractive index of the particle (complex), $F'=[F(\text{RH})]/F(80)]$, where $F=1-0.9\ln[1-(\text{RH}/100)]$ and RH $\equiv$ relative humidity in percent, $\theta_{\text{fov}}=$ field of view (half-angle) in the elevation plane, $A\equiv$ area of the receiving aperture, $D\equiv$ ground range, $r=(D^2/2R)[(1-l_T)^2-l_R^2]+D\sum(1-l_T)$, $\sum\equiv$ elevation angle of transmitter, $R\equiv$ effective radius of the earth, $l_{T,R}=(2h_{T,R}R/D^2)^{1/2}$, and $h_{T,R}\equiv$ height of the transmitter, receiver above sea level. The coefficients (c_j, b_j;$j = 1, 2, 3$) are shown in Figure 7.4 for various wind speeds and a relative humidity level of 80%.

For multiple forward scatter, the integrated pathloss is given by

$$\frac{P_R}{P_T} \approx \frac{2.515Ap(\gamma)p(0)c \exp\{-\tau_0 - \tau_1\}}{D_1} \tag{7.2.29}$$

where $p(\gamma) \equiv$ medium's scalar phase function evaluated at scattering angle γ and

$$\gamma \equiv \frac{D}{R} + \sin^{-1}\left(1 - \frac{h_T}{R}\right) + \sin^{-1}\left(1 - \frac{h_R}{R}\right)$$

$$\tau_0 = cZ_0,$$

$$\tau_1 = cD_1,$$

$$Z_0 = (2Rh_T)^{1/2} + R \tan \gamma - R(\sec \gamma - 1) \csc \gamma,$$

$$D_1 = (2Rh_R)^{1/2} + R(\sec \gamma - 1) \cot \gamma.$$

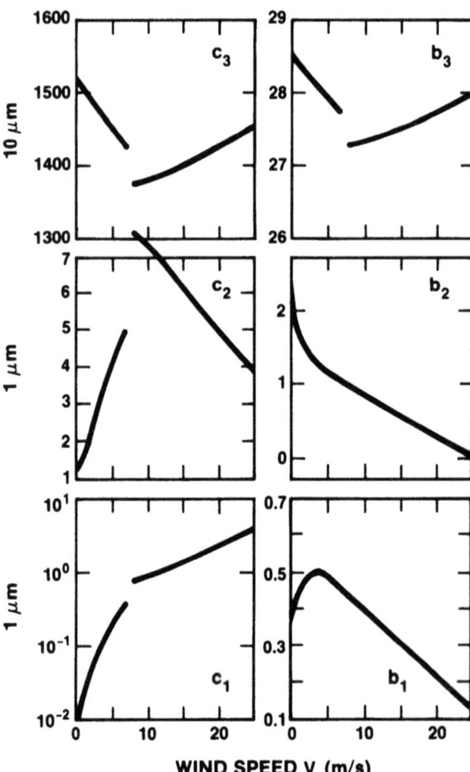

Figure 7.4. Wind speed dependence of size distribution coefficients.

If both the transmitter and receiver are close to the earth surface, Equation (7.2.29) reduces to

$$\frac{P_R}{P_T} \approx \frac{5.03\beta A}{D} p^2(0) \exp\{-\tau\} \tag{7.2.30}$$

with $\tau = cD$. Table 7.3 shows a comparison of predicted pathloss for both models with experimental measurements taken over a 63-km path between San Diego and Oceanside, California.[59,60] For this experiment, the measured pathloss were derived from the 514.5-nm line of a 1-W argon-ion laser and from the 532-nm and 1064-nm lines of a Q-switched Nd:YAG laser. The geometric horizon was 40 km from the transmitter for the former and 19 km for the latter two beams. No evidence of pulse broadening was observed. It is apparent that both models give reasonable predictive results, with the single-scatter model being the more accurated of the two. More experimental data are needed to better establish the validity of these models.

7.2.2.3. The Green's Function for Multiple Scattering

In classical electrodynamics, the general solution of the Maxwell equations for propagating electromagnetic fields in dielectric and free-space media is the Green's function.[77] This concept can easily be extended to include radiance propagation in multiple-scattering medium as well.[1,7,17,24,78] In this subsection we shall show this extension explicitly.

Yura, among others, has shown that the form of the Maxwell equations for particulate media in the small-angle scattering limit is essentially the same as that shown in Equation (7.2.13) for the radiative transfer equation under the same approximation.[19,25] Thus, for negligible absorption we have

$$\left[\frac{\partial}{\partial z} + \boldsymbol{\gamma} \cdot \frac{\partial}{\partial \mathbf{r}} + b\right] N = \iint_{-\infty}^{+\infty} d^2\boldsymbol{\gamma}^1\, \beta(|\boldsymbol{\gamma}.\boldsymbol{\gamma}'|) N(z; \mathbf{r}, \boldsymbol{\gamma}') \tag{7.2.31}$$

as the radiance transport equation, subject to the specification of N on the plane $z = 0$. Our problem is to find the Green's function $g(z, \mathbf{r}, \boldsymbol{\gamma}, 0, \mathbf{r}', \boldsymbol{\gamma}')$ for the radiance existing at the point $(\mathbf{r}, z_0)$ arriving in the transverse direction due to a point source at $(\mathbf{r}', 0)$ emitting flux in the direction $\boldsymbol{\gamma}'$. The complete solution to Equation (7.2.31) was derived by Arnush[22] and is given by

$$N(z_0; \mathbf{r}, \boldsymbol{\gamma}) = (2\pi)^4 \iiiint_{-\infty}^{+\infty} d^2\boldsymbol{\nu}\, d^2k \exp\{-i(\boldsymbol{\nu} \cdot \boldsymbol{\gamma} + \mathbf{k} \cdot \mathbf{r})\}$$
$$\times \tilde{N}(z, \boldsymbol{\nu} + z\mathbf{k}, \boldsymbol{\nu}) G_{z_0}(\boldsymbol{\nu}, \mathbf{k}) \tag{7.2.32}$$

Table 7.3. Comparison of Both Single-Scatter and Multiple-Forward-Scatter Models with San Diego-to-Oceanside Propagation Results (63-km OTH Range)

Date	Extinction coefficient (km^{-1})	Observed[a] pathloss (dB)	Single-scatter pathloss[b] (dB)	Relative difference (dB)	Multiple-scatter pathloss[c] (dB)	Relative difference (dB)
15 January 1976	0.115[d]	−96	−99	−3	−102	−6
21 January 1976	0.043[d]	−86	−84	+2	−87	−1
28 January 1976	0.126[d]	−102	−102	0	−105	−3
25 February 1976	0.148	−108	−107	+1	−110	−2

[a] Measurement of pathloss (P_R/P_T) at transmitter elevation angle of 3 mrad.
[b] Relative humidity = 80%; wind speed = 5 m/s; effective earth radius ratio = 1.2204.
[c] $p(0) = 16$; $\frac{4}{3}$ earth assumed.
[d] Calculated from a ducted beam.

where

$$\tilde{N}_0(z_0, \mathbf{k}, \boldsymbol{\nu}) = \iiiint\limits_{-\infty}^{+\infty} d^2r\, d^2\boldsymbol{\gamma} \exp\{i(\boldsymbol{\nu}\cdot\boldsymbol{\gamma} + \mathbf{k}\cdot\mathbf{r})\} N_0(z_0, \mathbf{r}, \boldsymbol{\gamma}) \tag{7.2.33}$$

$$G_{z_0}(\mathbf{k}, \boldsymbol{\nu}) = \exp\left\{-bz_0 + \int_0^{z_0} d\zeta\, Q(\boldsymbol{\nu} + \zeta\mathbf{k})\right\} \tag{7.2.34a}$$

$$= \exp\left\{-bz_0\left[1 - \int_0^1 du\, Q(\boldsymbol{\nu} + uz_0\mathbf{k})\right]\right\} \tag{7.2.34b}$$

and

$$Q(\boldsymbol{\nu}) = \iint\limits_{-\infty}^{+\infty} p(\boldsymbol{\gamma}) \exp\{i\boldsymbol{\nu}\cdot\boldsymbol{\gamma}\}\, d^2\boldsymbol{\gamma} \tag{7.2.35a}$$

$$= 2\pi \int_0^\infty d\gamma\, \gamma J_0(\gamma\nu) p(\gamma) \tag{7.2.35b}$$

is the Fourier transform of the scalar phase function.†

If we define

$$g(z_0, \mathbf{r}, \boldsymbol{\gamma}) = \frac{1}{(2\pi)^2} \iint\limits_{-\infty}^{+\infty} G_{z_0}(\mathbf{k}, \boldsymbol{\nu}) \exp\{-i(\boldsymbol{\nu}\cdot\boldsymbol{\gamma} + \mathbf{k}\cdot\mathbf{r})\}\, d^2\boldsymbol{\nu}\, d^2k \tag{7.2.36}$$

then substituting Equations (7.2.33) and (7.2.36) into Equation (7.2.32) and inverting the order of integration yields

$$N(z_0, \mathbf{r}, \boldsymbol{\gamma}) = \iint\limits_{-\infty}^{+\infty} N_0(0, \mathbf{r}', \boldsymbol{\gamma}') g(z_0, \mathbf{r} - \mathbf{r}' - z_0\boldsymbol{\gamma}', \boldsymbol{\gamma} - \boldsymbol{\gamma}')\, d^2r'\, d^2\boldsymbol{\gamma}' \tag{7.2.37}$$

which is the Green's function solution of the problem.[7,78] One can then compute g by the following procedure:

1. Find the Fourier transform of the volume scattering function [Equation (7.2.37)].
2. Integrate and exponentiate the Fourier transform contained in Equation (7.2.36).
3. Find the inverse transform [Equation (7.2.36)].

†The derivation of this solution will be discussed in more detail in the following section.

Unfortunately, this is a very hard to do analytically for most real-world scalar phase functions, and one must resort to either numerical[(26)] or approximate techniques for solutions.[(2,22,31)]

Lutomirski has developed[(7,78)] approximate expressions for g which can be applied to a number of communications and surveillance problems. In particular, he observed from Equation (7.2.34) that when the argument of Q was zero, the system function G became $G(0, 0)$, which equals unity. Conversely, when the argument of Q was very large, he found that $G(\infty, \boldsymbol{\nu}) = (G(\mathbf{k}, \infty) = \exp(-bz)$. Thus he was able to quantify the behavior in Equation (7.2.35) into the following series:

$$Q(\nu) = \int_0^\infty d\gamma\, 2\pi\gamma p(\gamma)\left[1 - \frac{\gamma^2\nu^2}{4} + \cdots\right] = 1 - \frac{\nu^2\gamma_0^2}{4} + \cdots \qquad \text{for } \boldsymbol{\nu}\gamma_0 \ll 1 \quad (7.2.38)$$

where

$$\gamma_0 = 2\pi \int_0^\infty d\gamma\, \gamma^3 p(\gamma), \qquad \left[2\pi \int_0^\infty d\gamma\, \gamma p(\gamma) = 1\right] \qquad (7.2.39)$$

is the mean square scattering angle. Specifically, he was able to write

$$G_{z_0}(\mathbf{k}, \boldsymbol{\nu}) = \begin{cases} \exp\left\{-\dfrac{bz_0\gamma_0^2}{4}\left(\boldsymbol{\nu}^2 + z_0\boldsymbol{\nu}\cdot\mathbf{k} + \dfrac{z_0^2k^2}{3}\right)\right\}, & \\ \qquad \text{for } \nu \ll 1/\gamma_0\sqrt{(bz)},\ k \ll 1/\gamma_0 z_0\sqrt{(bz_0)}, & \\ \exp\{-bz_0\}, \qquad \text{otherwise} & \end{cases} \Bigg\}\, bz_0 \gg 1$$

which yields

$$g(z_0, \mathbf{r}, \boldsymbol{\gamma}) = \begin{cases} \dfrac{12}{\pi^2(bz_0)^2\gamma_0^2}\exp\left\{-\dfrac{4}{bz_0\gamma_0^2}\left[\left|\boldsymbol{\gamma} - \dfrac{3\mathbf{r}}{2z_0}\right|^2 + \dfrac{3r^2}{z_0^2}\right]\right\} & \\ \qquad \text{for } \boldsymbol{\gamma} \gg \gamma_0(bz_0)^{1/2},\ r \gg z_0\gamma_0(bz_0)^{1/2} & (7.2.40a) \\ \exp\{-bz_0\}\boldsymbol{\delta}(\mathbf{r} - z_0\boldsymbol{\gamma})\boldsymbol{\delta}(\boldsymbol{\gamma}), \qquad \text{otherwise} & (7.2.40b) \end{cases}$$

as the Green's function for particulate multiple scattering. Equation (7.2.40a) generates the propagation of multiply scattered radiation in the limit of many scatterings, and Equation (7.2.40b) is the extracted component of the incident field as it traverses the particulate medium. Both of these equation assume a source located at $(\mathbf{z}, \mathbf{r}') = (0, 0)$ with $\boldsymbol{\gamma}' = (0, 0)$.

As with the previous models, Equation (7.2.40) is a first-order approximation of the complete multiple-scattering solution. In addition to the above small-angle scattering which occurs with the multiple-scattering medium, the propagating light field may experience particulate absorption and large-angle scattering. Both aspects rob energy from the propagating line-of-sight field in obvious ways, as least as far as a narrow field-of-view receiver system is concerned. However, all three aspects have a further effect on power loss in the forward direction. The increased path a scattered light field must traverse in order to exit a plane-parallel medium increases its probability of being absorbed and thus affects the potential received power in some nonlinear way. This potential extra loss can be significant in some situations of interest, say, in the marine channel, where a is reasonably large. We saw in Chapter 6 that optical absorption transformed irradiance decay from a linear degradation to an exponential falloff described by an effective attenuation rate called the diffuse attenuation coefficient, k_d. This coefficient accounts for the backward-scattered radiation as well as the "total" absorption loss. In the marine channel we found that k_d is approximately given by

$$k_d \approx a + \frac{b}{m'}$$

where m' is a constant between 6 and 50 depending on the water type involved. The conclusion is that Equation (7.2.40) can be a very effective tool for systems analysis if used properly, but it must be used with care. It essentially has the same use restrictions as we found for the mutual coherence function developed in Section 7.1. Unfortunately, these restrictions have not always been clearly stated, or completely understood for that matter. In the next subsection we will attempt to address these issues in sufficient analytical detail to give the reader a clear understanding of the power and limitations of this formulation.

7.2.2.4. The Arnush–Fante Multiple-Forward-Scattering Approximations

The first complete solution of the radiative transfer equation under the small-angle scattering approximation was performed by Bravo–Zhivotovskiy *et al.* using Fourier transforms and numerical integration techniques.[26] Specifically, they defined the Fourier transform of the radiance to be

$$\tilde{N}(z, \mathbf{k}, \boldsymbol{\nu}) = \iiiint_{-\infty}^{+\infty} d^2r\, d^2\boldsymbol{\gamma} \exp\{i(\boldsymbol{\nu}\cdot\boldsymbol{\gamma} + \mathbf{k}\cdot\mathbf{r})\} N(z, \mathbf{r}, \boldsymbol{\gamma}) \tag{7.2.41}$$

and transformed Equation (7.2.15) into the following form:

$$\left\{\frac{\partial}{\partial z} - \mathbf{k}\cdot\frac{\partial}{\partial\boldsymbol{\gamma}} + c[1 - \omega_0 Q(\boldsymbol{\nu})]\right\}\tilde{N}(z, \mathbf{k}, \boldsymbol{\nu}) \approx N_0(z, \mathbf{k}, \boldsymbol{\nu}) \tag{7.2.42}$$

using Equation (7.2.35). By assuming an initial narrow-beam radiance distribution, they were able to solve Equation (7.2.42) for the Fourier transform of the scattered radiance

$$N_s(z_0, \mathbf{k}, \boldsymbol{\nu}) = N_0(0, \mathbf{k}, \boldsymbol{\nu} + z_0\mathbf{k}) \exp\left\{-cz + \frac{b}{Q(0)} \int_0^{z_0} Q(|\boldsymbol{\nu} + \zeta\mathbf{k}|)\, d\zeta\right\} \tag{7.2.43}$$

where $\tilde{N}_0(0, \mathbf{k}, \boldsymbol{\nu})$ is the Fourier transform of the initial radiance distribution at $z = 0$. To obtain the scattered radiance, Equation (7.2.43) can be evaluated numerically to yield the desired results. Bravo-Zhivotovskiy *et al.* calculated the irradiance entering a circular aperture of radius r_1 as a function of scattering length for the empirically derived scalar phase function

$$p(\boldsymbol{\gamma}) = \frac{1}{2\pi\gamma_0|\boldsymbol{\gamma}|} \exp\left\{-\frac{|\boldsymbol{\gamma}|}{\gamma_0}\right\} \tag{7.2.44}$$

for seawater, where $\boldsymbol{\gamma}_0$ is the rms scatter defined by Equation (7.1.13) and is on the order of 0.10.(79) Table 7.4 summarizes their results in terms of the normalized log received power (P_R/P_∞). Figure 7.5 shows a comparison of these results with a set of experimental measurements reported by Ganich and Levin(27) and clearly illustrates good agreement between the two data sets, within the accuracy of the experiment. Other data comparisons were presented by Bravo-Zhivotovskiy *et al.* and showed similar predictive performance for their model.(26) With the theory clearly established by these Soviet researchers, other workers began investigating specific aspects and approximate forms of this model.

In 1972, Arnush developed a closed-form solution for the scattered radiance using the general transform solution given in Equation (7.2.31).(22)

Table 7.4. $-\mathrm{Log}\,(P_R/P_\infty)$ for Various Values of $(r_1/z\gamma_0)$ and bz

	$-\log(P_R/P_\infty)$ for $(r_1/z\gamma_0)$ =									
bz	10^{-2}	$3\cdot10^{-2}$	10^{-1}	$2\cdot10^{-1}$	$5\cdot10^{-1}$	1.0	1.5	2.0	2.5	3.0
1	0.425	0.379	0.319	0.250	0.143	0.069	0.037	0.021	0.014	0.010
2	0.851	0.757	0.629	0.491	0.277	0.139	0.078	0.043	0.026	0.016
4	1.693	1.502	1.214	0.995	0.549	0.280	0.157	0.092	0.055	0.037
6	2.523	2.217	1.726	1.327	0.784	0.415	0.241	0.147	0.090	0.059
8	3.322	2.854	2.135	1.635	0.979	0.538	0.325	0.204	0.130	0.084
10	4.015	3.329	2.421	1.863	1.140	0.646	0.409	0.260	0.171	0.112
12	4.478	3.615	2.609	2.024	1.269	0.742	0.475	0.314	0.211	0.143
14	4.710	3.777	2.742	2.146	1.374	0.827	0.542	0.367	0.252	0.174

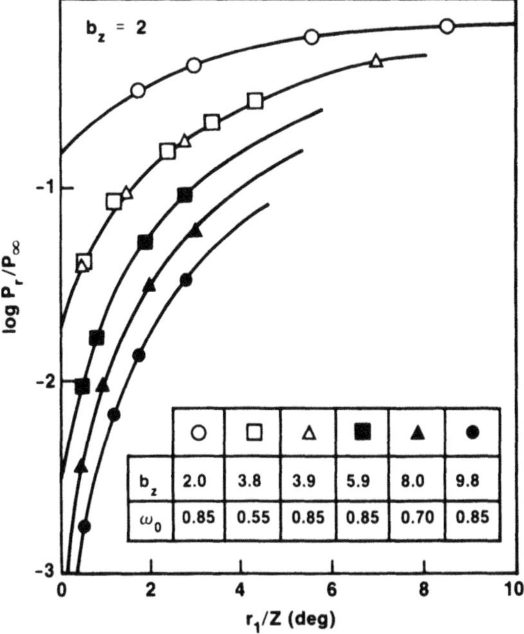

Figure 7.5. Comparison of the theoretical curves derived by Bravo-Zhivotovskiy *et al.*[26] with the experimental data of Ganich and Levin.[27]

Using the scalar phase function given in Equation (7.2.37), he was able to write Equation (7.2.43) as

$$\begin{aligned} N_s(z_0, \mathbf{k}, \boldsymbol{\nu}) &\approx \tilde{N}_0(0, \mathbf{k}, \boldsymbol{\nu} + z_0 \boldsymbol{k}) \exp\left\{-cz + \omega_0 c \int_0^{z_0} Q(\boldsymbol{\nu} + \zeta \mathbf{k})\, d\zeta\right\} \\ &\approx \tilde{N}_0(0, \mathbf{k}, \boldsymbol{\nu} + z_0 \mathbf{k}) \exp\left\{-\tau + b \int_0^{z_0} \frac{d\zeta}{(1 + \boldsymbol{\gamma}_0^2 |\boldsymbol{\nu} + \zeta \mathbf{k}|^2)^{1/2}}\right\} \end{aligned} \tag{7.2.45}$$

which reduces to

$$\tilde{N}_s \approx N_0(0, \mathbf{k}, \boldsymbol{\nu} + z_0 \mathbf{k}) \exp\left\{-(1 - \omega_0)\tau - \frac{\omega_0 \tau \mathbf{g}_0^2}{2}\left(\boldsymbol{\nu}^2 + z_0 \boldsymbol{\nu} \cdot \mathbf{k} + \frac{z_0^2 k^2}{3}\right)\right\} \tag{7.2.46}$$

by keeping only the low spatial and angular frequency components of the transfer function

$$\exp\left\{-\tau + \omega_0 c \int_0^{z_0} Q(\boldsymbol{\nu} + \zeta \mathbf{k})\, d\zeta\right\} \tag{7.2.47}$$

This last expression is the Green's function we derived in the previous subsection. Equation (7.2.46) comes about by noting that Equation (7.2.45) differs significantly from zero only when

$$|\boldsymbol{\nu} + \zeta \mathbf{k}|^2 \gamma_0^2 \ll 1$$

for ζ in the range $0 < \zeta < z$, given $bz \gg 1$ and $\gamma_0 \ll 1$. The impact of the above expansion is that Equation (7.2.46) is in a form which has a closed-form inverse Fourier transform. Specifically, this equation can be Fourier inverted to yield the expression for the scattered radiance

$$N_s(z, \mathbf{r}, \boldsymbol{\gamma}) \approx N_0' \exp\left\{-a_0^2 \boldsymbol{\gamma}^2 + a_1 \boldsymbol{\gamma} \cdot \mathbf{r} - \frac{r^2}{r_a^2}\right\} \tag{7.2.48}$$

where

$$N_0' = \frac{P_T \exp\{-(1-\omega_0)\tau\}}{\pi^2 (2\omega_0 \tau \gamma_0^2)(r_0^2 + \frac{1}{6}\omega_0 \tau \gamma_0^2 z^2)}$$

$$a_0^2 = \frac{1}{2\omega_0 \tau \gamma_0^2} + \frac{z_0^2}{4 r_2^2}$$

$$a_1 = z_0 / r_a^2$$

$$r_a^2 = r_0^2 + \tfrac{1}{6}\omega_0 \tau \gamma_0^2 z_0^2$$

The mean square radius of this distribution is given by

$$\overline{r^2} = \frac{1}{I(z)} \iint_{-\infty}^{+\infty} d^2 r\, r^2 I(z, r)$$

and computes to be

$$\overline{r^2} \approx \tfrac{1}{3}\omega_0 \tau \gamma_0^2 z_0^2$$

for r_0 small. The above expression for $\overline{r^2}$ is completely consistent with the reported literature[(22)] and was the expression used in the previous development for multiple forward scatter. Integrating Equation (7.2.48) for the power captured by a field-of-view, finite-aperture receiver, Arnush obtained

$$P_R(r_1) \approx P_T \exp\{-(1-\omega_0)\tau\}\left[1 - \exp\left\{-\left(\frac{3 r_1^2}{2\omega_0 \tau \gamma_0^2 z_0^2}\right)\right\}\right] \tag{7.2.49}$$

which gives

$$\frac{P_R}{P_\infty} \equiv \frac{P_R(r_1)}{P_R(\infty)} = \left[1 - \exp\left\{-\left(\frac{3r_1^2}{2\omega_0\tau\gamma_0^2 z_0^2}\right)\right\}\right] \tag{7.2.50}$$

for the normalized received power. A comparison was then made between Equation (7.2.50) and Table 7.3, and this is shown in Figures 7.6 and 7.7. Arnush concluded from these figures that Equation (7.2.49) agreed quite well with the numerical calculation of Bravo-Zhivotovskiy *et al.* over the range $0.01 < r_1/z_0\gamma_0 < 3$ for $bz > 8$ and that the agreement improved as $r_1/z_0\gamma_0$ increased. Karp recomputed Equation (7.2.49) for the received power captured by a spherical cap rather than a planar aperture and compared predictions from the derived expression with the Duntley tank data.[22] Figures 7.8 and 7.9 depict the results of the data comparison. Figure 7.10 shows the power measurement geometry used in these comparisons. Although the agreement is not perfect, it is remarkably close. Stotts showed that the radiance expression derived by Arnush was essentially independent of the particular form of the scalar phase function used in the derivation and his formulation could be extended to time-dependent scattered radiance

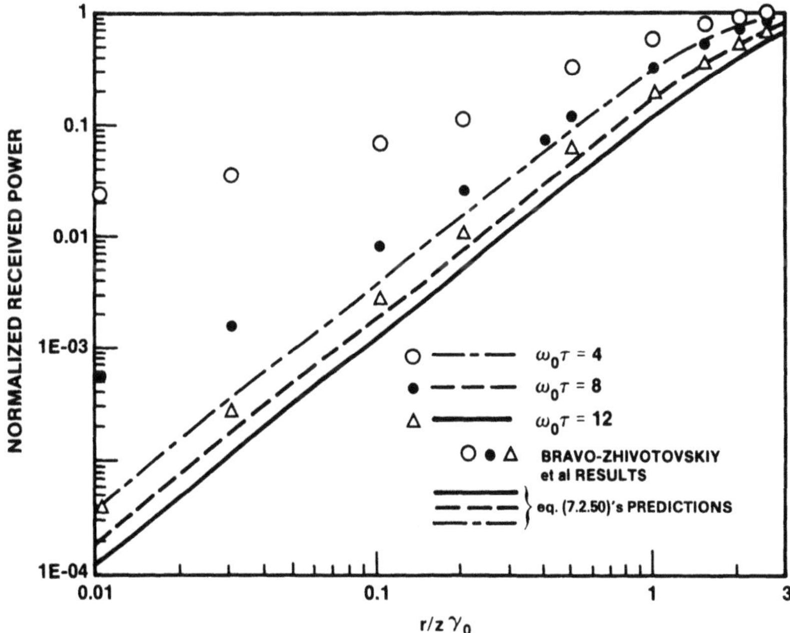

Figure 7.6. Comparison of the normalized received power under various scattering conditions and aperture sizes, set 1.

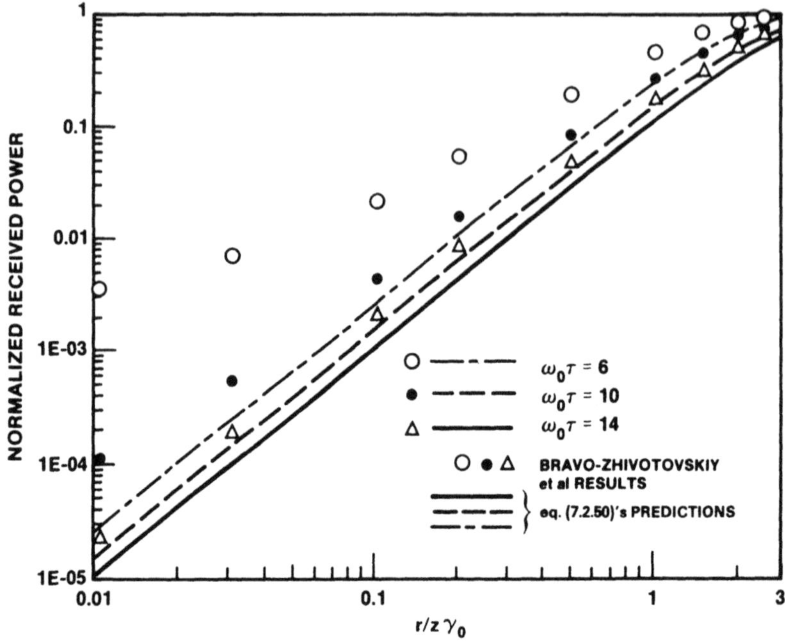

Figure 7.7. Comparison of the normalized received power under various scattering conditions and aperture sizes, set 2.

determinations.[52] Thus it appeared that a closed-form solution for the scattered radiance was available for use in general communications analysis.

Fante developed a similar result as Arnush,[22] but his approach was somewhat different.[2,31] Specifically, he used the highly peaked nature of the volume scattering function as justification for expanding the scattered radiance contained in the integrand of Equation (7.2.13) into a Taylor series and retaining only the first two terms in his development. The resulting solution was similar in form to Equation (7.2.46), and he was able to use this result to investigate the form of this expression for various correlation functions and to correct a result previously reported by Dolin.[80] In addition, Fante investigated the influence of turbulent plasma on received power, beam divergence, the mutual coherence function, and laser backscatter. References 2, 4, 6, and 12 summarize much of the work in this area based on the above formulations. Although these formulations appear to predict received power or irradiance distributions quite well, it was found in the mid-seventies that the above closed-form solutions were not always able to characterize with great accuracy the received radiance distribution emerging from a particulate medium.

During the summer of 1975, a number of sophisticated ocean optics experiments were conducted from a barge moored in the Southern California

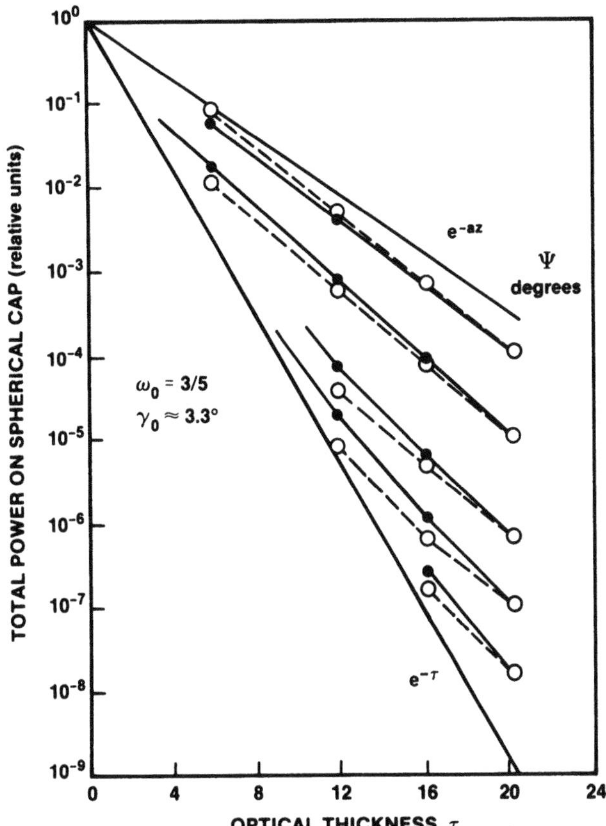

Figure 7.8. Received power on a spherical cap derived from the Arnush model as a function of optical thickness for several cap sizes, compared with the Duntley tank data for $\omega_0 = 3/5$ and $\gamma_0 = 3.3$.

bight north of Santa Catalina Island, and the results were summarized in a report authored by Driscoll *et al.*[30] The intent of these experiments was to better understand the propagation of solar and pulsed laser radiation through the air/sea interface and the marine channel. One set of experiments involved measuring the resultant radiance distribution after a collimated laser beam has traversed a volume of seawater. To obtain these measurements, a blue-green dye laser was submerged to depths of as much as 55 meters and directed upoward toward the surface where an underwater radiance scanner was located. Each device was fastened onto a separate optical bench, the two optical benches being perpendicular. Figure 7.11 illustrates the measurement geometry described. The axis of the laser was inclined 12.5° from the zenith in the plane of the experiment towards the

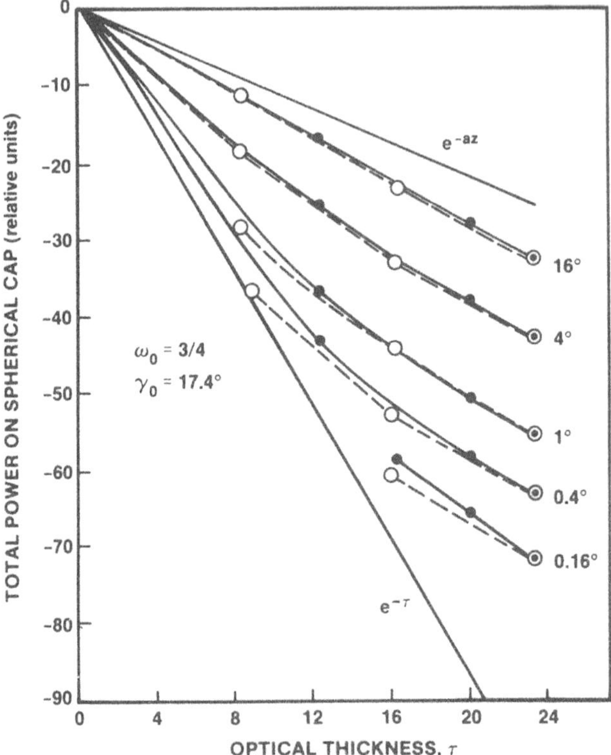

Figure 7.9. Received power on a spherical cap derived from the Arnush model as a function of optical thickness for several cap sizes, compared with the Duntley tank data for $\omega_0 = 3/4$ and $\gamma_0 = 17.4$.

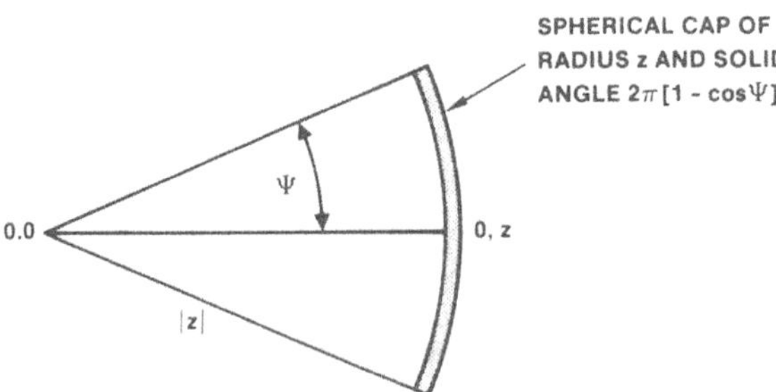

Figure 7.10. Measurement geometry for a spherical cap.

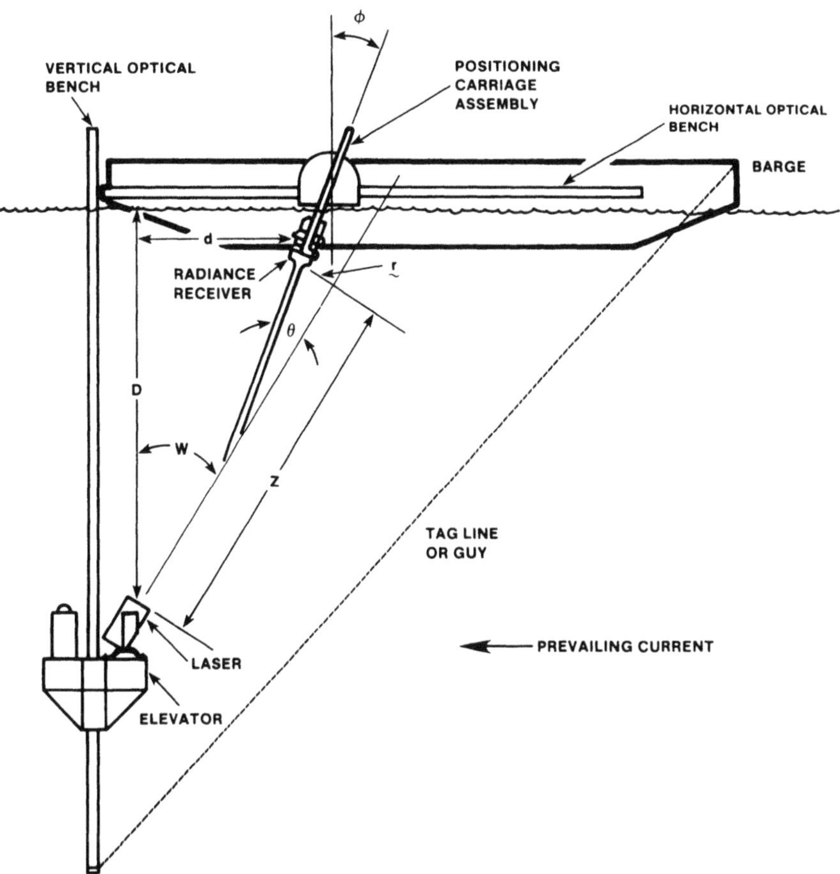

Figure 7.11. Experimental configuration of the laser radiance propagation measurements.

horizontal optical bench. The submerged radiance receiver traveled along this horizontal bench and could scan angle +80° from the nadir in the plane of the experiment. Tables 7.5 and 7.6 outline the operating characteristics of the laser transmitter and radiance receiver, respectively.[30,31] The radiance receiver could be interchanged with the laser source for measurements of solar radiance at depths. Figure 7.12–7.14 show measured laser radiance at source depths of 22.9 m, 38.1 m and 45.7 m, respectively, in the plane. Figure 7.15 depicts a typical water column profile for the experimental site, and Figure 7.16 compares a measured volume scattering function with a normalized form of the scalar phase function given in Equation (7.2.44). Figure 7.17 illustrates a typical normalized solar radiance distribution (radiant loss) as measured at a depth of 45.7 m on 24 June 1975 under somewhat cloudy conditions. Figure 7.18 shows the integrated value of this radiance

Table 7.5. Laser Specifications

Mechanical	
Canister size	41-cm diameter × 84-cm height (16 × 33 in.)
	46-cm diameter flange (18 in.)
Weight	109 kg (240 lb)
Displacement	114 kg (250 lb)
Depth limit	180 m (600 ft)
Electrical	
Input power	1 kW, 3 phase, 400 Hz, Y
Lamp energy	Up to 10 J
Operating parameters	
Pulse repetition rate	20 pulses/s
Peak pulse amplitude	5 kW (±6% rel, ±10% abs)
Pulse width	0.75 μs
Pulse energy	4 mJ/pulse (nom.)
Center wavelength	521.4 nm
Bandwidth	4.66 nm
Beamwidth (in water)	21.4 mrad (FWHM)

profile as a function of receiver field to view. Figures 7.19–7.21 compare the Arnush model with typical solar radiant loss measurements at depths of 15.2 m, 18.3 m, and 21.3 m, respectively.[30] Figures 7.22–7.24 compares the Arnush model with selected laser radiance measurements at depths of 25 m, 42.8 m, and 48.2 m, respectively.[31] It is clear from these graphs that the Arnush model does not predict the radiance (or radiant loss) well, except off-axis as seen in Figure 7.24. This tendency was also seen for some solar data.[30] This is not too surprising in the first comparison set since these data were taken in the $2.86 < bz_0 < 4.0$ range at small values of r_1/z_0. Driscoll *et al.* were able to get the theory to correlate better with the experimental data in some cases by applying a small correction term (see

Table 7.6. Underwater Radiance Receiver

Aperture	7.6-cm diameter, $f/1.3$
Field of view	±21.3 mrad (in water)
Minimum detectable signal	2×10^{-7} W cm^{-2} sr^{-1}
Maximum detectable signal (attenuator in optical path)	5 W cm^{-2} sr^{-1}
Optical bandwidth	100 Å centered at 5240 Å
Detector	RCA PF1023 5.08-cm diameter, 10-state photomultiplier
Detector electrical bandwidth	10 MHz
Output	0 to 10-V pulse to oscilloscope
	0 to 10-V dc averaged peak value to DVM

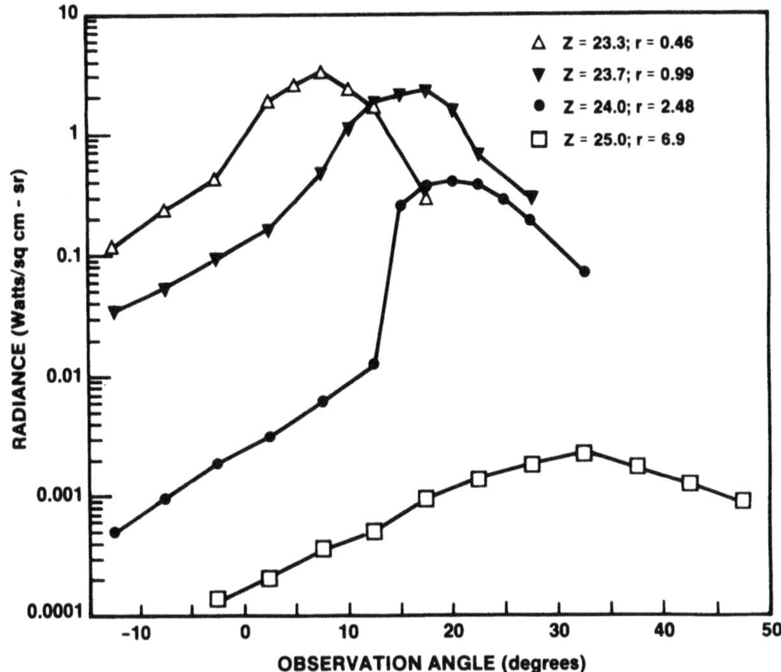

Figure 7.12. Measured laser radiance at a source depth of 22.9 m.

Appendix C of Ref. 30) to the Arnush model.[22] Figure 7.25 shows one example of this model improvement. But this correction did not always work.[30] On the other hand, the laser radiance results were reasonably close to the required validity range ($bz_0 = 4.73$ and $r_1/z_0\gamma_0 = 1.95$ in Figure 7.22, $bz_0 = 8.1$ and $r_1/z_0r_0 = 0.64$ in Figure 7.23, $bz_0 = 8.5$ and $r_1/z_0\gamma_0 = 1.08$ in Figure 7.24) and should be in better agreement with the experimental data shown in the first two figures. The possible conclusion one could draw from these comparisons is that the validity range of the radiance distribution near the optical axis is more stringent than that for the integrated effects of this quantity. That is, the closed-form solution of the radiative transfer equation in the small-angle scattering approximation may not be valid for small and medium-range scattering lengths near the optical axis and only its integrated valued over reasonable apertures can be used in that vicinity. let us investigate this point more closely.

From Fraunhoffer diffraction theory, it is known that the propagated electric field distribution in the far field is essentially the Fourier transform of the source field distribution.[81] This implies that near the optical axis, only the low-frequency spatial and angular frequency components of the transformed field are present. In other words, the contribution of radiance

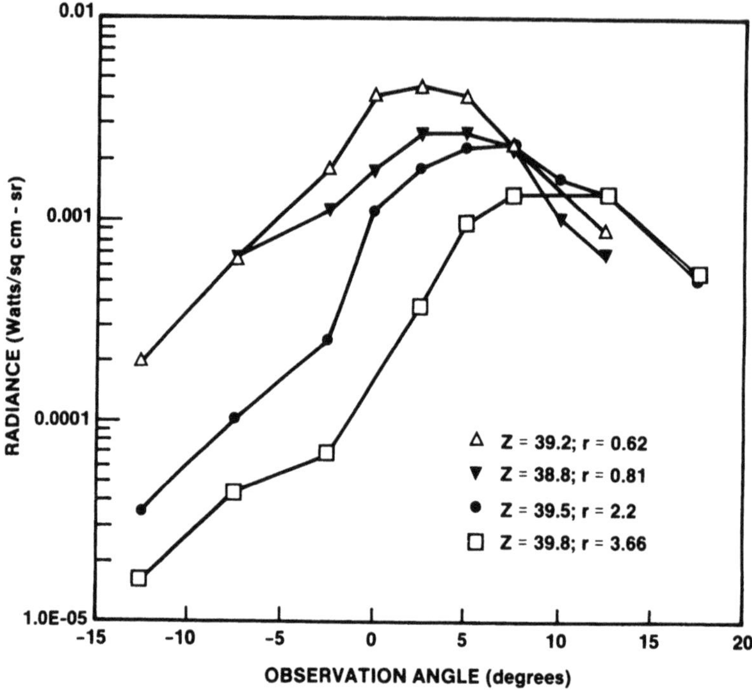

Figure 7.13. Measured laser radiance at a source depth of 38.1 m.

near the axis from high-frequency spatial and angular structure in the source is negligible. This fact is the basis of the Huygens–Fresnel theory and allows the closed-form solution of the Maxwell equations for electromagnetic wave propagation characterization in free-space and turbulent media, as seen in the preceding chapters. One other way of interpreting this fact is that the electromagnetic fields of the incident source distribution can be geometrically ray traced from the source to the receiver plane in both cases. From Chapter 5, we know that turbulence affects the refractive index of the atmosphere very slightly ($\Delta n \sim 10^{-5}$–10^{-6}) and this manifests itself as microradian or less deflections of the propagating electric field distributions. Although particulate scattering of light is highly forward-biased in direction in both the atmospheric and marine channels, the average scatter angle is in the degrees (even for multiple forward scatter) and one can no longer think of this type of scattering as ray traceable. More specifically, one must consider these marine and atmospheric channels as true generalized linear systems.

Both Arnush and Fante derived specific forms of the transfer function for a multiple-scattering medium, based on energy content considerations of this function. They reasoned that all of the forward-scattered radiance would be contained therein, and the weaker diffusive multiple scatter was

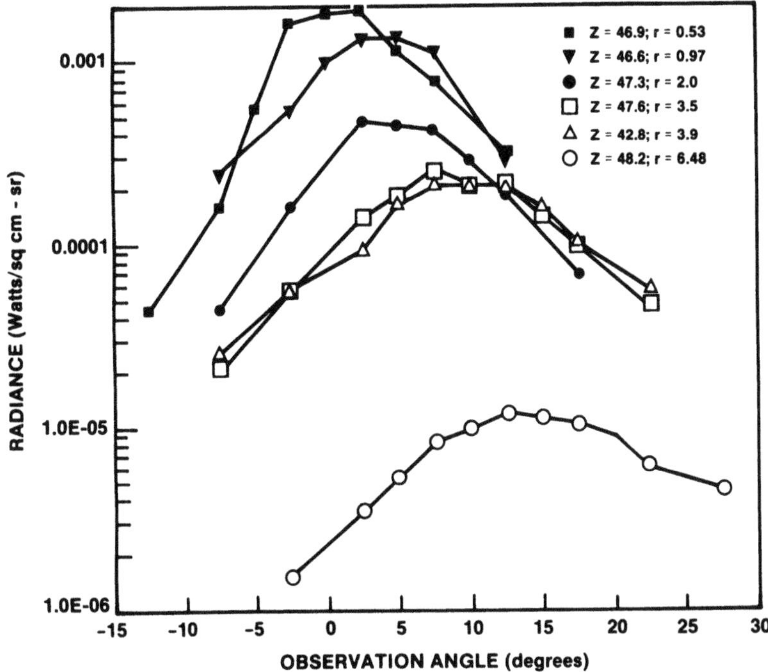

Figure 7.14. Measured laser radiance at a source depth of 45.7 m.

therefore neglected. However, from a linear systems point of view, just the opposite holds. In a generalized linear system, a set of input functions is mapped onto a set of output functions. In our case, this means that the particulate medium of interest will map initial solar or laser radiance distributions into the scattered radiance. Marechel, among others, has shown that the small details in any input function are contained in the high-frequency Fourier component of its transform, and its coarse details are contained in the low-frequency components.[81] For this input to be transferred with minimal loss in fidelity, the high-frequency Fourier components must be relatively unaltered during inversion. In more specific terms, it is the high-frequency components of the scattered transform radiance which must be retained if specific detail of the radiance distribution near the optical axis is desired. Stotts confirmed this hypothesis by expanding Equation (7.2.47) in the high spatial and angular frequency component limit to obtain the following scattered radiance distribution for a plane-wave source input[33]:

$$N_s \approx N_0'' e^{-\tau}[\delta(\gamma) + \omega_0 \tau p(|\gamma|)] \tag{7.2.51}$$

This expression is the solution of the general radiative transfer equation assuming only single scatter.[82] It is apparent that this expression reflects

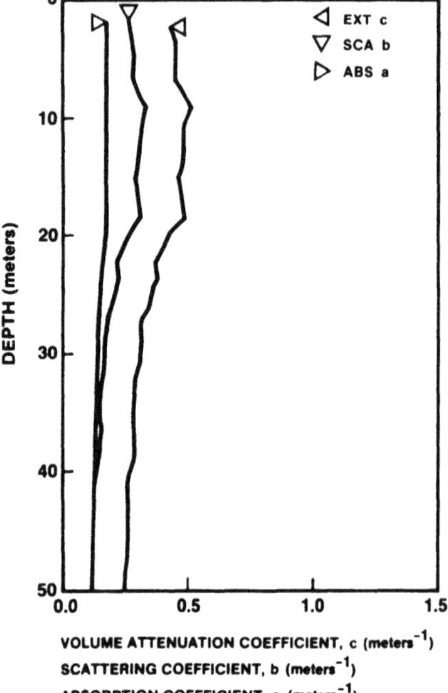

Figure 7.15. Ocean properties—520 nm. Santa Catalina Island. Latitude: 33-27.2 N. Longitude: 116-29.0 W. 23 June 1975. 1024 PDT.

the peak radiance distribution form suggested earlier for multiple forward scatter. More specifically, if we integrate Equation (7.2.51) over a field of view θ_{for} to obtain the scattered radiance, we find that

$$N_s \approx I_0'' e^{-\tau}[1 + \omega_0 \tau \Phi(\theta_{for})] \sim I_0'' \exp\{-(1 - \omega_0 \Phi)\tau\} \qquad (7.2.52)$$

for $\omega_0 \Phi \tau \ll 1$. Equation (7.2.52) is of the form suggested by Bucher(57) and derived earlier in Section 7.2.2.1. This implies that the high-frequency components of Equation (7.2.47) contain information on multiple forward scatter and radiance near the optical axis, and the low-frequency components describe diffusive multiple scatter.

In summary, Equation (7.2.47) contains all the information necessary to describe radiance propagation through a particulate multiple-scattering medium. In those situations where integrated effects or off-axis radiance values are required in receiver planes where $bz > 8$, approximation techniques can be applied with care to this equation to yield the desired results. Otherwise, numerical techniques must be employed for the radiance/irradiance calculation. Besides the numerical methods described earlier,(26) there are two other ways of solving the radiative transfer equation given in Equation (7.2.13) numerically. One was developed by Tam and Zardecki(32) and the other by Helliwell.(50) In addition, diffusion theory has been applied

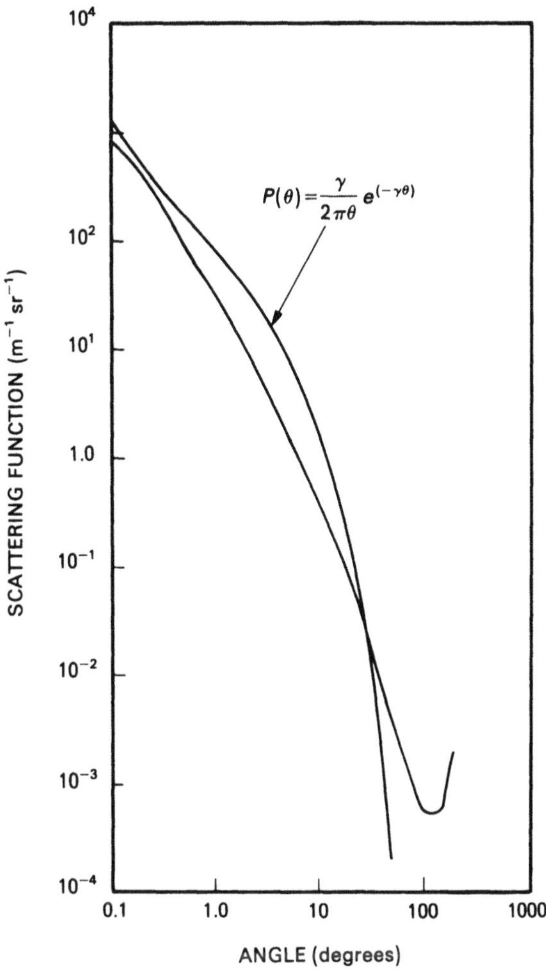

Figure 7.16. Volume scattering function. 520 nm. 23 July 1975. 2106 STA.CAT/21-SEA.

to the radiative transport equation for the case of light propagation through relatively thick or dense particulate media.[4,12]

7.2.2.5. Other Mathematical Approaches

Tam and Zardecki[32] began with an inversion equation similar to Equation (7.2.43); specifically, they had

$$N(z, r, \phi) = \frac{\pi \exp\{-kz\}}{\gamma_1^2 (2\pi)^4} \int d^2\zeta \int d^2\eta \, e^{\tilde{\Omega}(z)} \times \exp\left\{-\frac{(\boldsymbol{\zeta}+\boldsymbol{\eta} Z)^2}{4\gamma_1^2} + i(\boldsymbol{\zeta}\cdot\boldsymbol{\phi}+\boldsymbol{\eta}\cdot\mathbf{r})\right\} \tag{7.2.53}$$

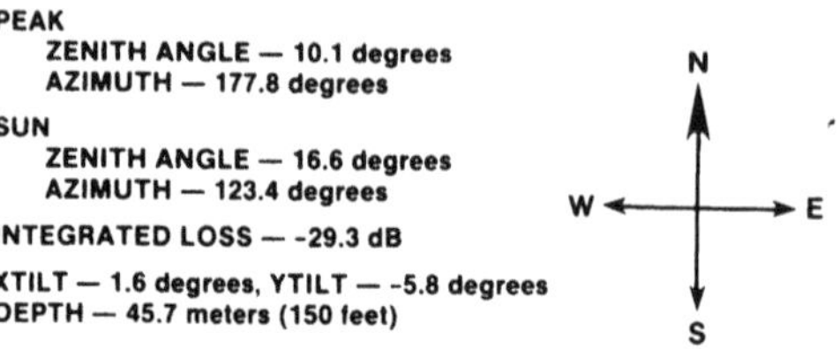

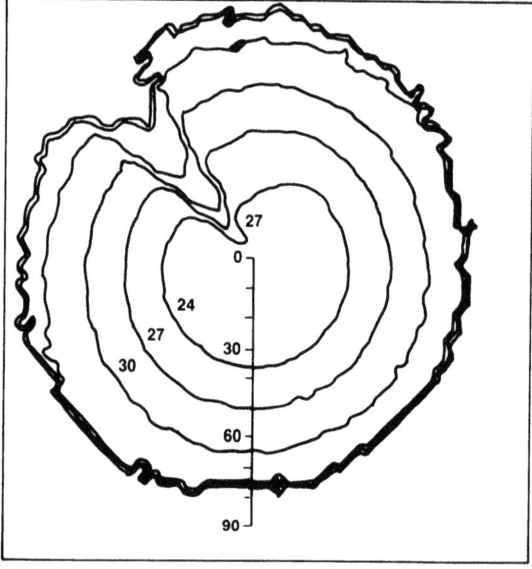

Figure 7.17. Radiant loss (dB/unit solid angle).

with

$$\Omega(z) = b \int_0^z \exp\left\{ -\frac{1}{4\gamma_0^2} |\boldsymbol{\zeta} + \boldsymbol{\eta}(z - z')|^2 \right\} dz' \tag{7.2.54}$$

and γ_1 being the angular width of the source at $z = 0$. To numerically integrate Equation (7.2.53), they first expanded Equation (7.2.54) and integrated over and term by term. This gave

$$N(z, r, \phi) = \frac{\pi}{\gamma_1^2} \frac{\exp_e\{-bz\}}{(2\pi)^4} \sum_{m=0}^{\infty} \frac{\sigma^m}{m!} I_m(z, r, \phi), \tag{7.2.55}$$

where

$$I_0(z, r, \phi) = 2(2\pi)^3 \gamma_1^2 z^{-2} \exp\left\{ -\frac{\gamma_1^2}{z^2}(x^2 + y^2) \right\} \delta\left(\phi - \frac{r}{z} \right) \tag{7.2.56}$$

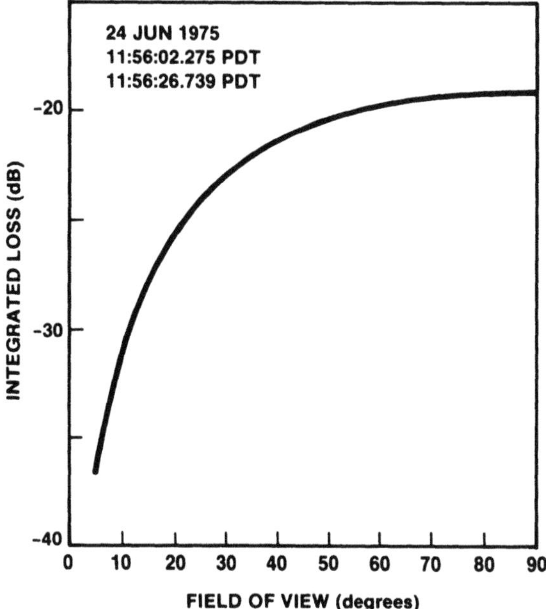

Figure 7.18. Integral of radiance.

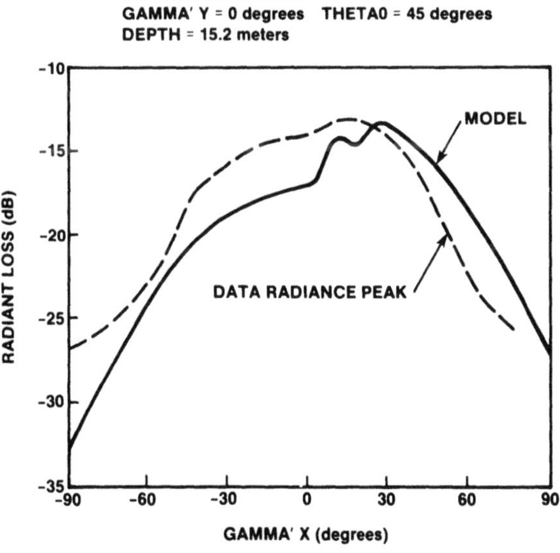

Figure 7.19. Radiance profile through sun angle. 1124 PDT. 24 June 1975.

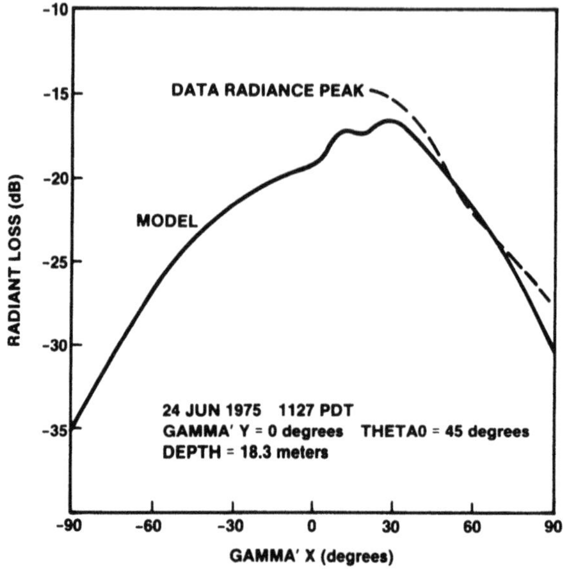

Figure 7.20. Radiance profile through sun angle. 1127 PDT. 24 June 1975.

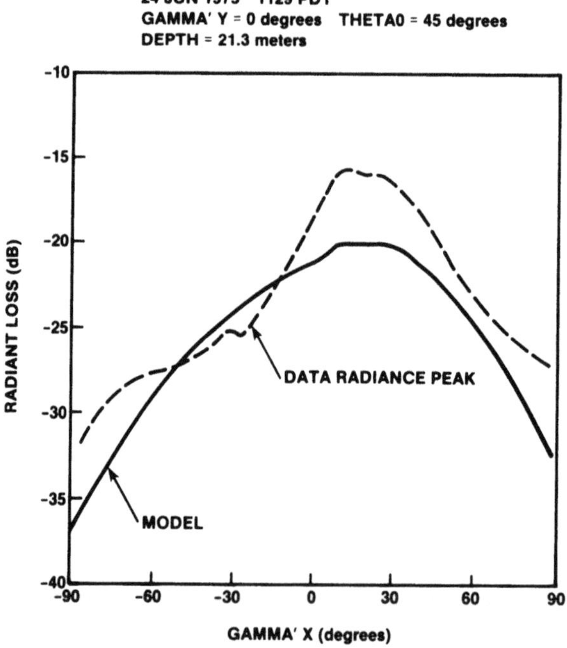

Figure 7.21. Radiance profile through sun angle. 1129 PDT. 24 June 1975.

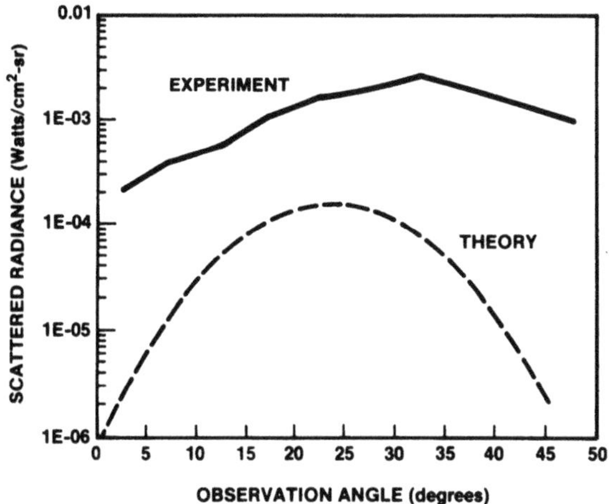

Figure 7.22. Comparison of theory with measured data for $\mathbf{r} = (6.9\ \text{m}, \pi)$ and $z = 25$ m.

and where, for $m \geqslant 1$,

$$I_m = (2\pi)^2 z^{m-2} \exp\left\{ -\frac{\phi^2}{(m/\gamma_0^2) + \gamma_1^{-2}} \right) \int_0^1 \cdots \int_0^1 dz_1' \cdots dz_m'$$

$$\times \Delta^{-1}(m) \exp\left\{ -\left[\frac{(m/4\gamma_0^2) + \gamma_1^{-2}}{\Delta(m)}\right]\left[\frac{r}{z} - \left(1 - \frac{1}{m}\sum_{j=1}^{m} z_j\right)\phi\right]^2\right\}, \tag{7.2.57}$$

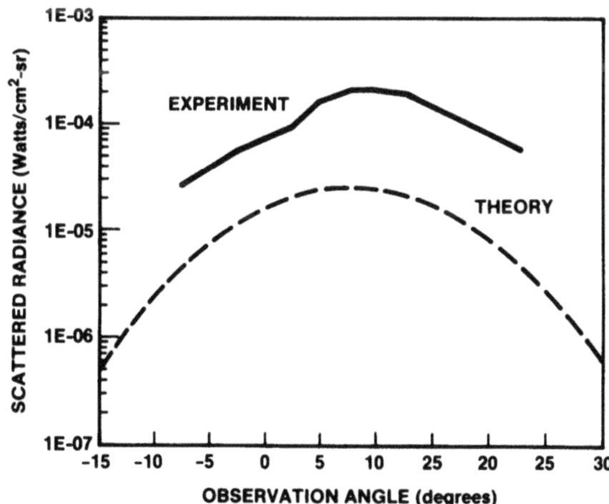

Figure 7.23. Comparison of theory with measured data for $\mathbf{r} = (3.9\ \text{m}, 0)$ and $z = 42.8$ m.

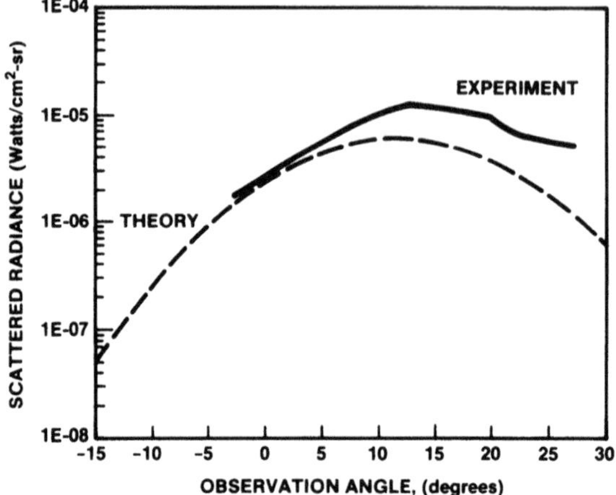

Figure 7.24. Comparison of theory with measured data for $\mathbf{r} = (6.48 \text{ m}, \pi)$ and $z = 48.2$ m.

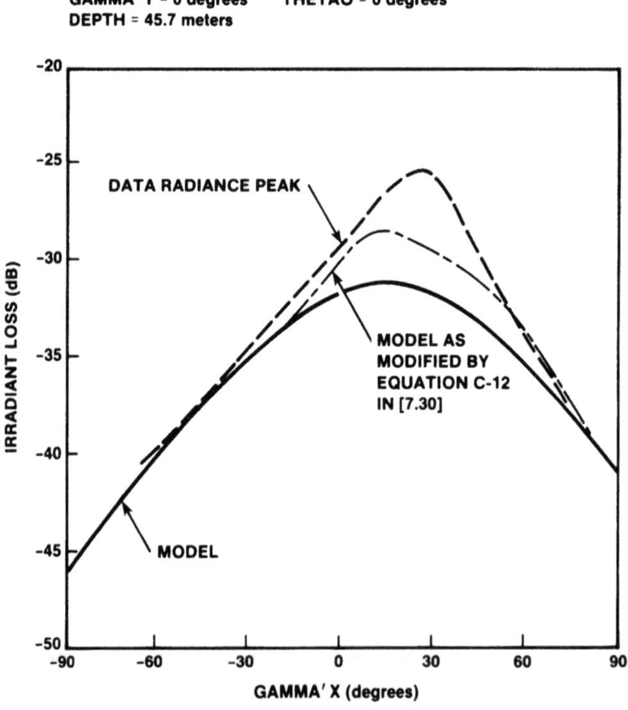

Figure 7.25. Radiance profile through sun angle.

with

$$\Delta(m) = \frac{1}{4\gamma_0^4}\left\{m \sum_{j=1}^{m} (1 - z_j)^2 - \left[\sum_{j=1}^{m} (1 - z_j')\right]^2\right\}$$
$$+ \frac{1}{4\gamma_0^4}\left[m + \sum_{j=1}^{m} (1 - z_j)^2 - 2 \sum_{j=1}^{m} (1 - z_j)\right] \qquad (7.2.58)$$

Figure 7.26 is an example radiance comparison of the "exact" answer given in Equation (7.2.56) with the approximate Arnush approach described above. It is clear that this technique gives the forward-scatter performance one expects from the analysis.

Helliwell,[50] on the other hand, approached the multiple-forward-scatter characterization problem from a more classical solution point of

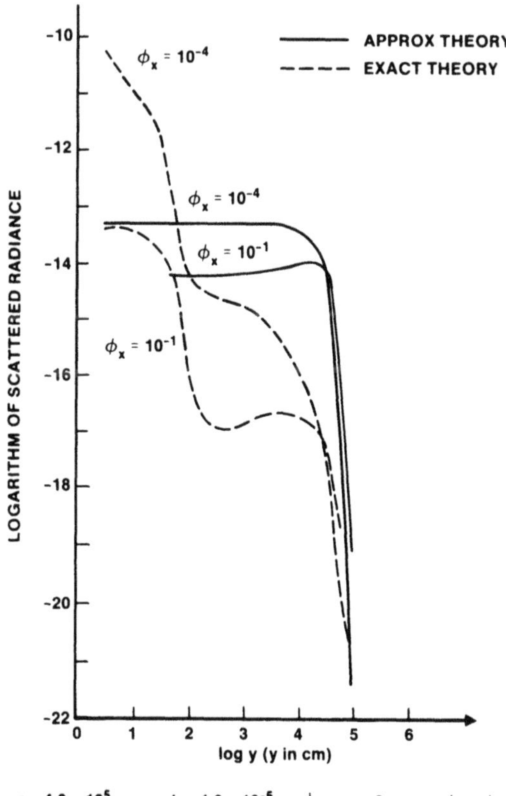

Figure 7.26. Scattered radiance versus transverse distance, as predicted by exact theory and the approximate theory based on the Arnush approximation.

view. Specifically, he reformulated the general radiative transfer equation given in Equations (7.2.11) and (7.2.12) into the following form:

$$\sin\theta\left[\mu\frac{\partial N}{\partial r}+\frac{1-\mu^2}{r}\frac{\partial N}{\partial\mu}\right]+\cos\theta\frac{\partial N}{\partial z}+[c(z)-K]N$$
$$=\exp\{Ks\}\frac{b(z)}{4\pi}\int_0^\pi\int_0^\pi p^*(\alpha)\exp\{-Ks'\}N(z,r,\phi',\theta')\sin\theta'\,d\theta'\,d\phi' \tag{7.2.59}$$

where

$$\mu=\cos\theta$$

and

$$S=\min\left[\frac{z-Z}{\cos\theta},\frac{r\mu+(R^2-r^2-(r\mu)^2)^{1/2}}{\sin\theta}\right]$$

where $Z=0$ for $0\leqslant\theta\leqslant\pi/2$ and $Z=z_0$ for $\pi/2<\theta\leqslant\pi$. In Equation (7.2.59), K is a constant to be derived, $P^*(\alpha)$ is the sum of p evaluated at $+\varphi'$ and p evaluated at $-\varphi'$ under the assumption $N(z,\mathbf{r},\theta,-\varphi)=N(z,\mathbf{r},\theta,\varphi)$ and $0\leqslant z\leqslant z_0$. The notation s' has been used in the equation since s is a function of θ and φ. A similar equation can be obtained for stratifical media. The boundary conditions for N are as follows: The solution at $r=0$ is obtained from Equation (7.2.59) first for $\mu=-1$ (the singularity at $r=0$ disappears) and then for $\mu\neq-1$ using $\partial\bar{L}/\partial\mu=0$ [the equation that Equation (7.2.59) reduces to for $r=0$]. With the above clearly defined, Helliwell was then able to apply a finite difference scheme similar to the s_n' method of Carlson.[83] This was accomplished by dividing all the dimensions into the following intervals (not necessarily equal):

$$0=r_0<r_1<r_2<\cdots<r_{\mathrm{NI}}=R \tag{7.2.60a}$$

$$0=z_0<z_1<z_2<\cdots<z_{\mathrm{NJ}}=D \tag{7.2.60b}$$

$$0=\phi_{\mathrm{NK}}<\phi_{\mathrm{NK}-1}<\cdots<\phi_1<\phi_0=\pi \tag{7.2.60c}$$

$$0=\theta_0<\theta_1<\theta_2<\cdots<\theta_{\mathrm{NL}}=\pi \tag{7.2.60d}$$

This allows the integral in Equation (7.2.59) to be approximated by the relation

$$\int_0^{\pi}\int_0^{\pi} p^*(\alpha)N(r_i, z_j, \phi', \theta')\sin\theta' d\theta' d\phi'$$

$$\approx \sum_{k'=0}^{NK}\sum_{l'=0}^{NL} N(r_i, z_j, \phi_{k'}, \theta_{l'})\frac{1}{4\pi}\int_A\int p^*(\alpha)\sin\theta'\, d\theta'\, d\phi'$$

$$= \sum_{k'=0}^{NK}\sum_{l'=0}^{NL} N_{i,j,k',l'}P_{k,l,k',l'} \tag{7.2.61}$$

The integration for $k', l' = k, l$ is done analytically and for $k', l' \neq k, l$ is approximated with a simple two-dimensional trapezoid rule with a fine $(\Delta\theta', \Delta\varphi')$ mesh. The angular region A over which the integral in Equation (7.2.61) is carried out is a suitable region around (θ_l, φ_h).

For $k = 0$, $(\varphi = \varphi_0 = \pi$ and $\mu = -1)$ the equation to be solved for $N_{k=0,l}$ is

$$-\sin\theta_l\frac{\partial N_{k=0,l}}{\partial r} + \cos\theta_l\frac{\partial N_{k=0,l}}{\partial z} + [c(z) - K]N_{k=0,l}$$

$$= \exp\{Ks\}b(z)\sum_{k'=0}^{NK}\sum_{l'=0}^{NL} N_{ijk'l'}P_{0lk'l'} \tag{7.5.62}$$

For $k = 1, \ldots, NK$, an implicit differencing for $\partial/\partial\mu$ is used and the equation to be solved for $\bar{L}_{k,l}$ is

$$\sin\theta_l\left[\mu_k\frac{\partial N_{k,l}}{\partial r} + \frac{1-\mu_k^2}{r}\frac{N_{k,l}}{\mu_k - \mu_{k-1}}\right] + \cos\theta_l\frac{\partial N_{k,l}}{\partial z} + [c(z) - K]\bar{L}_{k,l}$$

$$= \exp\{Ks\}b(z)\sum_{k'=0}^{NK}\sum_{l'=0}^{NL} N_{ijk'l'}P_{klk'l'} - \sin\theta_l\left[\frac{1-\mu_k^2}{r}\frac{(-N_{k-1,l})}{\mu_k - \mu_{k-1}}\right]$$

(7.2.63)

At the point (r_i, z_j), the difference approximation for $\partial/\partial r$ and $\partial/\partial z$ depends upon the direction of the radiance, that is, on φ_k and θ_l. In Figure 7.27, the point (r_i, z_j) is illustrated with its eight neighboring points. If φ_k and θ_l are such that the projection of the direction vector onto the (r, z) plane is from direction I, then points 1, 6, and 7 are used to approximate the derivatives. That is

$$\frac{\partial N_{kl}}{\partial_r} = \frac{N_{ujkl} - N_{i-1,jkl}}{r_i - r_{i-1}} \tag{7.2.64}$$

$$\frac{\partial N_{kl}}{\partial z} = \frac{N_{i-1,j+1,kl} - N_{i-1,jkl}}{z_{j+1} - z_j} \tag{7.2.65}$$

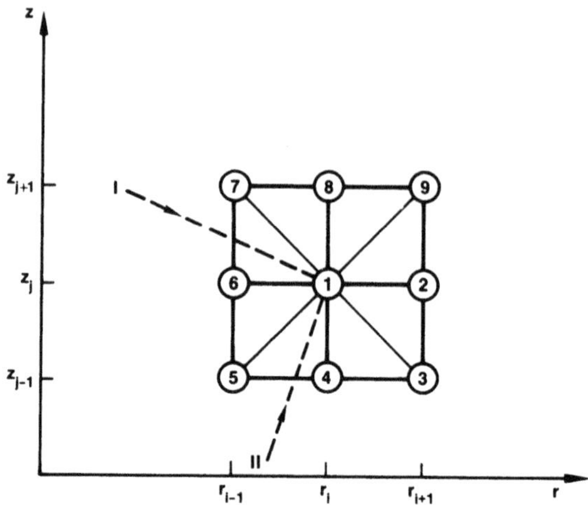

Figure 7.27. Schematic drawing to illustrate Carlson's scheme.

Likewise, if the projection is from II then points 1, 4, 5 are used. So,

$$\frac{\partial N_{kl}}{\partial r} = \frac{N_{i,j-1,kl} - N_{i-1,j-1,kl}}{r_i - r_{i-1}} \tag{7.2.66}$$

$$\frac{\partial N_{kl}}{\partial z} = \frac{N_{ijkl} - N_{i,j-1,kl}}{z_j - z_{j-1}} \tag{7.2.67}$$

Similar expressions hold for each of the other six possible triangles into which the projected direction may fall. For each pair (θ_l, φ_k), the resulting difference equation can be easily solved for N_{ijkl} in terms of previously determined or specified values of N. All that needs to be done is to start at the appropriate boundary and sweep through the (r_i, z_j) grid in the correct direction.

Since the above procedure requires knowledge of N for solution of Equation (7.2.59), Helliwell pointed out that this equation must be solved iteratively. Although somewhat computationally intensive, the finite difference method yields satisfactory results. Figure 7.28 compares predictions derived by Helliwell with the underwater radiance distributions reported by Tyler,[(84)] assuming a source at 4.24 m. It is obvious that this method produces results in reasonable agreement with Tyler's data. Figure 7.29 compares finite difference predictions with those results previously shown in Figure 7.24 and illustrates better agreement than the Arnush model. Unfortunately, it did not do as well with other data sets from that experiment, and no explanation was apparent.[(50)]

When the density of particulates is relatively high (greater than one percent) or the path length is sufficiently large that the propagation is

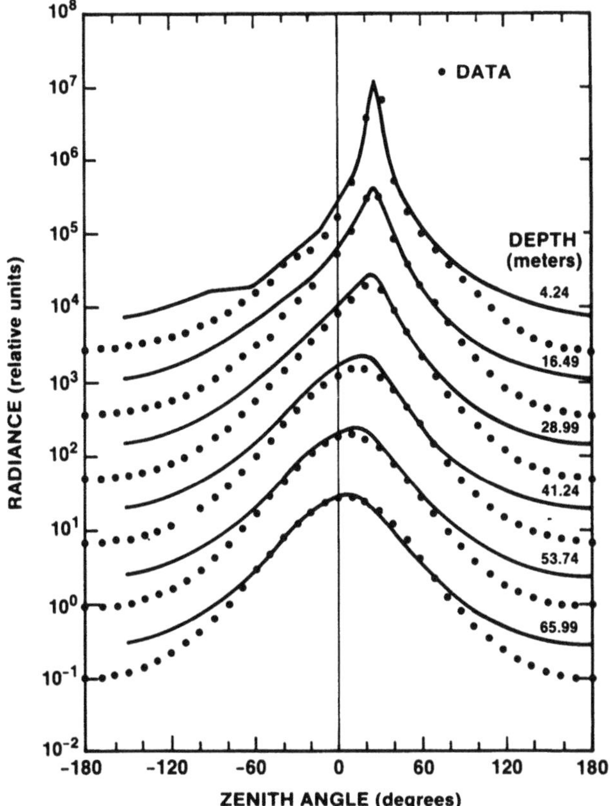

Figure 7.28. Calculated in-water radiance distribution with source at 4.24 m equal to the data at 4.24 m.

multiple scatter dominated, diffusion theory (as noted before) can be called upon to quantify the radiation transfer. Recent work by Furutsu[85,86] and Ishimaru[(87)] applies the diffusion approximation to the space-time radiative transport equation to yield solutions for pulse propagation through thick clouds. These results have shown good agreement with some experimental data.[(88)] The two approaches are equivalent in the CW (monochromatic) case but yield different pulse solutions due to the inclusion of different time-dependent derivatives.[(89)] References 88 and 89 review these works and their difference in fine details. However, both require certain specific conditions to be true for the medium:

1. Particle sizes must be much larger than the wavelength of interest (parabolic-equation validity assumption); and
2. Angular distribution of the radiant intensity must be nearly isotropic.

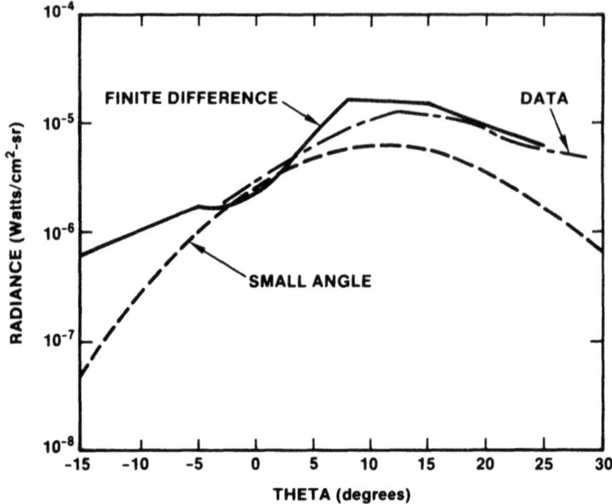

Figure 7.29. Comparison of theories with data for $\mathbf{r} = (6.48\text{ m}, \pi)$ and $z = 48.2$ m.

As noted by Ishimaru,[89] more work must be done to overcome these areas of concern for optical communications analysis. One way of overcoming these problems is through Monte Carlo simulation, the topic of the next section.

7.2.2.6. Monte Carlo Simulation

Monte Carlo simulation is a computer-based technique which can be applied to any problem where one knows the probability for each step in a sequence of events and desires the statistics of all possible outcomes. Many researchers, most notably Bucher,[61] Gordon *et al.*,[46] Plass and Kattawar,[47,48] and Lerner and Summers,[44] have contributed significantly to our understanding of optical radiance and irradiance propagation through particulate media using this technique. In this section, we shall review the Monte Carlo method and some of the results produced relevant to the aforementioned channels. The works of Bucher[61] and Kattawar[45] are drawn upon heavily for this section.

In any Monte Carlo simulation, one must have an adequate supply of uniformly distributed random numbers between 0 and 1. Today, most researchers utilize pseudorandom number generators in their computers to create that supply. These generators are actually computer algorithms. The simplest and most widely used number generators have the form

$$X_i = kX_{i-1} + \delta \quad (\text{modulo } M)$$

where M is equal to $2^{(\text{bit length}-1)}$ for a binary computer, with the first bit being the sign bit and k, δ, and x_i integers between 0 and $M - 1$. Clearly, we need to divide the resulting sequence by M to yield a random number chain of values always between 0 and 1. The period of the above pseudorandom number generator ranges between $2^{(\text{bit length}-2)}$ to M, with this latter value occurring if

1. M and δ have no common divisors
2. $k = 1$ (modulo p) for every prime factor p of M
3. $k = 1$ (modulo 4) if M is a multiple of 4

This sequence is then used with the cumulative probabilities of each event involved in the simulation to produce, after a large number of runs, the relative frequencies of all the possible outcomes. Let us illustrate this for the case of an optical beam traversing a multiple-scattering medium.

Assume we have a particulate medium which is isotropic and homogeneous in nature and is shaped as a plane-parallel slab of infinite width and physical thickess z. Its scattering coefficient is b. The fundamental principle of Monte Carlo simulation is as follows: Let r be a random number which is uniformly distributed on $0 < r < 1$ and let $p(x)dx$ be the probability of x lying between x and $x + dx$ with $\Delta_1 \leqslant x \leqslant \Delta_2$. Here

$$\int_{\Delta_1}^{\Delta_2} p(x)\, dx = 1$$

and

$$\begin{aligned} r = P(x) &\equiv \text{cumulative probability} \\ &= \int_{\Delta_1}^{x} p(x')\, dx' \end{aligned}$$

Given a random number r, x is uniquely determined and falls with frequency $p(x)dx$ in the interval (x, dx). In our case, the probability of light interacting with a particle in the interval z'' and $z'' + dz''$ is given by

$$P(z'')\, dz'' = \exp\{-bz''\}\, dz'' \tag{7.2.68}$$

Hence

$$r = \int_0^{z'} \exp\{-bz''\}\, dz'' = 1 - \exp\{-bz'\} \tag{7.2.69}$$

or

$$z' = \frac{-1}{b} \ln(1 - r) \tag{7.2.70}$$

Since the sequence $\{1 - r_i\}$ is uniformly distributed like the chain $\{x_i\}$, one can replace Equation (7.2.70) with

$$z' = \frac{-1}{b} \ln (r) \tag{7.2.71}$$

in the simulation without any loss in generality or accuracy. What it does do is eliminate one of the computation steps, thereby reducing the computing times of each simulation run. When one is propagating 50,000 to 100,000 photons, this reduction can be significant. In any event, we move the photon the distance z' and assume an interaction when a particle has occurred. As we have assumed no absorption, a scattering event must be the result. From our previous work, it is apparent that the scalar phase functions are azimuthally symmetric as well as forward-biased. Thus, for the scattering event we need two random numbers. The first goes to choosing the rotation angle φ' from its uniformly distributed set between 0 and 2π. The second to determining the polar angle θ' for the inverse relation of

$$r'' = \int_0^{\theta'} p_a(\theta'') \sin \theta'' \, \lambda\theta''$$

where $p_a(\theta)$ is the azimuthally averaged scalar phase function of the medium. Given these two scatter angles, θ' and φ', we select another random numbers and use it to determine a new propagation distance

$$z''' = \frac{-1}{b} \ln (r'')$$

We then move the photon from its location $(0, 0, L')$ a distance L'' along the ray direction (θ', φ'). We then determine a new ray direction (θ'', φ'') and propagation distance z'''' using the above techniques and move the photon to its new interaction position. This process is contained until the photon exits the scattering medium through one of its planar boundaries. This computational procedure is repeated with as many photons as necessary to generate the desired statistics and accuracy. Light exiting the top (incident) boundary is indicative of the backscattered irradiance. Light leaving the bottom boundary indicates the transmitted irradiance. Generally, for each of the exiting photons, we record its total travel distance, exit angle, and radius of exit relative to the straight-through direction, or any other information necessary to generate the desired statistical measures.

It should be clear from the above example that if a medium is optically thin, i.e., $1/b \gg z$, then most of the incident radiation will exit with very few interactions. (The optical thickness generally indicates the average number of interactions light will incur traversing a multiple-scattering medium.) As the probable error of the simulation is inversely proportional to the square root of the number of photons used, a reasonably good simulation in this situation will be obviously quite wasteful and costly in terms of computer time. To counter this problem, many researchers employ the method of forced collisions and introduce a statistical weight into their simulation. That is, an incident photon is no longer considered a single entity, but rather a group of photons. For example, in the above simulation, each photon incident on the medium is multiplied by an initial weight, say W_0. Normally, the fraction of light escaping a first collision and passing through the medium is given by $\exp\{-\omega_0\tau\}$, where ω_0 again is the optical scattering thickness of the medium and τ is the optical thickness of the medium. Therefore, the amount of light still within the medium is proportional to $(1 - e^{-\omega_0\tau})$. We will now force a collision by rewriting our probability density function for distance as

$$p'(t)\,dt = \frac{e^{-t}\,dt}{1 - e^{-\omega_0\tau}}$$

where t is normalized to the mean free path, D (i.e., normalized to the mean free path between collisions). Hence we have

$$r' = \frac{1 - e^{-t'}}{1 - e^{-\omega_0\tau}} \tag{7.2.72a}$$

or

$$r' = \frac{1}{b}\ln\left[1 - r'(1 - e^{-\omega_0\tau})\right] \tag{7.2.72b}$$

replacing Equations (7.2.70) and (7.2.71) in the simulation. To remove the "bias" from the computation, we need to multiply the weight W_0 by $(1 - e^{-\tau})$, thereby giving the fraction of light remaining after the interaction. This approach can also be utilized to take into account particulate and molecular absorption, or absorption by a surface. In particular, each incident photon is assigned a statistical weight that is adjusted by the albedo at each collision to account for the finite probability of absorption. The path of each weighted photon is followed until this weight drops below a predetermined level.[(45)] With this basic material on Monte Carlo simulation behind us, we will now turn to some of the results generated using this technique. For our purposes here, we shall review the work of Bucher for nonabsorbing particulates[(61)] and Lee *et al.* for absorbing particulates,[(90,91)] both studies pertaining solely to optical communications analysis. However, the reader is well advised to consult those references previously cited for additional information and understanding in the general area of optical communications and surveillance.

In most system applications, optimum performance depends upon both the amount and form of the received signal involved. For example, a small amount of light confined to a relatively short time interval is much more detectable in background-limited communications than if the same flux level is spread over a larger time span. Thus specific knowledge of the degrading effects incurred from a signal beam traversing a particulate medium is mandatory to a successful system design. Generally, a collimated optical pulse will experience, in the above beam spreading, dispersion in angle of arrival, attenuation, degradation of spatial coherence, and dispersion in the time and frequency of the waveform modulating it.[61] The four major effects are:

1. Angular spreading
2. Spatial spreading
3. Multipath time spread
4. Total transmission.

Figure 7.30 depicts the first three effects graphically.[61] Danielson *et al.* investigated the steady-state transmission and reflection of a plane wave from a nonabsorbing cloud possessing a Henyey–Greenstein scalar phase function.[92] This function has the form

$$p(\theta) = \frac{(1 - g^2)}{2(1 + g - 2g \cos \theta)^{1.5}} \tag{7.2.73}$$

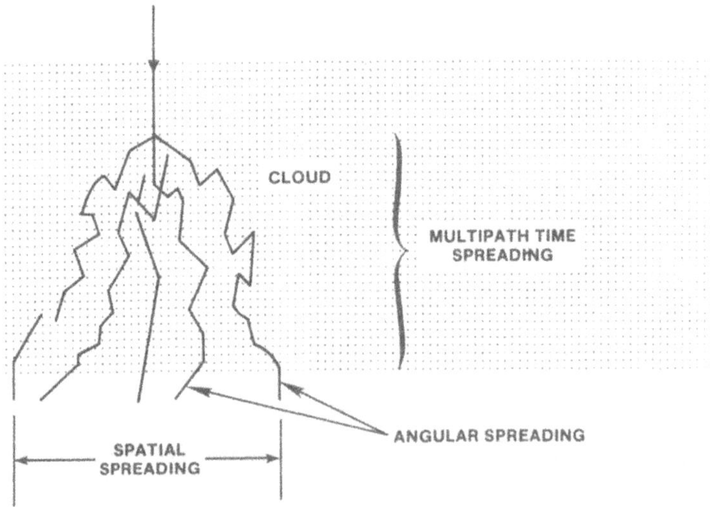

Figure 7.30. Parameters of light propagation in a nonabsorbing particulate medium.

where g is the asymmetry factor defined previously. Plass and Kattawar have also performed extensive research in this area using a variety of different phase functions.[48] However, the above analyses were directed more towards the physics of radiation transfer, rather than towards engineering optical systems. Bucher recognized this void and investigated the more complex question of atmospheric communications using various scattering functions $p(\gamma)$.[61] In particular, he found that angular and spatial spreading, total transmission, and multipath time spreading become independent of the phase function used to create them when the diffusion thickness, defined as

$$\tau_d = \tau[1 - \overline{\cos\theta}] \tag{7.2.74}$$

exceeds a value of 2 to 3. Let us look at this in more detail, beginning with angular spreading.

Angular spreading is the apparent angular broadening of a collimated beam due to scattering within the medium. In his cloud simulation, Bucher observed a sharply peaked (within 5° of the normal) distribution of exit angles centered around the straight-through axis within a radius of 0.8 times the cloud's physical thickness. In addition, the total irradiance contained in this distribution was much more than that in the direct beam. Figure 7.31 illustrates this latter facet, as a function of the cloud's optical thickness. It is apparent from this figure that the peaked distribution is decreasing at a slower rate (~2.6 dB/τ versus 4.34 dB/τ) than the direct beam is. Bucher attributed this additional irradiance to a strong forward-scattering behavior. Experimental verification of this hypothesis has been reported,[59,60] as noted earlier.

In those simulations involving clouds with $\tau_d > z$, Bucher found the distribution of exit angles to be nearly uniform in brightness about the cloud normal. That is, he observed that the light rays emerging from a given section of cloud bottom tended to be concentrated within a 45° cone about the normal. However, when a projected area correction factor of $1/(\cos\varphi_E)$ with φ_E the angle of exit, was applied, the brightness of the cloud's bottom boundary was fairly uniform at all viewing angles.

Spatial spreading is the dimensional increase of a beam's finite cross section and is a direct result of the angular spreading described above. Typically, its statistical distribution has a long tail with some rays having an exit radius several times the average. Thus Bucher chose to characterize spatial spreading in terms of the average radius $\bar{r}$ and also the radius r_c in which half of the transmitted rays exit the medium. Figure 7.32 shows $\bar{r}/D_d$ as a function of τ_d for the various scattering functions Bucher used in his analysis. Figure 7.33 gives r_c/D_d as a function of τ_d. The parameter D_d is

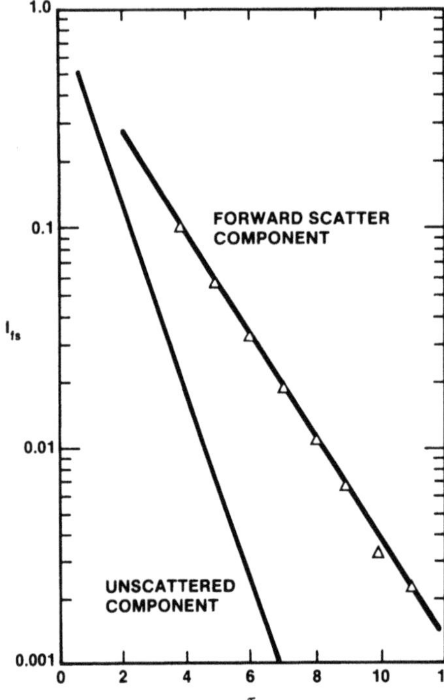

Figure 7.31. Intensity of the forward scatter and unscattered components.[61]

called the diffusion distance and is given by

$$D_d = \bar{l}/[1 - \overline{\cos\theta}] \tag{7.2.75}$$

where $\bar{l}$ represents the mean free path ($=1/b$ in this case) and $\overline{\cos\theta}$ again

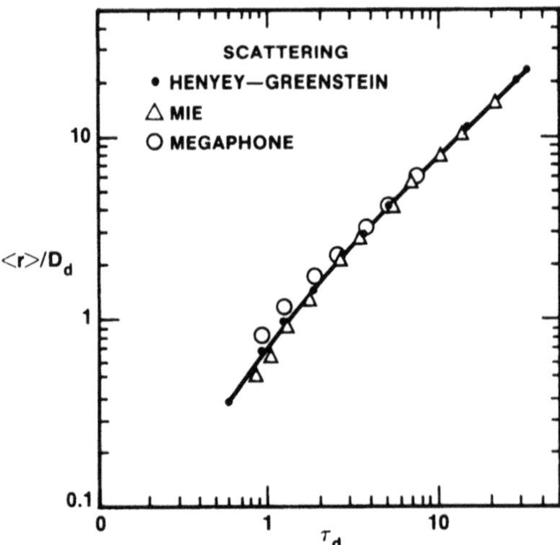

Figure 7.32. $\langle r\rangle/D_d$ as a function of τ_d. The 90% confidence limits on all data points are ±5% or better.[61]

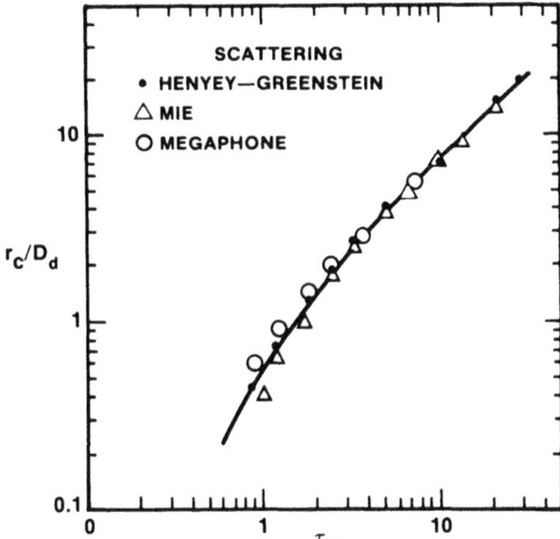

Figure 7.33. r_c/D_d as a function of τ_d. The 90% confidence limits on all data points are ±5% or better.[61]

the average cosine. Figure 7.34 depicts $\overline{\cos\theta}$ as a function of the normalized Mie parameter for nonabsorbing spheres. For the scattering medium modeled, $\pi o/2$ ranged from 17.05 to 21.54, thereby yielding a value of ~0.81 for $\overline{\cos\theta}$. This implies that $\tau_d \approx 5.26\tau$. It is apparent from Figures 7.30 and 7.31 that the three phase functions behave similarly for $\tau_d > 3$. Curve fitting the above data, Bucher found that the two radii could be expressed as

$$\bar{r} = 0.92 D_d(\tau_d)^{0.92} \tag{7.2.76}$$

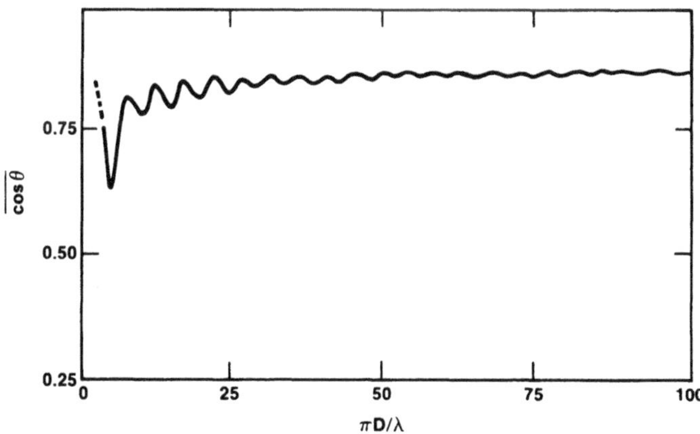

Figure 7.34. Cos θ as a function of the Mie scattering parameter $\pi D/\lambda$.

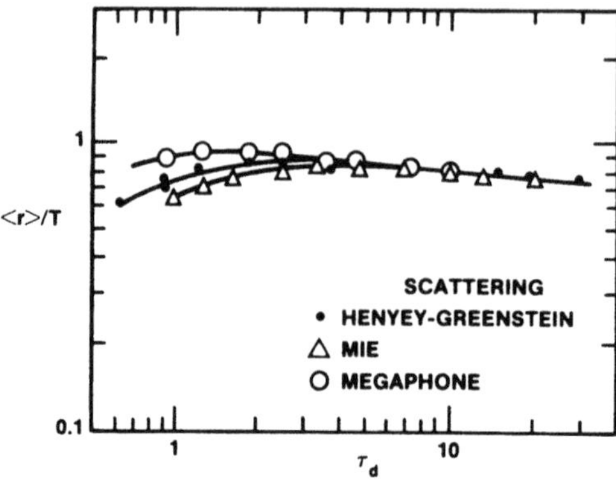

Figure 7.35. Plot of $<r>/T$ as a function of τ_d showing the saturation in spatial spreading as a cloud of fixed physical thickness T becomes optically thicker.[61]

and

$$r_c = 0.78 D_d (\tau_d)^{0.93} \tag{7.2.77}$$

respectively. Recalling the relationship between the physical thickness and mean free path, the above equations can be rewritten as

$$\bar{r} = 0.92 z (\tau_d)^{-0.08} \tag{7.2.78}$$

and

$$r_c = 0.78 z (\tau_d)^{-0.07} \tag{7.2.79}$$

respectively. With this normalization, we see that the normalized spatial spreading saturates and even decreases as a cloud of fixed physical thickness becomes more optically dense. This is shown clearly in Figure 7.35 for $\bar{r}/z$ and in Figure 7.36 for r_c/z.

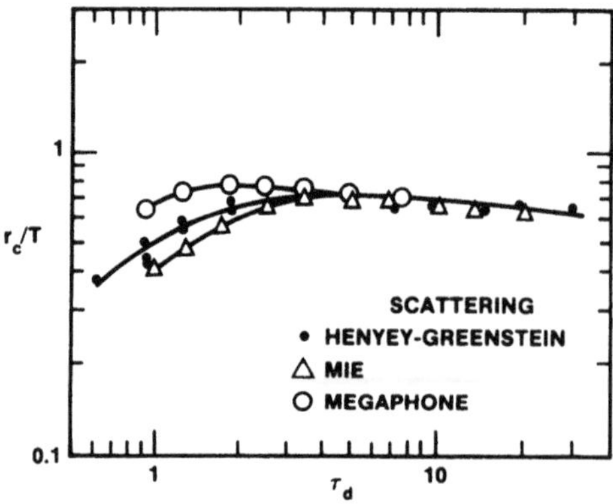

Figure 7.36. Plot of r_c/T as a function of τ_d showing the saturation in spatial spreading as a cloud of fixed physical thickness T becomes optically thicker.[61]

Danielson *et al.* showed in their 1969 work[92] that the transmittance for thick clouds was inversly dependent on the medium's diffusion thickness τ_d. Bucher verified this result through Monte Carlo simulation and, in fact, developed analytical expressions for this effect. In particular, it can be shown that the optical transmittance for thick clouds is given by

$$T_c \approx \frac{A(\varphi_s)}{1.42 + \tau_d} \tag{7.2.80}$$

where

$$A(\phi_s) = 1.69 - 0.5513\phi_s + 2.7173\phi_s^2 - 6.9866\phi_s^3 + 7.1446\phi_s^4 - 3.4249\phi_s^5 + 0.6155\phi_s^6 \tag{7.2.81}$$

In the above, φ_s is the signal zenith angle (relative to the cloud top normal) in radians. Figure 7.37 gives the cloud transmittance as a function of optical thickness for $\overline{\cos\theta} = 0.875$ and various zenith angles. It is obvious that the cloud transmittance is reduced approximately 6 dB for $\varphi_s = 0$ to $\varphi_s = 90°$. It is interesting to note that Equation (7.2.81) follows $(\cos\varphi_s)^{1/2}$ for $\varphi_s < 85°$.

The last engineering parameter Bucher chose to investigate was multipath time spreading in clouds. In characterizing their effect, he found it necessary to use three statistical measures: the mean pulse width $\bar{t}$, its standard deviation γ_t, and the width of the most intense portion of the time

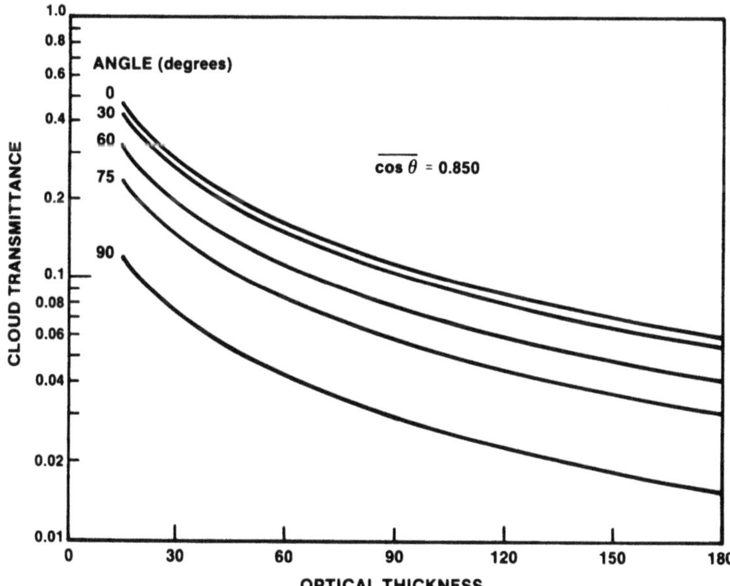

Figure 7.37. Cloud transmittance as a function of optical thickness for various zenith angles.

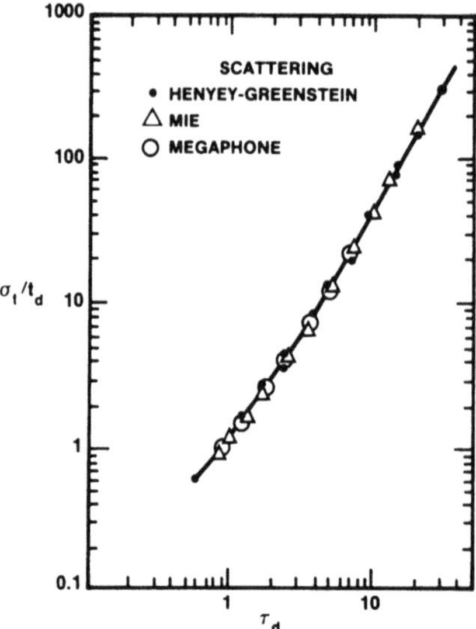

Figure 7.38. σ_t / D_d as a function of τ_d. The 90% confidence limits on all data points are ±5% or better.[61]

distribution. This latter parameter, which is the multipath time spread L, is defined as the shortest range of multipath time values which encompasses 63% of the total transmitted light. Figures 7.38–7.40 give σ_t, $\bar{t}$, and L, respectively, as a function of diffusion depth τ_d. The statistical distributions of time yielding these results were typically of the form $t\,e^{-kt}$. It is apparent

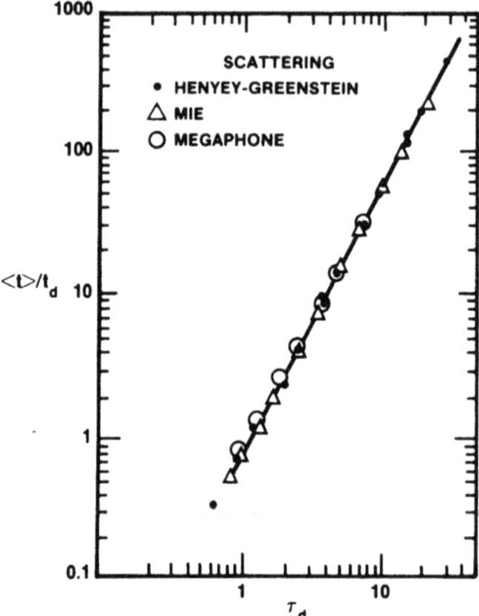

Figure 7.39. $\langle t \rangle / Dd$ as a function of τ_d. The 90% confidence limits on all data points are ±5% or better.[61]

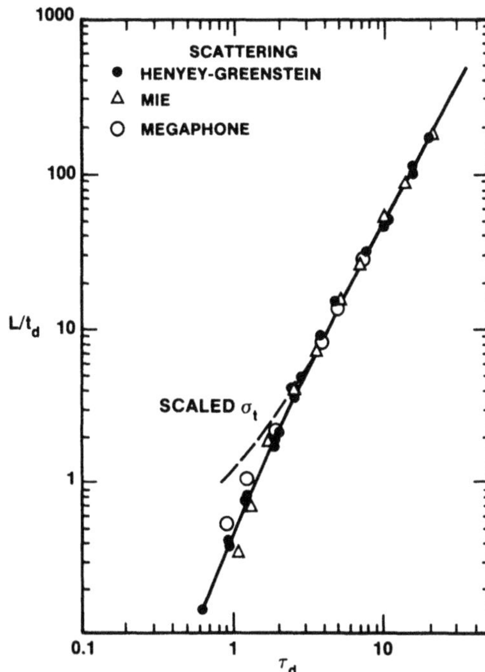

Figure 7.40. L/D_d as a function of τ_d. The 90% confidence limits on all data points are ±5% or better.[61]

that all three scattering functions again yield similar results when τ_d exceeds 2. The normalization factor t_d is given as

$$t_d = \frac{D_d}{c_0} \tag{7.2.82}$$

where c_0 is the speed of light in air. Curve fitting these results, Bucher found that

$$\bar{t} = 0.62 t_d (\tau_d)^{1.94} \tag{7.2.83}$$

$$\sigma_t = 0.64 t_d (\tau_d)^{1.81} \tag{7.2.84}$$

$$L = 0.74 t_d (\tau_d)^{1.83} \tag{7.2.85}$$

for $\tau_d > 3$. Clearly, the above can be rewritten as

$$\bar{t} = 0.62 z (\tau_d)^{0.94} \tag{7.2.86}$$

$$\sigma_t = 0.64 z (\tau_d)^{0.81} \tag{7.2.87}$$

$$L = 0.74 z (\tau_d)^{0.83} \tag{7.2.88}$$

When absorption becomes involved in the multiple-scattering process, the angular spreading decreases due to the increased probability of light being absorbed due to the additional path induced by particulate scattering. Lee *et al.* have assessed the influence of particulate absorption on multipath time spread and cloud transmittance.[90,91] In particular, they found that

$$\bar{t} = \frac{1}{c_0} \frac{0.34 D' \tau_d^{1.9}/(1-g)}{0.49 + (1-\omega_0)\tau_d^{1.1}/(1-g)} \tag{7.2.89}$$

and

$$\sigma_t = \frac{1}{c_0} \frac{0.58 D \tau_d^{1.5}/(1-g)}{0.49 + (1-\omega_0)\tau_d^{1.1}/(1-g)} \tag{7.2.90}$$

for the pulse broadening incurred in absorbing media, and

$$T_c = \frac{A(\phi_s)}{1.42 + \tau_d} \frac{2k(\tau + g_1)\, e^{-k(\tau+g_1)}}{1 - e^{2k(\tau+g_1)}} \tag{7.2.91}$$

for cloud transmittance where

$$k' = [3(1-g)(1-\omega_0)]^{1/2} \tag{7.2.92}$$

$$g_1 = \frac{1.42}{(1-g)} \tag{7.2.93}$$

Equations (7.2.90) and (7.2.91) are illustrated graphically in Figures 7.41 and 7.42, respectively. It is apparent from these plots that pulse broadening is more sensitive to the effects of absorption than cloud transmittance is. However, both show that a little absorption can significantly affect light propagation for optical thickness greater than 20. This point was brought out in our apparent properties discussion in Chapter 6.

In terms of water propagation, Lee *et al.* investigated propagation in the diffusion regime.[91] Recall that diffusive multiple scattering creates radiance distributions of the form

$$N(z, \theta) = S_1 N(\theta) \exp\{-kz\} \tag{7.2.94}$$

for downwelling radiance. Using typical Jerlov IB/II water characteristics and volume scattering functions, these researchers were able to develop a representative diffusion pattern for these situations. Specifically, they found that

$$N(\theta) \simeq \exp\{-\theta^2/\sigma^2\}/\pi\sigma^2 \tag{7.2.95}$$

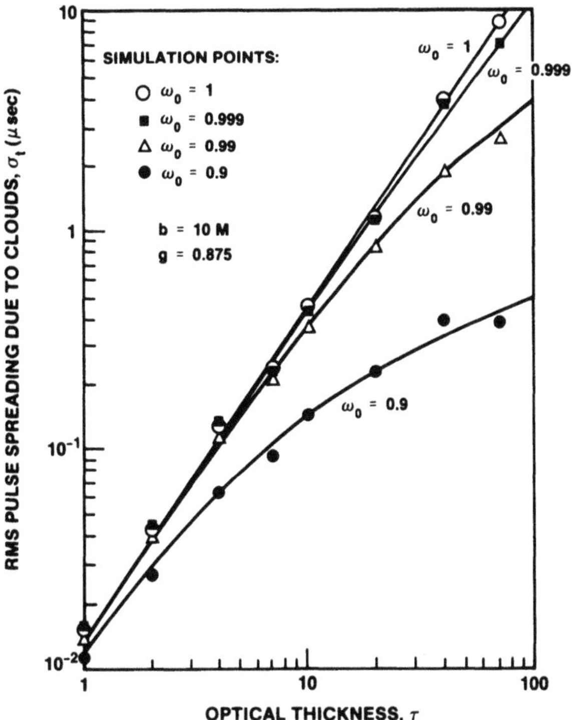

Figure 7.41. Pulse spreading by clouds.

where

$$\sigma = 0.6539 \text{ radians } (37.8°)$$

To check the validity of this expression, we calculate the normalized received power for an underwater receiver at depth. This yields

$$f_{sg}(\theta^w_{fov}) = \frac{2\pi \int_0^{\theta^w_{fov}} N(\theta') d\cos\theta'}{2\pi \int_0^{\pi/2} N(\theta') d\cos\theta'}$$

$$\approx 1.077311 I_0 - 0.178888 I_1 + 0.008198 I_2 \qquad (7.2.96)$$

where

$$I_0 = 1 - \exp\{-(\theta^w_{fov})^2/0.435336\}$$

$$I_1 = -(\theta^w_{fov})^2(1 - I_0) + 0.435204 I_0$$

$$I_2 = -(\theta^w_{fov})^4(1 - I_0) + 0.7580704 I_1$$

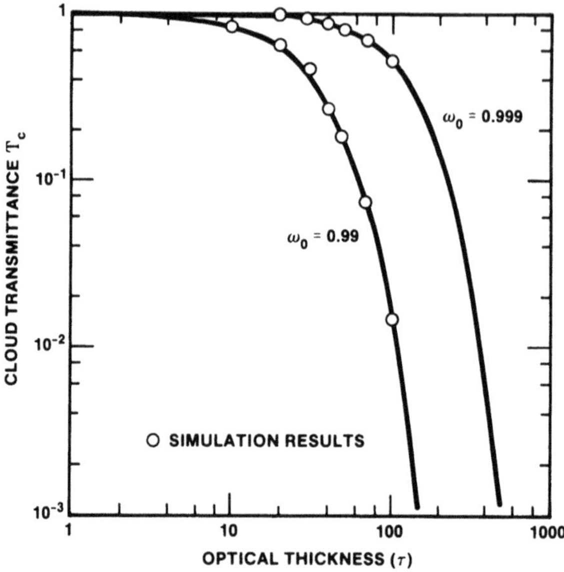

Figure 7.42. Cloud transmittance versus optical thickness.

and θ_{fov}^{w} is the receiver field of view in water. Figure 7.43 is a comparison of Equation (7.2.96) with a representative set of data taken from Ref. 30. It is apparent that the comparison is quite good.

One issue that we have not discussed is the angle of incidence effect on diffuse propagation. Again from the above simulation work, it has been

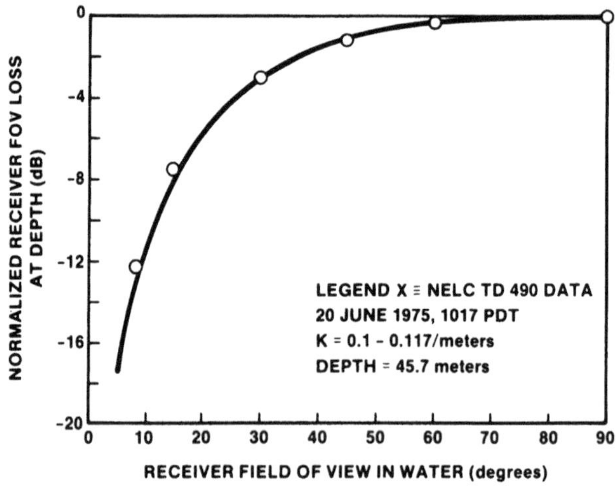

Figure 7.43. Comparison of normalized receiver field-of-view loss with experimental data.

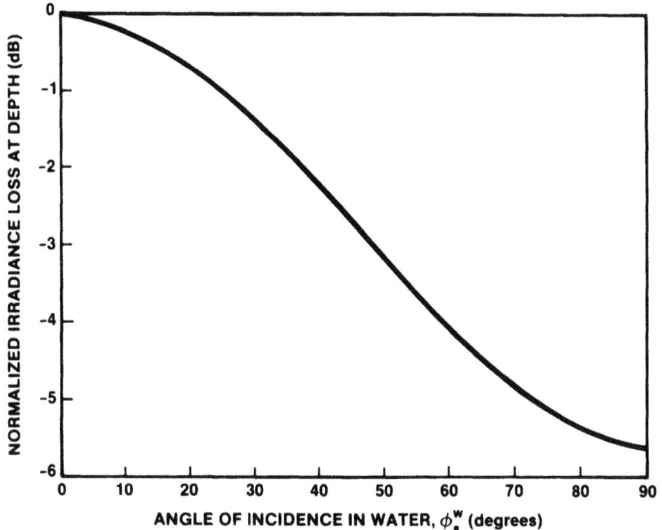

Figure 7.44. Normalized irradiance loss at depth.

found that

$$S_1 \propto f_w(\theta_S^2) \tag{7.2.97}$$

with

$$f_w(\theta_S^w) = 1.0 - 9.72 \times 10^{-4}\theta_S^w - 4.28 \times 10^{-4}(\theta_S^w)^2 \\ + 6.04 \times 10^{-6}(\theta_S^w)^3 - 2.4 \times 10^{-8}(\theta_S^w)^4 \tag{7.2.98}$$

and θ_S^w is the incidence angle in water given in degrees. Figures 7.44 illustrates normalized irradiance loss as a function of incidence angle. Figure 7.45 compares this equation, when inserted in Equation (7.2.94), with experimental lake data taken by Baker and Smith.[(93)] Again we see a very good fit between the data and predictions.

Let us now investigate the coupling medium between the atmospheric and marine channels, the air/sea interface.

7.3. The Air/Sea Interface

Based on the mathematical techniques one can draw upon to analyze radiance propagation in the marine and atmospheric channels, three basic

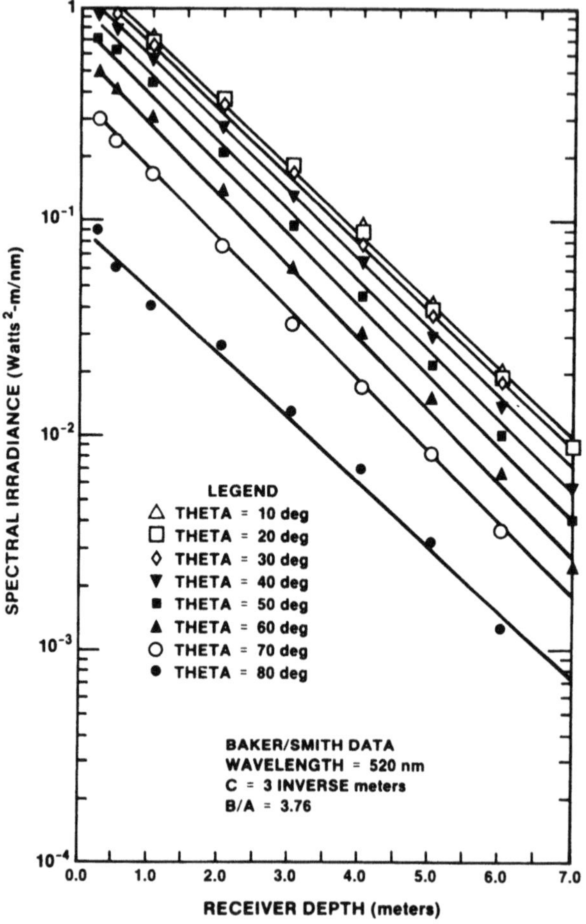

Figure 7.45. Comparison of Monte Carlo-based radiance model with experimental data.

methods for characterizing the transferal of radiance from the atmosphere to the ocean emerge, and these will be reviewed in this section. Two are based on geometrical optics; the other on irradiance transmittance.

7.3.1. Geometrical Optics Approach to Channel Coupling

In this subsection, a two-dimensional geometrical optics model will be developed to determine the effect of surface irregularities on beam spreading, pointing, and scintillation when traversing the boundary.[(3)] To do so, we consider the following model of an element of surface (Figure 7.46). A ray of light, γ degrees off the normal, impinges upon the surface whose

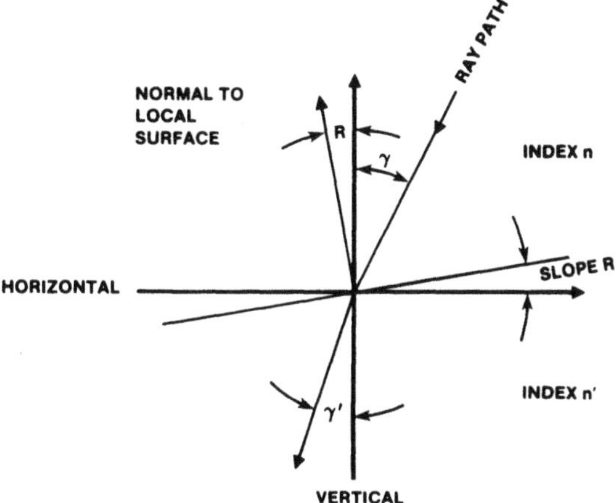

Figure 7.46. Model of an element of the surface.

local slope is R degrees from the horizontal. By Snell's law

$$n' \sin(\gamma + R) = n \sin(\gamma + R) \tag{7.3.1}$$

and

$$\gamma' = \sin^{-1}\left[\frac{n}{n'} \sin(\gamma + R)\right] - R \tag{7.3.2}$$

which is valid for both positive and negative slopes. γ and γ' are always taken with respect to the true vertical. If R is a random variable, the statistics of the slope must also be factored in. We do this in the following manner. Given a sample R from the set of possible slopes, γ' is well defined. That is, the probability of γ' conditioned upon γ and R is

$$p(\gamma'/\gamma, R) = \delta\left(\gamma' - \left\{\sin^{-1}\left[\frac{n}{n'} \sin(\gamma + R)\right] - R\right\}\right) \tag{7.3.3}$$

We arrive at the angular distribution of γ', given γ, by averaging over the variable R. Thus,

$$\begin{aligned} p(\gamma'/\gamma) &= \int_{-\infty}^{\infty} dR p_R(R) p(\gamma'/\gamma, R) \\ &= \int_{-\infty}^{\infty} dR p_R(R) \cdot \delta\left(\gamma' - \left\{\sin^{-1}\left[\frac{n}{n^1} \sin(\gamma + R)\right] - R\right\}\right) \end{aligned} \tag{7.3.4}$$

where $p_R(R)$ is the probability density of R.

Rigorously, this is merely the change of variables in the density $p_R(R)$ from R to γ' using

$$R = \tan^{-1}\left[\frac{\frac{n}{n'}\sin\gamma - \sin\gamma'}{\cos\gamma' - \frac{n}{n'}\cos\gamma}\right] = R(\gamma', \gamma) \tag{7.3.5}$$

Thus,

$$\begin{aligned} p(\gamma'/\gamma) &= P_R[R(\gamma', \gamma)]\left|\frac{dR(\gamma', \gamma)}{d\gamma'}\right| \\ &= P_R\left\{\tan^{-1}\left|\frac{\frac{n}{n'}\sin\gamma - \sin\gamma'}{\cos\gamma' - \frac{n}{n'}\cos\gamma}\right|\right\} \\ &\quad\times\left|\frac{\frac{n}{n'}\cos(\gamma-\gamma') - 1}{1 + \frac{n}{n'}^2 - 2\frac{n}{n'}\cos(\gamma-\gamma')}\right| \end{aligned} \tag{7.3.6}$$

Knowing $p(\gamma'/\gamma)$, we can compute the average spreading and offset of a ray incident at the angle γ. This becomes

$$\begin{aligned} \text{Average offset} &= \int \gamma' p(\gamma'/\gamma)d\gamma' = \overline{\gamma'} \\ &= \int\left\{\sin^{-1}\left[\frac{n}{n'}\sin(\gamma+R)\right] - R\right\}p_R(R)\,dR \\ &= \int \sin^{-1}\left[\frac{n}{n'}\sin(\gamma+R)\right]p_R(R)\,dR - \bar{R} \end{aligned} \tag{7.3.7}$$

Defining $\overline{\gamma^2}$ as

$$\begin{aligned} \overline{\gamma^2} &= \int \gamma'^2 p(\gamma'/\gamma)\,d\gamma' \\ &= \int\left\{\sin^{-1}\left[\frac{n}{n'}\sin(\gamma+R)\right] - R\right\}^2 p_R(R)\,dR \end{aligned} \tag{7.3.8}$$

the rms spread becomes

$$\Delta = (\overline{\gamma'^2} - \overline{\gamma'}^2)^{1/2} \tag{7.3.9}$$

There are some practical limitations to these results which require modification (Figure 7.47).

A ray of light with zenith angle γ will never intercept a wave whose slope is greater than $2\pi - \gamma$ because of wave obscuration. However, the ray will still penetrate the interface with probability one. Consequently, the limits of integration for R are set at $[-2\pi, 2\pi - \gamma]$, and the density $p_R(R)$ should be modified to that of

$$\frac{p_R(R)}{\int_{-\pi/2}^{\pi/2-\gamma} P_R(R)\,dR} = \begin{cases} p_R(R) & \text{for } -\pi/2 \leqslant R \leqslant \pi/2 - \gamma \\ 0, & \text{elsewhere} \end{cases} \tag{7.3.10}$$

The results in Equations (7.3.6) to (7.3.8) would then be modified by replacing $PR^{(R)}$ with Equation 7.3.10. In general, the results presented can be simplified by only considering those values of $(\gamma + R) < 45°$. This corresponds to the major operational requirements and gives good engineering insight into the behavior of a ray going through the air/sea interface. For this case

$$p(\gamma'/\gamma) \approx p_R\left[\frac{\frac{n}{n'}\gamma - \gamma'}{1 - \frac{n}{n'}}\right]\left|\frac{1}{1 - \frac{n}{n'}}\right|$$

$$\overline{\gamma'} \approx \frac{n}{n'}\gamma + \left(1 - \frac{n}{n'}\right)\bar{R}$$

$$\Delta^2 \approx \left|1 - \frac{n}{n'}\right|^2 \operatorname{Var}[R] \tag{7.3.11}$$

Notice that $|1 - (n/n')| < 1$ for the air/sea interface (index of water = 1.33, index of air = 1), and consequently the ray spreading is appreciably less than the slope distribution of the ocean. In addition, the surface adds the contribution $[1 - (n/n')]\bar{R}$ to the normal bending due to Snell's law.

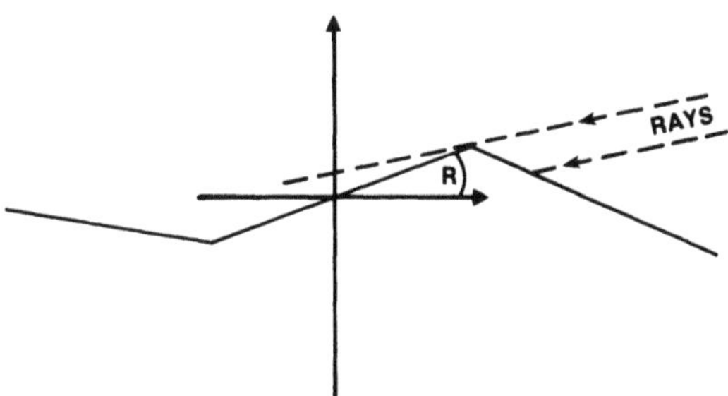

Figure 7.47. Modifications of the model.

In all cases, the model used represents an optical beam of zero cross section and zero divergence. The existing beam also has zero cross section and zero divergence but is being steered by the roughness of the surface. If this surface is the ocean, $'(\gamma'|\gamma)$ represents the average time history of the beam direction γ', with $p(\gamma'|\gamma)\, d\gamma$ the probability that it is pointing within a $d\gamma'$ interval of the γ' direction at any instant in time. This apparent beam wander would cause severe scintillation in an operating system. As the cross section of the beam increases, the refracting surface can no longer be considered locally flat and different portions of the beam are refracted at different angles. Consequently, it would be possible to average out the beam wander in the direction of the mean by spreading the beam over a larger portion of the surface. If the area of the beam is A, and the correlation length of the surface statistics is L, then there are approximately $A/\pi(L/2)^2$ identically distributed independent paths similar to diversity paths. If we further assume a depth z such that the beam cross section is greater than A, i.e., $(bz\gamma_0^2/3)7^A$, then all of the paths will overlap at the receiver. This can be analyzed in the following manner. First we notice that the probability of having the beam within an rms deviation about the mean is

$$\int_{\overline{\gamma'}-\Delta}^{\overline{\gamma'}+\Delta} p(\gamma'/\gamma)\, d\gamma' \tag{7.3.12}$$

which for the Gaussian density becomes 0.68. Thus, even if we had no time variations, the beam would be within a deviation of the mean only 68% of the time. Now suppose we pick N independent, identically distributed paths to the receiver and transmit $(1/N)$th of the power p_i in each path. Since the paths are identically distributed, the average direction of the sum is still γ'. Now, however, the variance of the sum becomes Δ^2/N, or a standard deviation for the sum of $\Delta/\bar{N}^{1/2}$ about the mean $\overline{\gamma'}$. If, for example, we set $N = 25$ and assume the central limit is approximately valid, the probability that the beam is within $\pm\Delta$ is now 0.999994. Since the correlation length is approximately the separation between independent spatial Nyquist samples, we see that

$$N \sim \frac{A}{\pi(L/2)^2} \tag{7.3.13}$$

and can be used accordingly. Further, it can be shown that the scintillation will reduce the average signal-to-noise ratio by the factor

$$\frac{1}{1+(\Delta^2/N)} \tag{7.3.14}$$

A verification of these results and the relationship in Equation (7.3.13) was shown qualitatively in Refs. 7 and 30.

Finally, we can interpret the function $p(\gamma'|\gamma)$ as a beam spreading factor. Thus, the output beam after traversing the surface will be

$$f(\gamma', r) = \int d\gamma \gamma p(\gamma', \gamma) f(\gamma, r) \tag{7.3.15}$$

or an average over all input ray directions weighted by the relative intensity. Notice that we have not restricted the results with regard to which medium corresponds to air and which to water. When going from air to water, set $n = 1$, $n' = 1.33$, and when going from water to air, set $n = 1.22$ and $n' = 1$. Then the computation of the beam moments after traversing the surface yields

$$\overline{\gamma'} = \frac{\int \gamma' f(\gamma', r)\, d\gamma'}{\int f(\gamma', r)\, d\gamma'}$$

$$\mathrm{Var}\,[\gamma'] = \frac{\int (\gamma' - \overline{\gamma'})^2 f(\gamma', r)\, d\gamma'}{\int f(\gamma', r)\, d\gamma'} \tag{7.3.16}$$

The results derived in this section were performed for a one-dimensional surface. To extend them to a three-dimensional surface is straightforward if we restrict ourselves to Cartesian coordinates. The variable R would then become the pair $R = (x, y)$, and the two-dimensional results would carry over to each of the orthogonal coordinates. The interpretation would then be one of projecting the true slope distribution onto the Cartesian coordinate system. Although simple in theory, the actual computations are difficult. If we use the linearization implicit in Equation (7.3.11), this problem is greatly simplified.

7.3.2. Irradiance Transmittance through the Air/Sea Interface

We first consider the transmission across a calm surface. Standard references show the following Fresnel transmission coefficient T_F for propagation from a medium (air) where the refractive index is unity to one where the refractive index is n ($\frac{4}{3}$ for water):

$$T_F(\theta_0) = 1 - A\left[\frac{(n^2 - \sin^2\theta_0)^{1/2} - \cos\theta_0}{(n^2 - \sin^2\theta_0)^{1/2} + \cos\theta_0}\right]^2 - B\left[\frac{(n^2 - \sin^2\theta_0)^{1/2} - n^2\cos\theta_0}{(n^2 - \sin^2\theta_0)^{1/2} + n^2\cos\theta_0}\right]^2, \tag{7.3.17}$$

where θ_0 is the angle of incidence with respect to the normal to the surface.[85] Thus

For the transverse electric (TE) mode: $A = 1$, $B = 0$ (7.3.18a)

For the transverse magnetic (TM) mode: $A = 0$, $B = 1$ (7.3.18b)

For unpolarized light: $A = B = \frac{1}{2}$ (7.3.18c)

The transmissivity of Equation (7.3.17) is plotted as a function of θ_0 for the three modes in Figure 7.48 for $n = \frac{4}{3}$. For incidence angles $0 < \theta_0 < 60°$:

1. Neglecting polarization effects produces a maximum error of ~7%.
2. The transmission coefficient for unpolarized light remains constant at $T_F = 0.98$ within approximately 2%.

If we now allow the surface to have slopes, then from Figure 7.48 it is clear that as long as the angle between the (local) normal to the surface and the incident radiation remains smaller than 60°, the transmissivity through the surface will remain approximately 98% (in the absence of wave occlusion effects). Losses at the air/sea interface are significant only for large incident angles with steep slopes or high winds. Calculations quantifying such losses have been made by Gordon[94] using the empirical relation

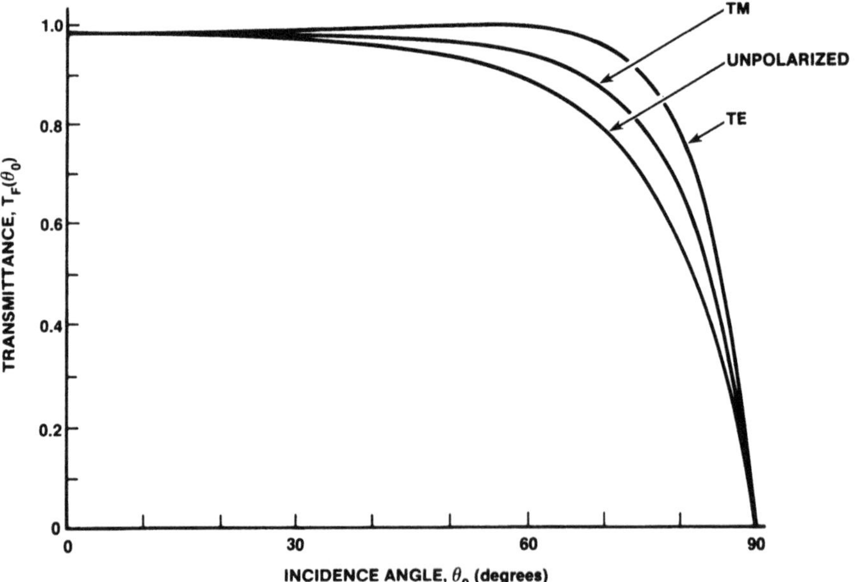

Figure 7.48. Fresnel transmission coefficient versus incidence angle. TE: transverse electric; TM: transverse magnetic.

suggested by Cox and Munk[95] from open-ocean measurements of slope versus wind speed:

$$\sigma_s = 0.0403\ V_w^{1/2} \qquad \text{(upwind)} \tag{7.3.19a}$$

and

$$\sigma_s = 0.0315\ V_w^{1/2} \qquad \text{(crosswind)} \tag{7.3.19b}$$

where V_w is the wind speed in knots, measured at a 41-ft altitude (the anemometer height normally used by ships at sea), and σ_s is the rms value (radians) of the approximately normal slope distribution for capillary waves. Her results are shown in Table 7.7 and Figure 7.49. It is apparent from Figure 7.49 that $T_{a/s}$ is fairly insensitive to wind speed. Using 8 m/s as the representative data set, $T_{a/s}$ can be expressed as a function of incident angle φ_s by

$$T_{a/s}(\phi_S) = 0.97186 - (0.11761 \times 10^{-2})\phi_S - (0.65066 \times 10^{-4})\phi_S^2 - (0.211011 \times 10^{-5})\phi_S^3 + (0.19292 \times 10^{-7})\phi_S^4 \tag{7.3.20}$$

Table 7.7. $T_{a/s}$ Time-Averaged Downlink Air/Sea Interface Transmittance

Signal zenith angle in air	Wind speed								
	0	1.03	2.06	4.12	7.21	10.3	13.4	16.5	19.6 m/s
	0	2	4	8	14	20	26	32	38 knots
0	0.979	0.977	0.976	0.974	0.970	0.967	0.963	0.960	0.956
5	0.975	0.974	0.972	0.970	0.966	0.963	0.959	0.956	0.952
10	0.964	0.962	0.961	0.959	0.955	0.951	0.948	0.944	0.941
15	0.945	0.944	0.943	0.940	0.936	0.933	0.929	0.926	0.922
20	0.920	0.918	0.917	0.914	0.910	0.907	0.903	0.899	0.896
25	0.887	0.885	0.884	0.881	0.877	0.873	0.870	0.866	0.863
30	0.847	0.845	0.844	0.841	0.837	0.833	0.829	0.826	0.822
35	0.800	0.798	0.797	0.794	0.790	0.786	0.782	0.779	0.775
40	0.747	0.745	0.743	0.741	0.736	0.733	0.729	0.725	0.722
45	0.687	0.685	0.684	0.681	0.677	0.673	0.669	0.666	0.663
50	0.620	0.619	0.617	0.615	0.611	0.608	0.605	0.602	0.599
55	0.548	0.546	0.545	0.543	0.540	0.538	0.536	0.534	0.532
60	0.469	0.468	0.468	0.466	0.465	0.464	0.464	0.464	0.463
65	0.385	0.385	0.385	0.386	0.387	0.389	0.391	0.393	0.395
70	0.295	0.298	0.299	0.303	0.310	0.315	0.321	0.325	0.329
75	0.203	0.209	0.214	0.244	0.236	0.247	0.255	0.262	0.268
80	0.113	0.126	0.136	0.153	0.172	0.186	0.197	0.206	0.213
85	0.0361	0.0610	0.0751	0.0969	0.119	0.135	0.148	0.157	0.165
90	0	0.0265	0.0390	0.0594	0.0809	0.0961	0.108	0.117	0.124

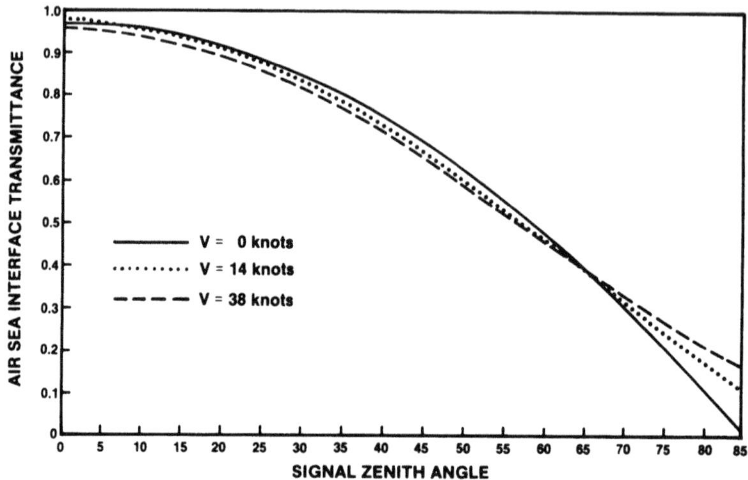

Figure 7.49. Graph of time-averaged downlink air/sea interface transmittance.

In Equation (7.3.20), φ_S is expressed in degrees.

For thin-cloud scenarios, the air/sea transmittance is given by

$$T'_{a/s} = \int_0^{\pi/2} T_{a/s}(\phi_S) f_w(\theta_S^w) \cos\phi_S \, d\phi_S \tag{7.3.21}$$

where

$$\theta_S^w = \sin^{-1}\left(\frac{1}{1.33}\phi_S\right) \tag{7.3.22}$$

7.3.2.1. Light Transmittance through Sea Foam

An additional effect on air/sea transmissivity is due to the interaction of the wind and the sea surface: the appearance of sea surface foam or whitecaps, as well as thin foam streaks in the direction of the wind. Sea foam is readily apparent to anyone flying at low altitude over the ocean on a windy day; it appears in blotches of the order of meters that are highly reflective and therefore would present a considerable transmission loss to a beam passing through them.

The difficulty in quantifying the significance of foam is the variability with both wave spectrum and wind speed. An empirical model for the fraction of ocean area covered (whitecaps and foam streaks) as a function of wind speed has been developed.[96] Assuming 100% reflectivity for the foam, the model can be used to infer an area-averaged interface transmissivity. That result, taken from Ref. 97, is illustrated in Figure 7.50. The

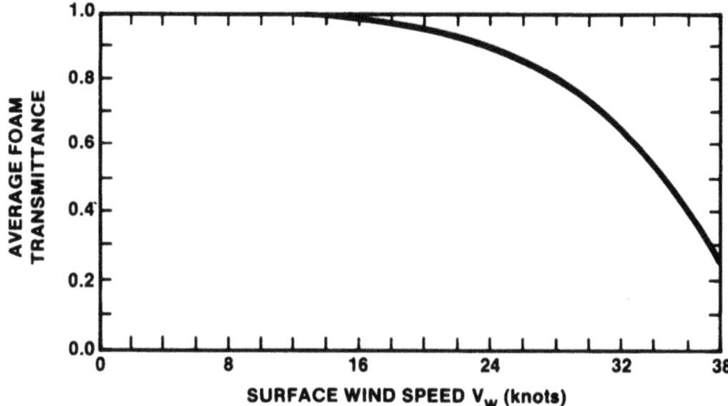

Figure 7.50. Average transmittance due to wind-driven foam and streaks.

(one-way) transmissivity is reduced by 3 dB at a wind speed of approximately 32 knots. Of the two wind-driven effects, the loss due to foam is greater than that due to capillary wave slopes. The total air/sea interface transmittance is the product of this term and the Fresnel transmittance $T_{a/s}$. However, foam is certainly subject to large and rapid variations, and caution should be exercised in inferring better than a gross estimate of the magnitude of this effect from the assumed model for wind velocities in excess of approximately 25 knots.

7.3.3. A Green's Function Approach to Air/Sea Interface Radiance Coupling

Lutomirski and Snead[7] have developed a technique which, in principle, allows the calculation of the resulting radiance distribution for an arbitrary specification of the wave slope as a function of position on the surface.

Figure 7.51 shows the principal planes at which the radiance is computed. The radiance just above the surface is represented as the product of two terms: an irradiance term I dependent on position in that plane $I(r')$, and a ray term $R(k)$ dependent on the distribution of ray angles at that point. If the initial irradiance distribution is Gaussian over a plane inclined at angle θ_0 with respect to the zenith (the primed system), then we can represent the radiance f over a horizontal plane just above the surface by

$$N^+ = I^+R^+ \tag{7.3.23}$$

where

$$I^+(x, y) = \frac{1}{2\pi\sigma_s^2} \exp\left\{ -\frac{y^2 + x^2 \cos^2 \theta_0}{2\sigma_s^2} \right\} \cos \theta_0 \tag{7.3.24}$$

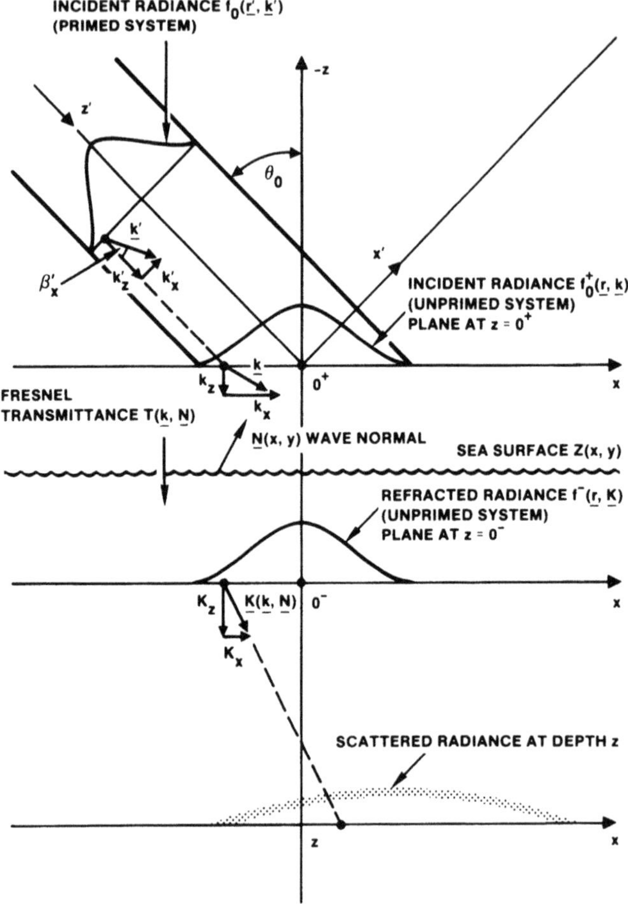

Figure 7.51. Principal planes at which radiance is computed.

and

$$R^+ = \delta(k_y)\delta(k_x - \sin\theta_0) \tag{7.3.25}$$

In Equations (7.3.24) and (7.3.25), we use Cartesian coordinates x, y, z, with z denoting depth in the water, so k_y, k_x are the projections of unit vectors along the x- and y-directions, respectively; σ_s is the standard deviation of the Gaussian spot; and δ is the Dirac delta function.

If we use $P(x, y)$ to denote the upward unit vector normal to the ocean surface at the point x, y, then, in terms of the local slope components m_x, m_y, we have

$$\mathbf{p} = \frac{-m_x\mathbf{e}_x - m_y\mathbf{e}_y - \mathbf{e}_z}{(1 + m_x^2 + m_y^2)^{1/2}} \tag{7.3.26}$$

Using Snell's law, it can be shown that the transformation of the unit ray vectors above the surface k to below-surface values $\boldsymbol{\gamma}$ is given by

$$\mathbf{k} = n\{\boldsymbol{\gamma} - (\mathbf{N} \cdot \boldsymbol{\gamma})\mathbf{N} - \mathbf{N}[(\mathbf{N} \cdot \boldsymbol{\gamma})^2 - (1 - 1/n^2)]^{1/2}\} \tag{7.3.27}$$

where $n = \frac{4}{3}$. To first order in m_x, m_y, Equations (7.3.26) and (7.3.27) combine to yield

$$k_x = n(\gamma_x - \mu m_x) \tag{7.3.28}$$

$$k_y = n(\gamma_y - \mu m_y) \tag{7.3.29}$$

The downlink coupling factor

$$\mu(\theta_0) = [1 - (\sin \theta_0/n)^2]^{1/2} - (1/n) \cos \theta_0$$

is plotted in Figure 7.52. Note that $\mu(0°) = \frac{1}{4}$ is a reasonable approximation out to 20°.

The radiance just beneath the ocean surface is then

$$N^- = I^- R^- \tag{7.3.30}$$

where

$$I^- = T_f(\theta_0) I^+ \tag{7.3.31}$$

(T_f being the Fresnel transmission coefficient) and

$$R^- = \delta(\gamma_y - \mu m_y)\delta(\gamma_x - \mu m_x - \sin \theta_0/n) \tag{7.3.32}$$

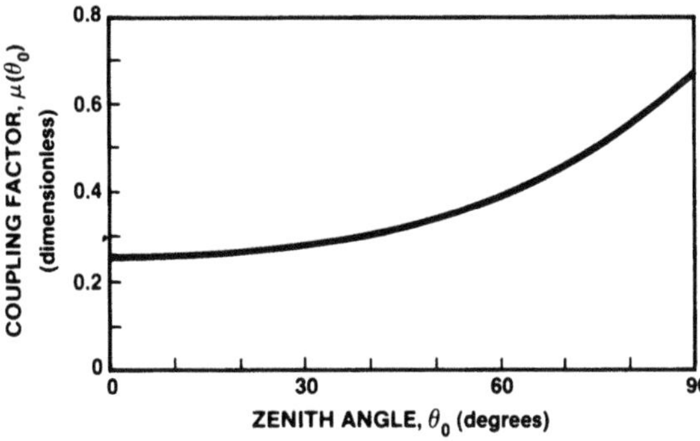

Figure 7.52. Downlink coupling factor between ray vector and wave slope.

is the ray angle distribution at (x, y). The slope components m_x, m_y will vary with position (x, y) on the surface. After the radiance just beneath the surface N is specified, the radiance over a plane at depth z beneath the surface may be calculated using the Green's function g, derived in Section 7.2.2.3. The form of the Green's function, which allows us to compute the differential configuration from N^- to the radiance at a point (x, y) at depth z, $(\mathbf{r}, z)$, and ray direction $\mathbf{K}$, is shown in Section 7.2.2.3 to be

$$dN_z(\mathbf{r}, \boldsymbol{\gamma}) = N(\mathbf{r}', \boldsymbol{\gamma}')g(\mathbf{r} - \mathbf{r}' - z\boldsymbol{\gamma}', \boldsymbol{\gamma} - \boldsymbol{\gamma}')\, d^2\mathbf{r}'\, d^2\boldsymbol{\gamma}' \quad (7.3.33)$$

and the total radiance of $\mathbf{r}$, z in the direction $\mathbf{K}$ is shown to be

$$N_z(\mathbf{r}, \boldsymbol{\gamma}) = \int\int N(\mathbf{r}', \boldsymbol{\gamma}')g(\mathbf{r} - \mathbf{r}' - z\boldsymbol{\gamma}', \boldsymbol{\gamma} - \boldsymbol{\gamma}')\, d^2\mathbf{r}'\, d^2\boldsymbol{\gamma}' \quad (7.3.34)$$

In principle, Eq. (7.3.34) completes the solution to the downlink problem in the small-angle-scattering approximation. The extent to which the magnitudes of the surface slope m_x, m_y are known, and vary with position (x, y), can be inserted into Equation (7.3.30) for N^-, and the extent to which the volume scattering functions of the water are known can be used to determine the Green's function g. The radiance (W/m^2/sr) at the point $(\mathbf{r}, z)$ in the distribution $\boldsymbol{\gamma}$ can be found by integrating over all incident directions and surface positions $\mathbf{r}$. Further, the radiance (W/m^2) at the point $(\mathbf{r}, z)$ can be found by integrating N_z over all $\boldsymbol{\gamma}$. In particular, integrating both sides of Equation (7.3.34) over $\mathbf{K}$ yields

$$I_z(\mathbf{r}) = \int N_z(\mathbf{r}, \boldsymbol{\gamma})\, d^2\boldsymbol{\gamma} = \int\int \bar{N}(\mathbf{r}', \boldsymbol{\gamma}')G(\mathbf{r} - \mathbf{r}' - z\boldsymbol{\gamma}')\, d^2\mathbf{r}'\, d^2\boldsymbol{\gamma}', \quad (7.3.35)$$

where

$$G(\mathbf{r} - \mathbf{r}' - z\boldsymbol{\gamma}') = \int g(\mathbf{r} - \mathbf{r}' - z\boldsymbol{\gamma}', \boldsymbol{\gamma} - \boldsymbol{\gamma}')\, d^2\boldsymbol{\gamma} \quad (7.3.36)$$

is a pseudo-Green's function that directly relates the irradiance at $(\mathbf{r}, z)$ to the radiance $N^-(\mathbf{r}', \boldsymbol{\gamma}')$ just beneath the surface at depth.

The advantage of this formulation, as in all Green's function-type solutions, is the separation of the geometry of the problem (which is the spatial extent and form of the radiance distribution N^-) from the propagation problem (which is defined by either g or G, which in turn determines how elements of N^- contribute to the radiance or irradiance, respectively).

7.4. System Considerations for the Optical Scatter Channel

It is clear from the previous discussions that light from a point source, after passing a few scattering lengths through water or a cloud, will subtend a solid angle to the receiver (Ω_s) of approximately one steradian about the zenith. This has two implications. First, it will be impossible to spatially discriminate between the source and the sun since both will appear at the same location with the same profile. Consequently, frequency filtering will be very important, but will be aggravated by the large field of view required. Second, we see that a field of view of 1 sr implies A/λ^2 spatial degrees of freedom so that we can only seriously consider a noncoherent communication system.

A further complication to the design problem results whenever we have to penetrate a cloud. Although clouds have negligible absorption losses, we have seen that there is a large multipath time spread imposed on the signal due to the large number of scatterings. We have seen that this pulse stretching can be anywhere from one to one hundred microseconds, thereby limiting the coherence bandwidth to under a megacycle—most likely even under ten kilocycles. Thus, to get large bandwidths one would have to resort to multiple frequencies, which is exceedingly difficult in an incoherent system.

We will take as a baseline for the design a satellite-to-submarine communication system operating in the blue-green region of the spectrum (5000 Å), which must also be able to penetrate clouds.

7.4.1. Modulation Format Selection

It has been shown that M-ary pulse-position modulation (PPM) is the optimum modulation format for background-limited optical communications.[98,99] Figure 7.53 illustrates synchronous PPM and its important relations. In a basic time period called the interpulse interval, one determines in which one of M slots, each of width ΔT seconds, the received pulse resides. For this discussion, we will ignore pulses which overlap more than one slot, i.e., pulse broadening by the optical scatter channel is assumed less than ΔT, with perfect bit synchronization. The difference between the interpulse interval and $M\Delta T$, the word time, is called the dead time and is denoted $N\Delta T$. In most situations, N is not an integer. This latter time is usually associated with the time required for the laser to recharge or some other such system consideration. $\log_2 M = k$ is the number of bits sent per PPM word or pulse. Figure 7.54 shows the binary word (011) in an octal PPM format. It is important to point out that this type of modulation assumes a mechanism for achieving time synchronization and has an average data rate equal to [Equation (3.7.8)]

$$\frac{\log_2 M}{(M + N)\Delta T} \tag{7.4.1}$$

Once the PPM coding scheme has been selected, a value for M must be

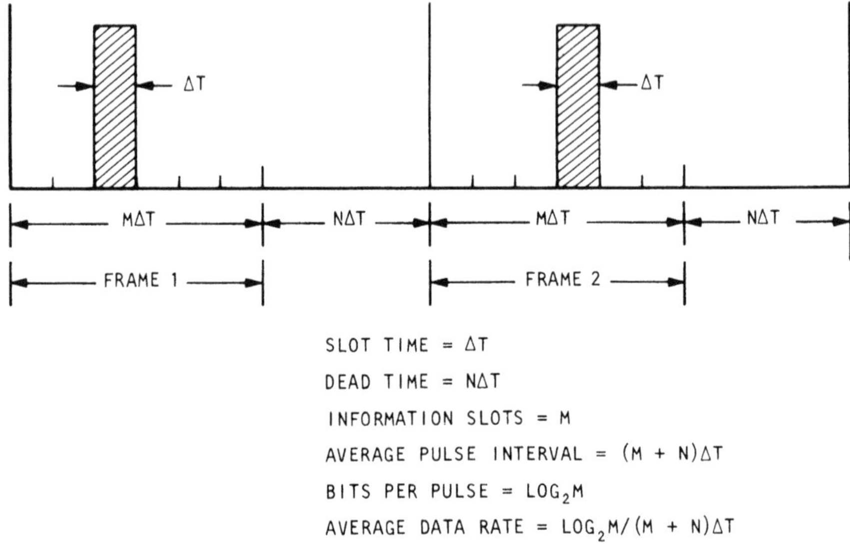

Figure 7.53. Pulse-position modulation (PPM) and its important relationships.

chosen. We will show here how one picks M to maximize the average data rate.

To maximize the data rate requires us to maximize the quantity shown above. To do this we set

$$T_D = N\Delta T$$

$$\Delta T = \alpha T_P$$

where we have defined the communication slotwidth in terms of the laser dead time. Here α is a system constraint which is imposed upon the design.

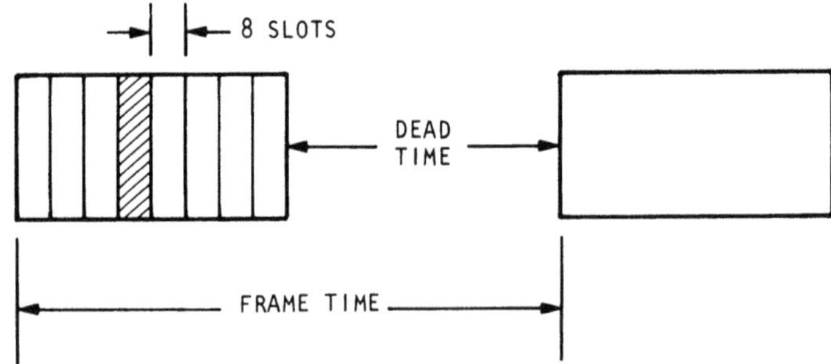

Figure 7.54. The binary word (001) in 8-ary PPM.

The pulse repetition rate R_P of the system can then be written as

$$R_P = \frac{1}{T_D(1+\alpha M)} \qquad \text{for } \alpha = \frac{\Delta T}{T_D}$$

$$R_{P_{\max}} = \frac{1}{T_D} \qquad \text{for } \alpha = 0 \tag{7.4.2}$$

This gives an average data rate equal to

$$\begin{aligned} R &= R_P \log_2 M \\ &= R_{P_{\max}} \left[\frac{\log_2 M}{1+\alpha M}\right] \end{aligned} \tag{7.4.3}$$

or in terms of the maximum pulse rate

$$\frac{R}{R_{P_{\max}}} = \frac{\log_2 M}{1+\alpha M} \equiv \beta < k \tag{7.4.4}$$

To maximize $R/R_{P_{\max}}$, Equation (7.4.4) is differentiated and set equal to zero. This yields

$$(1+\alpha M)\log_2 e = \alpha M \log_2 M$$

or

$$\alpha = \frac{\log_2 e}{M \log_2 (M/e)}$$

Figure 7.55 shows α as a function of the optimum word length in bits per pulse. Figure 7.56 gives the maximum normalized average data rate β as a function of word length in bits per pulse. It is apparent from this last figure that the normalized data rate moves to its asymptotic maximum value (equal to the number of bits per pulse sent) for k greater than 6. This means that one could potentially send as much information as one wanted by adjusting the slotwidth to the appropriate interval. Unfortunately, neither the optical scatter channel nor existing electro-optical and electronic technologies allow an arbitrary slotwidth to be employed. As indicated earlier, the former is really the limiting factor. Thus we see that given a slotwidth ΔT and a dead time T_D, the maximum data rate is

$$R_{\max} = R_{P_{\max}} \left[\frac{\log_2 M_{\text{opt}}}{1+\alpha M_{\text{opt}}}\right] \tag{7.4.5}$$

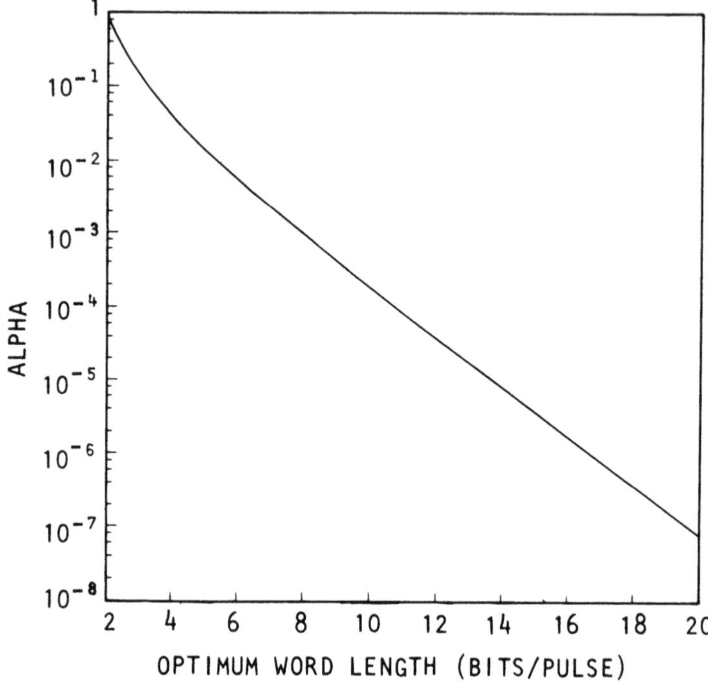

Figure 7.55. Ratio of slotwidth to system dead time as a function of optimum word length.

which can be determined from Figures 7.55 and 7.56. To transmit at a lower data rate, one can use a lower $R_{P_{\max}}$ or a smaller value of M. To operate at a higher data rate would require either a higher value for $R_{P_{\max}}$ or the use of multiple frequencies to increase M.

7.4.2. Noise Considerations

If we assume that the primary source of noise is the sun, which can be approximated as a blackbody with temperature $T = 6000$ K, we can use the earlier discussions to shown that the noise spectral density is

$$N_{\text{ob}} = \lambda^2 N_{\text{BB}}(f) = \frac{hf}{e^{hf/kT} - 1} \tag{7.4.6}$$

where $N(f)$ is the spectral radiance. Our optical bandwidth B would be determined from the multipath spread ΔT to be

$$B = \frac{1}{2\Delta T} \tag{7.4.7}$$

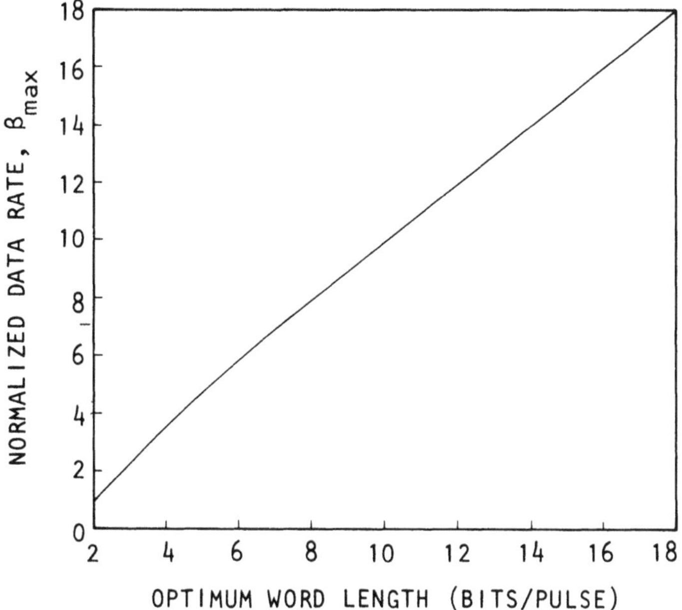

Figure 7.56. Maximum data rate as a function of optimum word length.

In practice, however, due to limitations in filter technology the actual optical system bandwidth B will be many times larger than B and consequently will allow more noise to enter than desirable. The ratio B_s/B would be the ratio of excess noise over minimum and can be several orders to magnitude. It has also been shown that the optimum way to process the current at the output of a photodetector is to count electrons. We can determine the average noise photoelectrons released as the incident noise energy E_n multiplied by the detector quantum efficiency and divided by hf. Thus, in general, we have, using Equation (7.4.6),

$$K_n = \frac{\eta En}{hf} = \frac{\eta}{hf} N_{\text{ob}}(B_s\,\Delta T)\frac{\Omega_{\text{fov}}}{\lambda^2} AL = \frac{\eta}{hf} N(f) AB_s\,\Delta T \Omega_{\text{fov}} L; \quad E_n = P_n\,\Delta T$$

where we have also included the spatial degrees of freedom as outlined in Chapter 1. Similarly, we see that the number of signal photoelectrons produced due to an incident signal energy E_s is

$$K_s = \frac{\eta E_s}{hf} \qquad E_s = P_{\text{ave}}/R_p$$

We can simplify K_n by recognizing that the scattering channel causes the noise source to be diffused to one steradian (in water) so that we can approximate $N(f)$ as $W(f)/\pi$ where $W(f)$ is the spectral irradiance.

7.4.3. Signal-to-Noise Ratio and Bit Error Probability

The most stressing part of the satellite/submarine link occurs during the daytime and is due to the presence of sunlight in the receiver field of view. We will concentrate our example on this case. For a more general treatment, see Refs. 3 and 5. The sunlight causes a high background noise level, which in turn causes a large value of K_n. For this case, we can treat the photoelectron count as a Gaussian random process and the signal count K_s as additive. It is then straightforward to show[100] that the probability of making an error in demodulating the PPM symbol, P_E [Equation (3.7.6), can be written as

$$P_E = 1 - \frac{1}{\sqrt{2\pi}} \int_{-\infty}^{\infty} \exp\{-(x - \text{SNR})^2/2\}(1 - \text{erfc}\,[x])M^{-1}\,dx \qquad (7.4.8)$$

where

$$\text{erfc}\,(x) = \frac{1}{\sqrt{2\pi}} \int_{x}^{\infty} \exp\{-t^2/2\}\,dt \qquad (7.4.9)$$

which can be approximated as[100]

$$P_E \cong \frac{(M-1)}{(\pi\text{SNR})^{1/2}} \exp\{-\text{SNR}/4\} \qquad (7.4.10)$$

with

$$\frac{S}{N} = \text{SNR} = \frac{K_s^2}{K_n} \qquad (7.4.11)$$

This is plotted in Figure 7.57. Since we are interested in bit error rate, P_B, we can relate P_B to P_E by[101]

$$P_B = \frac{2^{K-1}}{2^K - 1} P_E \qquad (7.4.12)$$

and recompute the curves as shown in Figure 7.58. Notice that as we go to a larger value of M, the signal-to-noise ratio required per bit decreases for the same performance. For the Gaussian channel, this ratio approaches -1.59 dB as $M \to \infty$. For the smaller values of M, one can also get a reduced

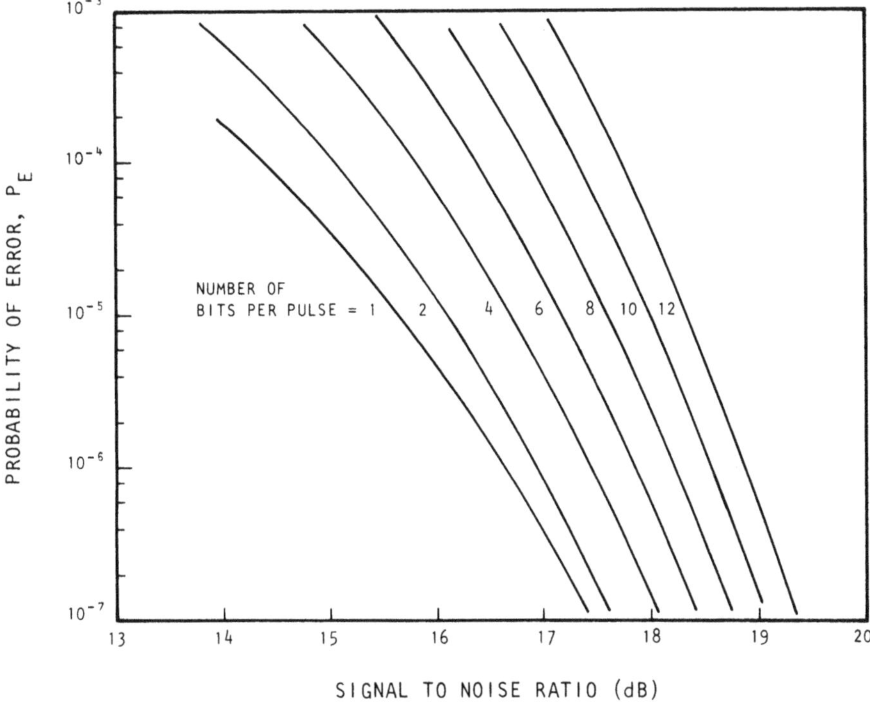

Figure 7.57. Detection probability of symbol error versus SNR.

per bit signal-to-noise ratio by using codes such as Reed–Solomon codes. These codes can yield several dB of improvement with modest complexity. However, one needs a minimum of 6 to 6.5 dB per bit signal-to-noise ratio for these codes so that one would only get 3 dB improvement if, say, $k = 8$ is selected.

7.4.4. Link Losses

Once we have determined the minimum required signal count K_s, and the expected noise count K_n, we can evaluate the required laser transmission system versus the desired system goals. Thus, for example, if we have a laser transmitter energy E_s and a repetition rate R_P, we have shown how we would optimize the transmission rate and determine the system performance. It is most convenient in optical systems, because of the high antenna gains, to consider the projected spot size of the beam in the receiver plane. We have shown that the antenna gain G is related to the projected solid angle Ω_T by

$$G = \frac{4\pi}{\Omega_T}$$

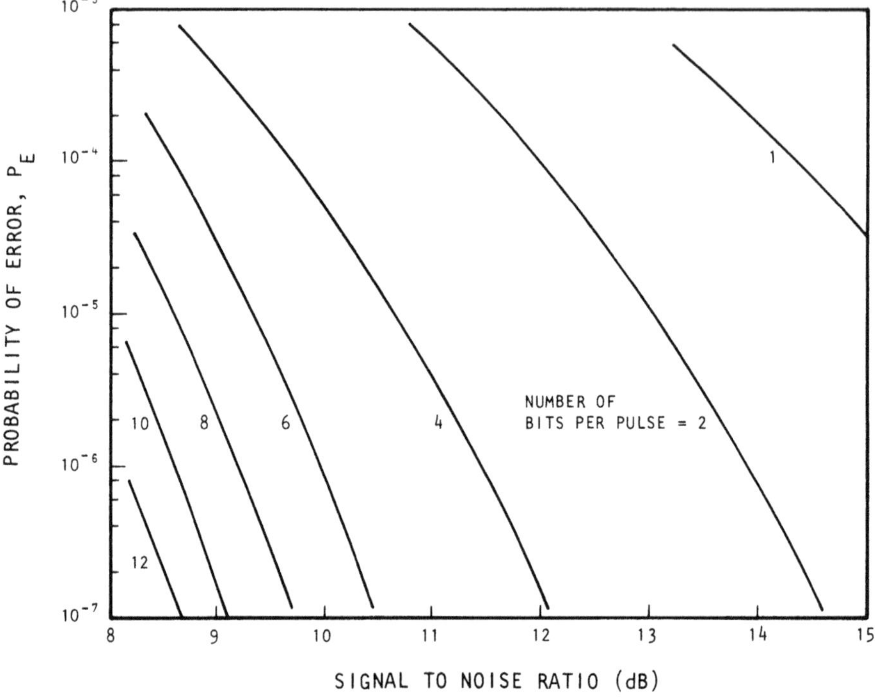

Figure 7.58. Detection probability of bit error versus SNR.

and we can usually consider the planar angle θ as $\Omega_T^{1/2}$. Thus the spot size at range R can be found to be

$$\text{Spot diameter} = \theta R = D$$

We normally wish to have as large a gain (small spot size) as possible. However, in a scattering environment there is usually a minimum spot size allowed, and this can be attributed to the spatial impulse response of the scattering medium. For clouds, the emerging spot out, D_{out}, is approximated by Equation (7.4.18) to be

$$D_{\text{out}} \simeq (D_{\text{in}}^2 + 4.7z^2\tau^{0.16})^{1/2}$$

where z is the cloud thickness. There is also a minimum spot size determined by the water, but this is at least an order of magnitude less than that for the cloud, so that for this parameter the cloud is the determining factor.

The transmission loss L will affect both the laser beam and the sunlight. For design purposes, we will follow the approach used in Ref. 3. We will determine the necessary power required and work backward to establish an available system margin or available "link loss L" that can be tolerated. This will then be used to trade off for depth penetration into the water or spot size D for greater coverage. For the background-limited condition, the word signal-to-noise ratio is equal to

$$\frac{S}{N} = \frac{\eta(P_{ave}/R_P)^2}{hf(P_n\,\Delta T)} \tag{7.4.13}$$

using Equation (7.4.11). Here P_{ave} is the average transmitter power. Assuming that both the signal and background noise are each attenuated by L,[3] we have

$$L = \left(\frac{R_p D_{out}^2}{4P_{ave}L''}\right)^2 \frac{\pi^2 W(f) hfB_s\,\Delta T\left(\frac{S}{N}\right)}{\eta A} \tag{7.4.14a}$$

$$= \left(\frac{R_p D_{out}^2}{4P_{ave}L''}\right)^2 \frac{\pi^2 W(f) hf\rho\left(\frac{S}{N}\right)}{A} \tag{7.4.14b}$$

where $\Omega_{fov} = 1$, P_{ave} is the average transmitter power, A is the receiver collector area, and L'' is the transmitter/receiver component losses. For the nominal set of numbers in Table 7.8, a set of generic curves were derived[3] and are shown in Figure 7.59. (The quantum-limited case is also included.) These curves are plotted as a function of the parameter ρ, where

$$\delta = \frac{B_s\,\Delta T}{\eta} \tag{7.4.15}$$

and B_s is the optical bandwidth in angstroms, ΔT is the pulsewidth in microseconds, and D_{out} is the spot size in kilometers.

Table 7.8. Parameters for the Standard Receiver

Pulse rate, R_p	100/second
Signal-to-noise ratio, S/N	20 (13 dB)
Average transmitter power, P_{ave}	1 (0 dBW)
Optical losses, L''	1 (0 dB)
Quantum efficiency, η	1 (0 dB)
Receiver aperture, A	1 m^2 (0 dB)

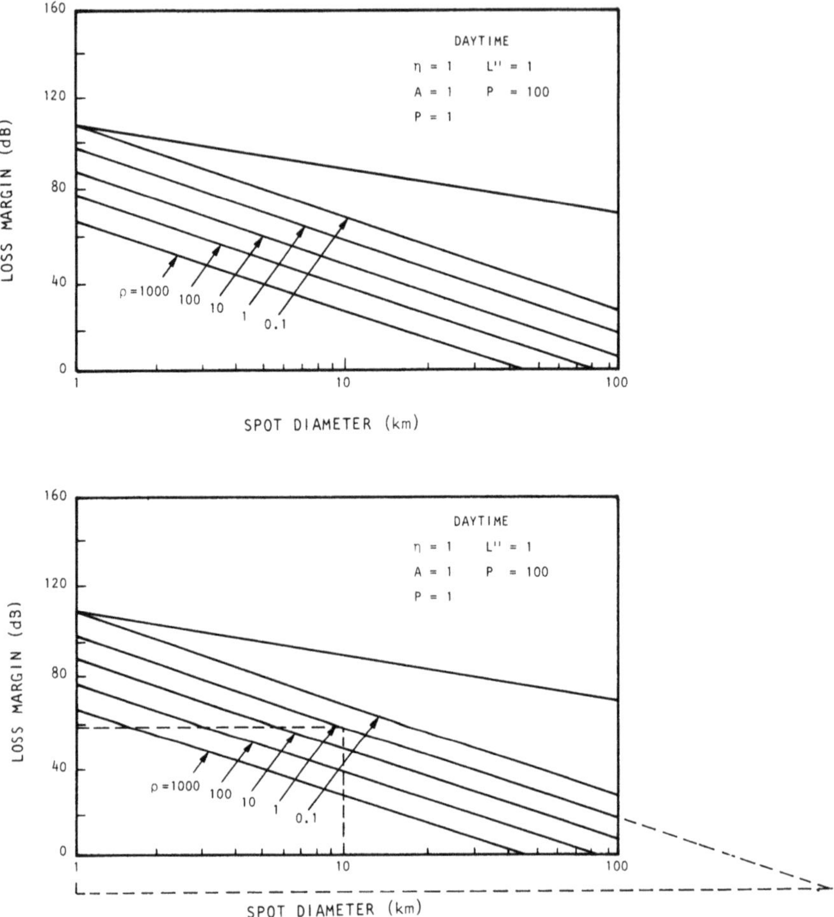

Figure 7.59. (a) Standard performance curve—daytime; (b) example system performance—daytime.

7.4.5. System Design

Let us suppose that we are to design a satellite-to-submarine communication system that is to operate at 100 bits per second with 95% availability in the Pacific Ocean. From Figure 7.60 we see that we will have to contend with optical thicknesses of about 150. From Figure 7.61 we also see that this will cause a pulse broadening to about 70 μs.[†] There are two contending lasers: the excimer laser with a dead time of 5 ms and the doubled NdYAG

[†]Notice that highest data rate possible would be $1/(7 \times 10^{-5}) = 1.43 \times 10^4$ bits/s for on-off binary modulation for $T_o = 0$.

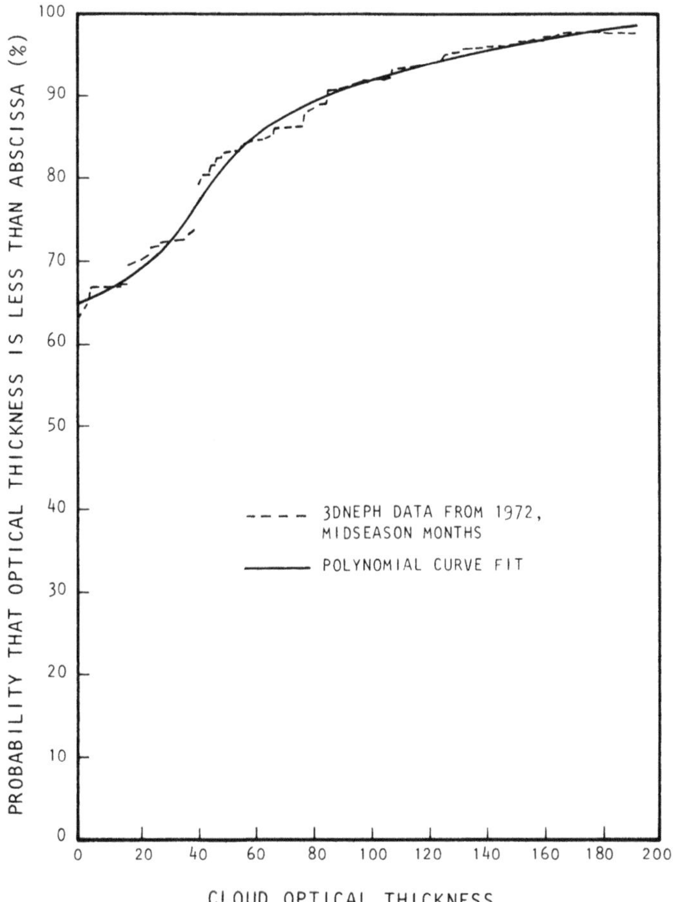

Figure 7.60. Probability of cloud optical thickness being less than abscissa off eastern coast of Hawaii.

with a dead time of 18 ms. This yields an α of 0.0039 for the NdYAG and an α of 0.014 for the excimer. Using Figure 7.55, this gives us $K_{\text{opt}} = 6.5$ for the NdYAG and $K_{\text{opt}} = 5$ for the excimer. Using Figure 7.56, this then gives us $\beta_{\max} = 6.5$ for the NdYAG and $\beta_{\max} = 4.5$ for the excimer. Thus the maximum data rates are 361 bits/s for the NdYAG and 900 bits/s for the excimer. Since we are only interested in a data rate of 100 bits/s we have a design option. Since power is of prime importance on a satellite, we will choose the largest word size M and the smallest repetition rate R_P. The result is that both lasers then yield the same design: 10 pulses per second at $k = 10$.† From Figure 7.58 we see that the per bit signal-to-noise

†$R = R_P \log_2 M$.

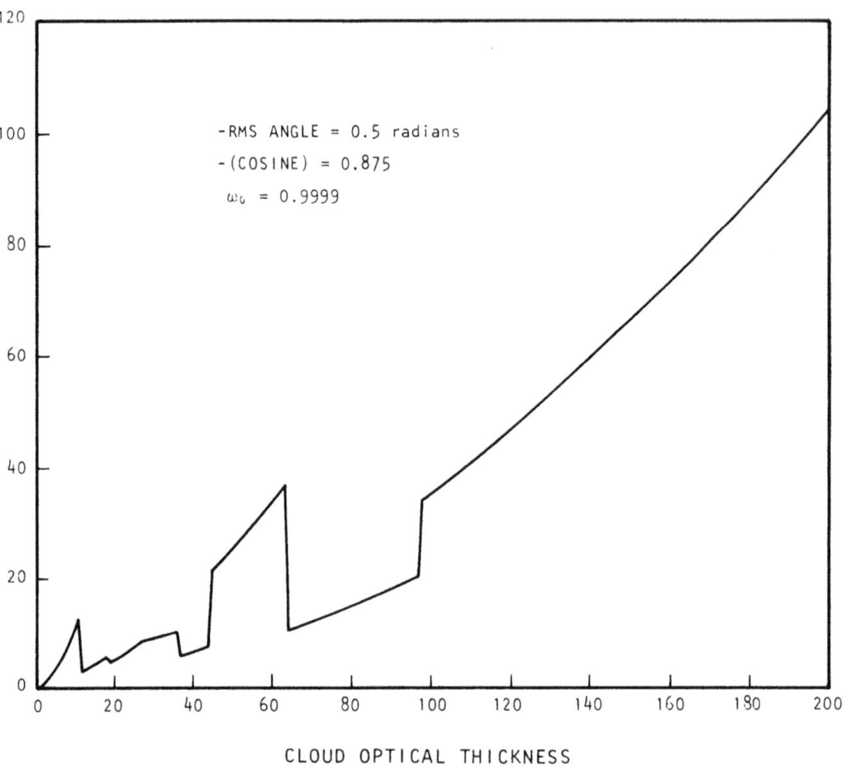

Figure 7.61. Multipath time spread; Ciany *et al.* climatology model.

ratio is about 9 dB so that we can get an additional 3 dB improvement using a Reed–Solomon code. This means that instead of needing a per pulse signal-to-noise ratio of 19 dB, we would only need 16 dB.

For the sun directly overhead and illuminating our cloud layer, the environmental loss is equal to

$$L \approx T_a T_c T_{a/s} T_w f_{sg}(\theta_{fov}^w) \tag{7.4.16}$$

where T_a is the atmospheric transmittance to the top of the cloud [Equation (7.2.8)],

$$T_a = \exp\{-0.3573\} \tag{7.4.17}$$

T_c is the cloud transmittance [Equation (7.2.80)], $T_{a/s}$ is the air/sea interface transmittance [Equation (7.3.22)], T_w is the irradiance transmittance to depth [Equation (7.2.94)], given by

$$T_w = \exp\{-kD\} \tag{7.4.18}$$

Table 7.9. Example System Design

	Margin relation to standard[a]	
	10 watt	100 watt
$R_p = 10$	−20dB	−20dB
$S/N = 16$ dB	3dB	3dB
P_{ave}	−20dB	−40dB
$L'' = 0.5$	6dB	6dB
T_{as}	+1.55dB	+1.55dB
T_e	13.05dB	13.05dB
$T_{a/s}$	+2.05dB	+2.05dB
T_w	17.36dB	17.36dB
f_{sg}	3.0dB	3.0dB
Loss relative to standard	+6.0dB	−14dB

[a] Notice that a gain is treated as a negative loss.

with k the diffuse attenuation coefficient (inverse meters) and D the receiver depth (meters), f_{sg} is the finite field-of-view receiver collection factor [Equation (7.2.96)], and $\theta_{fov}^{w} = 30°$.

If we assume a nominal surface wind speed of around 16 knots, we have the link budget shown in Table 7.9 for a 10-W and 100-W laser system, respectively, communicating with a receiver at a depth of 100 M in Jerlov I water. For the receiver, we will assume that we have a 1-m aperture and a spectral bandwidth of 0.01 Å. Thus $\rho = (70)(0.01)/1 = 0.7$.

Using Figure 7.59, we can locate the +6-dB and −14-dB points on the ordinate, and follow these lines to their intersections on the $\rho = 0.7$ curve. The spot diameters for +6-dB and −14-dB relative losses are 175 km and 500 km, respectively.

If, on the other hand, we were only interested in a 10-km spot size, we would trace 10 km up until it intersected the $\rho = 0.7$ line. This would then result in about 60 dB (80 dB) more loss margin, which could be used to gain a higher availability, greater depth, or higher data rate.

References

1. R. F. Lutomirski, Atmospheric degradation of electro-optical system performance, *Appl. Opt.* **17**, 3915-3921 (1978).
2. R. L. Fante, Electromagnetic beam propagation in turbulent media, *Proc. IEEE* **63**, 1669-1662 (1975).

3. S. Karp, Optical communications between underwater and above-surface (satellite) terminals, *IEEE Trans. Commun.* **COM-24**, 66–81 (1976).
4. A. Ishimaru, Theory and application of wave propagation and scattering in random media, *Proc. IEEE* **68**, 1030–1061 (1977).
5. L. B. Stotts and P. J. Titterton, Link Models for Space/Air-to-Subsurface Optical Communications Analysis, International Telemetering Conference, ITC/USA/80, San Diego, California, October 14–16, 1980.
6. R. L. Fante, Electromagnetic beam propagation in turbulent media: an update, *Proc. IEEE* **68**, 1424–1443 (1980).
7. R. L. Lutomirski and D. E. Snead, Green's Function Calculation of the Effects of the Air/Sea Interface on Optical Propagation, in: *Special Topics in Optical Propagation, AGARD Conference Proceedings No. 300*, pp. 3-1–3-8, Technical Editing and Reproduction Ltd., London (1981).
8. S. Chandrasekhar, *Radiative Transfer*, Clarendon, Oxford (1960) [Reprinted by Dover Books, New York (1960)].
9. R. W. Preisendorfer, *Hydrologic Optics*, U.S. Department of Commerce, National Oceanic and Atmospheric Administration, Environmental Research Laboratories (1976).
10. R. W. Preisendorfer, *Radiative Transfer on Discrete Spaces*, Pergamon Press, New York (1965).
11. H. C. Van de Hulst, *Multiple Light Scattering*, Vol. 1, Acadmic Press, New York (1980).
12. A. Ishimaru, *Wave Propagation and Scattering in Random Media*, Vols. 1 and 2, Academic Press, New York (1978).
13. N. G. Jerlov and E. S. Nielsen (eds.), *Optical Aspects of Oceanography*, Academic Press, New York (1974).
14. N. G. Jerlov, *Marine Optics*, Elsevier Oceanography Series No. 14, Elsevier, Amsterdam (1976).
15. Naval Blue-Green Single-Pulse Downlink Propagation Model, Naval Ocean Systems Center, San Diego, California, Technical Report 387 (January 1, 1979).
16. R. L. Fante, Wave propagation in random media: a systems approach, in: *Progress in Optics* (E. Wolf, ed.), Vol. XXII, Chapter 6, Elsevier (in preparation).
17. R. F. Lutomirski and H. T. Yura, Propagation of a finite optical beam in an inhomogeneous medium, *Appl. Opt.* **10**, 1654 (1971).
18. D. A. de Wolf, Coherence of a light through an optically dense tubid layer, *Appl. Opt.* **17**, 1280–1285 (1978).
19. H. T. Yura, A Multiple Scattering Analysis of the Propagation of Radiance through the Atmosphere, URSI Commission F-sponsored Conference on Propagation in Non-ionized Media, La Baule, France (1977).
20. R. M. Gagliardi and S. Karp, *Optical Communications*, Wiley-Interscience, New York (1976).
21. W. H. Wells, Loss of resolution in water as a result of multiple small-angle scattering, *J. Opt. Soc. Am.* **59**, 686 (1969).
22. D. Arnush, Underwater light-beam propagation in the small-angle scattering approximation, *J. Opt. Soc. Am.* **62**, 1109 (1972).
23. R. L. Fante, Propagation of electromagnetic waves through turbulent plasma using transport theory, *IEEE Trans. Antennas Propagat.* **AP-21**, 750–755 (1973).
24. R. F. Lutomirski, The Irradiance Distribution in a Scattering Medium, Pacific Sierra Research Corporation, PSR Note 73 (May, 1975).
25. H. T. Yura, Aerospace Corporation, private communications.
26. D. M. Bravo-Zhivotovsky, L. S. Dolin, A. G. Luchmin, and V. A. Sarelyev, Structure of a narrow light beam in sea water, *Atmos. Oceanic Phys.* **5**, 160–167 (1969) [translated by P. A. Kaehn].

27. P. Y. Ganich and I. M. Levin, Extinction of the brightness of self-luminous objects in a scattering medium, *Bull. Acad. Sci. (USSR), Atmos. Oceanic Phys.* **4**, (1968).
28. S. Karp, Optical Communications between Underwater and Above-Surface (Satellite) Terminals, Naval Electronics Laboratory Center Technical Document, unclassified, TD 430 (June 1, 1975).
29. S. Q. Duntley, Underwater Lighting by Submerged Lasers, Visibility Laboratory, Scripps Institute of Oceanography Technical Report, SIO REF 71-1 (June 1, 1971).
30. R. G. Driscoll, J. N. Martin, and S. Karp, OPSATCOM Field Measurements, Naval Electronics Laboratory Center Technical Document, unclassified, TD490 (June 1, 1976).
31. R. D. Anderson and L. B. Stotts, Underwater measurements between off-axis radiance compared with various analytical treatments of the radiative transfer equation, *J. Opt. Soc. Am.* **72**, 738-746 (1982).
32. W. G. Tam and A. Zardecki, Laser beam propagation in particulate media, *J. Opt. Soc. Am.* **69**, 68 (1979).
33. L. B. Stotts, Limitations of approximate Fourier techniques in solving radiative transfer problems, *J. Opt. Soc. Am.* **69**, 1719 (1979).
34. R. L. Fante, Range of validity of the quadratic approximation for propagation through a random distribution of large aerosol particles, *Appl. Opt.* **21**, 9-11 (1982).
35. R. P. Bocker, Naval Ocean Systems Center, private communication.
36. S. L. Valley (ed.), *Handbook of Geophysics and Space Environments*, McGraw-Hill, New York, Chapter 7, Table 7-4, pp. 7-23 (1965).
37. A. J. LaRocca, Atmospheric absorption, in: *The Infrared Handbook* (W. L. Wolfe and G. J. Zissis, eds.), The Environmental Institute of Michigan, Ann Arbor (1978).
38. W. K. Pratt, *Laser Communications*, John Wiley and Sons, New York.
39. A. Deepak (ed.), *Inversion Methods in Atmospheric Remote Sounding*, Academic Press, New York (1977).
40. J. Lenoble, Standard Procedures to Compute Atmospheric Radiative Transfer in a Scattering Atmosphere, International Association of Meteorology and Atmospheric Physics (IAMAP), Radiation Commission, National Center for Atmospheric Research, Boulder, Colorado (July, 1977).
41. S. A. W. Gerstl and A. Zardecki, Discrete-ordinates finite-element method for atmospheric radiative transfer and remote sensing, *Appl. Opt.* **24**, 81-93 (1985).
42. A. Zardecki, S. A. W. Gerstl, and J. F. Embury, Application of the 2-D discrete-ordinate method to multiple scattering of laser radiation, *Appl. Opt.* **22**, 1346-1353 (1983).
43. G. N. Plass, G. W. Kattawar, and F. E. Catchings, Matrix operator theory of radiative transfer; Part 1, Rayleigh scattering, *Appl. Opt.* **12**, 314-329 (1973).
44. R. M. Lerner and J. D. Summers, Monte Carlo description of time- and space-resolved multiple forward scatter in natural water, *Appl. Opt.* **21**, 861-869 (1982).
45. G. W. Kattawar, Monte Carlo methods in radiative transfer, in: Multiple Light Scattering in Atmospheres, Oceans, Clouds and Snow, Institute for Atmospheric Optics and Remote Sensing, Short course No. 420, Williamsburg, Virginia, December 4-8, 1978.
46. H. R. Gordon, O. B. Brown, and M. M. Jacobs, Computed relationships between the inherent and apparent optical properties of a flat homogeneous ocean, *Appl. Opt.* **29**, 417-427 (1976).
47. G. N. Plass and G. W. Kattawar, Monte Carlo calculations of radiative transfer in the Earth's atmosphere-ocean system; Part 1, Flux in the atmosphere and ocean, *J. Phys. Ocean.* **2**, 139-145 (1972).
48. G. N. Plass and G. W. Kattawa, Monte Carlo calculations of light scattering in clouds, *Appl. Opt.* **7**, 415-419 (1968).

49. W. E. Meador and W. R. Weaver, Two-function approximations, in: Multiple Light Scattering in Atmospheres, Oceans, Clouds and Snow, Institute for Atmosphere Optics and Remote Sensing, Short course No. 420, Williamsburg, Virginia, December 4–8, 1978.
50. W. S. Helliwell, A finite difference solution to the radiative transfer equation for in-water radiance, *J. Opt. Soc. Am.* **2**, 1325–1330 (1985).
51. H. M. Heggestad, Optical communications through Multiple Scattering Media, Massachusetts Institute of Technology, Research Laboratory for Electronics Technical Report 472 (November, 1968).
52. L. B. Stotts, The radiance produced by laser radiation traversing a particulate multiple scattering medium, *J. Opt. Soc. Am.* **67**, 815–816 (1977).
53. A. Ishimaru and S. T. Hong, Two frequency mutual coherence function, coherence bandwidth and coherence time of millimeter and optical waves in rain, fog and tubulence, *Radio Science* **11**, 551–559 (1976).
54. A. Ishimaru and S. T. Hong, Multiple scattering effects on coherent bandwidth and pulse distortion of a wave propagating in a random distribution of particles, *Radio Science* **10**, 637–644 (1975).
55. K. Furutsu, Multiple scattering of waves in a medium of randomly distributed particles and derivation of the transport equation, *Radio Science* **10**, 29–44 (1975).
56. P. H. Levine, Megatek Corporation, private communication.
57. E. A. Bucher, Propagation models for optical communications through fog and clouds, *Proc. Nat. Electron. Conf.* **29**, 180–185 (1975).
58. A. Gordon, Practical approaches to underwater multiple-scattering problems, *Proc. Soc. Photo-Opt. Instrum. Eng.* **64**, 84–93 (1975).
59. L. B. Stotts, Atmospheric, Space and Underwater Optical Communications, National Science Foundation Grantee-Users Meeting on Optical Communications, Pittsburg, Pennsylvania, June 5–7, 1978.
60. L. B. Stotts, Satellite, Surface and Subsurface Optical Communications, International Telemetering Conference, ITC/USA/'78, Los Angeles, California, November 14–16, 1978.
61. E. A. Bucher, Computer simulation of light pulse propagation for communication through thick clouds, *Appl. Opt.* **12**, 2391–2400 (1973).
62. L. B. Stotts, Closed form expression for optical pulse broadening in multiple scattering media, *Appl. Opt.* **17**, 504–505 (1978).
63. M. A. Millbach, Computer Simulation of Light Propagation through a Scattering Medium, Masters thesis, Navy Postgraduate School, Monterey, California (June, 1978).
64. E. A. Bucher and R. M. Lerner, Experiments on light pulse communication through atmospheric clouds, *Appl. Opt.* **12**, 2401–2414 (1973).
65. R. A. Elliot, Wave Propagation in Particulate Media, Oregon Graduate Center, Annual Summary Report, Contract No. N0014-79-c-0897 (May 31, 1981).
66. J. C. Matter and R. G. Bradley, Optical pulse propagation through clouds, *Appl. Opt.* **20**, 554–563 (1981).
67. W. H. Paik, M. Tebyani, D. J. Epstein, R. S. Kennedy, and J. H. Shapiro, Propagation experiments in low-visibility atmospheres, *Appl. Opt.* **17**, 899–905 (1978).
68. J. S. Ryan and A. I. Carswell, Laser beam broadening and depolarization in dense fog, *J. Opt. Soc. Am.* **68**, 900–908 (1978).
69. R. S. Kennedy and J. H. Shapiro, Multipath Dispersion in Low Visibility Optical Communication Channels, Rome Air Development Center Technical Report, RADC-TR-77-73 (February, 1977).
70. W. S. Ross, W. P. Jaesar, J. Nakai, T. T. Nguyen, and J. H. Shapiro, Atmospheric optical propagation—an integrated approach, *Opt. Eng.* **21**, 775–785 (1982).
71. J. A. Curcio and L. F. Drummeter, Jr., Experimental Observations of Forward Scattering of Light in the Lower Atmosphere, Naval Research Laboratory, Technical Report No. NRL 6152 (September 30, 1985).

72. G. T. Ruck, Feasibility of Non-line-of-sight Laser Communications, Battelle Memorial Institute, Columbus, Ohio, Report No. BAT-171-A (December 15, 1964).
73. M. King and S. Kainer, Some parameters of a laser-type beyond-the-horizon communication link, *Proc. IEEE* **53**, 137 (1965).
74. Division 6, Quarterly Technical Summary, Space Communications, MIT Lincoln Laboratory, Cambridge, Massachusetts (March 15, 1969), pp. 10–12, DDC AD-851886.
75. R. S. Kennedy, Communication through optical scattering channels: An introduction, *Proc. IEEE* **58**, 1651 (1970).
76. P. H. Levine and M. E. O'Brien, ELOS Meteorology Sensitivity Study, Megatek Final Report No. R2005-099-F-1, Contract No. N00123-75-C-0328, Task MEG-TA-009 (November 15, 1977).
77. J. D. Jackson, *Classical Electrodynamics*, Second Edition, John Wiley and Sons, New York (1975).
78. R. F. Lutomirski, D. E. Snead, and W. L. Woodie, The Marine Boundary Layer Optical Communication Link, Pacific Sierra Research Technical Report, PSR Report 811 (July, 1978), Appendix A.
79. D. Bauer and A. Morel, Etude aux petits angles des l'indicatrix des diffusion de la lumière par les equx de mer, *Ann. Geophys.* **23**, 122 (1967).
80. L. Dolin, Propagation of a narrow light beam in a medium with strongly anisotropic scattering, *Radiophys. Quantum Electron.* **9**, 40–47 (1966).
81. J. W. Goodman, *Introduction to Fourier Optics*, McGraw-Hill, New York (1968).
82. M. A. Boc and A. Deepak, Multiple scattering corrections to the solar aureole, in: *Proceedings of the Third Conference on Atmospheric Radiation*, Davis, California, June 28–30, 1978, pp. 12–13, American Meteorological Society, Boston (1978).
83. R. D. Richtmyer and K. W. Morton, *Difference Methods for Initial-Value Problems*, Interscience Publishers, New York (1967).
84. J. E. Tyler, Radiance distribution as a function of depth in an underwater environment, *Bull. Scripps Inst. Oceanog.* **7**, 363 (1960).
85. K. Furutsu, Diffusion equation derived from space–time transport equation, *J. Opt. Soc. Am.* **70**, 360–366 (1980).
86. S. Ito and K. Furutsu, Theory of light pulse propagation through thick clouds, *J. Opt. Soc. Am.* **70**, 366–374 (1980).
87. A. Ishimaru, Diffusion of a pulse in densely distributed scatters, *J. Opt. Soc. Am.* **68**, 1045–1050 (1978).
88. S. Ito, Comparison of diffusion theories for optical pulse waves propagated in discrete random media, *J. Opt. Soc. Am. A* **1**, 502–505 (1984).
89. A. Ishimaru, Difference between Ishimaru's and Furutsu's theories on pulse propagation in discrete random media, *J. Opt. Soc. Am. A* **1**, 506–509 (1984).
90. G. M. Lee, G. M. Ciany, G. Schroeder, and J. Fenier, Availability models for space-to-earth optical communication links, Appendix A, in: S. Karp, A Test Plan for Determining the Feasibility of Optical Satellite Communications through Clouds at Visible Frequencies, Naval Ocean Systems Center Technical Note, TN279 (July 1, 1978).
91. G. M. Lee, C. M. Ciany, and C. Tranchita, McDonnell-Douglas Astronautics, private communication.
92. R. E. Danielson, D. R. Moore, and H. C. Van de Hulst, The transfer of visible radiation through clouds, *J. Atmos. Sci.* **26**, 1078–1087 (1969).
93. K. S. Baker and R. C. Smith, Quasi-inherent characteristics of the diffuse attenuation coefficient for irradiance, *Proc. Soc. Photo-Opt. Instrum. Eng. Ocean Optics VI* **208**, 60–63 (1969).
94. J. Gordon, Direction Radiance (Luminescence) of the Sea Surface, Scripps Institution of Oceanography, SIO Ref. B9-20 (October, 1969).
95. C. Cox and W. Munk, Statistics of the sea surface derived from sun glitter, *J. Mar. Res.* **13** (2), 63 (1954).

96. H. R. Gordon, Albedo of the ocean-atmospheric system: Influence of the sea foam, *Appl. Opt.* **16**, 2257–2260 (1976).
97. Naval Blue-Green Single Pulse Downlink Propagation Model, Naval Ocean Systems Center, San Diego, California TR 387 (January 1, 1979).
98. S. Karp and R. M. Gagliardi, The design of a pulse-position-modulated optical communication system, *IEEE Trans. Commun. Tech.* **COM-17**, 670–676 (December, 1969).
99. R. M. Gagliardi and S. Karp, M-ary Poisson detection and optical communications, *IEEE Trans. Commun. Tech.* **COM-17,** 208–216 (1969).
100. C. W. Helstrom, *Statistical Theory of Signal Detection*, Pergamon Press, New York (1968).
101. A. J. Viterbi, *Principles of Coherent Communications*, McGraw-Hill, New York (1966).

Appendix A

Generalized Radiometry

The traditional radiometric concepts are, as was pointed out in Section 1.2, usually assumed to be valid for fields generated by spatially incoherent sources. The question which naturally arises is can these laws be applied to radiation generated by highly coherent sources, such as the laser? Indeed, is there a relationship between the directional properties of a source and its degree of coherence? Recent research has shown that there is a very close connection between the degree of source spatial coherence and the angular distribution of radiant flux. One would suspect that this must be the case after comparing the radiant intensity patterns generated by blackbody and laser sources of identical physical size. The blackbody radiant intensity is given by $J(\theta) = J(0) \cos \theta$, which exhibits a broad angular dependence. The laser, however, can have almost all of its energy directed within an angular cone of divergence on the order of 10^{-4} radians for a 2-mm diameter beam. One of the primary differences between the laser and the blackbody source is the high degree of spatial coherence exhibited by the laser.

Within the last 10 years principles of generalized radiometry have been developed which treat sources of arbitrary states of coherence and clarify the class of sources for which the traditional radiometry is applicable. The first attempt at deriving radiometric quantities from the theory of partial coherence was made by Walther[(1)] in 1986 with his introduction of the generalized radiance. Walther's work formed the nucleus for a series of papers, starting in 1971, by Marchand and Wolf[(2)] which developed the mathematical structure underlying the generalized radiometry. A subsequent paper by Moran[(3)] provided an alternative derivation of the generalized radiance, established the conditions under which the generalized and traditional radiance functions may be regarded as identical, and clarified, to some extent, the connections between the generalized radiometry and transport theory in random media. We now proceed to derive the generalized radiance function following the approach given by Moran.

A.1. The Generalized Radiance

We begin our development of the Generalized radiance by considering the propagation geometry shown in Figure 2.1. The propagation of the coherence function in vacuum over a distance Z is governed by the well-known equation[4,5]

$$\Gamma(\mathbf{r}_{s1}, \mathbf{r}_{s2}, Z) = \left(\frac{k_0}{2\pi}\right)^2 \iint_{-\infty}^{\infty} \frac{\Gamma(\mathbf{r}_{a1}, \mathbf{r}_{a2}, Z=0)}{\mathbf{r}_{sa1}\mathbf{r}_{sa2}}$$

$$\times \cos(\theta_1)\cos(\theta_2)\, e^{-ik_0(\mathbf{r}_{sa1}-\mathbf{r}_{sa2})}\, d^2\mathbf{r}_{a1}\, d^2\mathbf{r}_{a2} \quad \text{(A.1.1)}$$

where $\Gamma(\mathbf{r}_{s1}, \mathbf{r}_{s2}, Z)$ is the mutual coherence function between the points $\mathbf{r}_{s1}$ and $\mathbf{r}_{s2}$ in the plane σ_Z, r_i is the distance from $\mathbf{r}_{a1i}$ to $\mathbf{r}_{a2i}$, $k_0 = 2\pi/\lambda$, λ is the mean wavelength of the radiation, and θ_i is the angle between the Z axis and the line connecting the points $\mathbf{r}_{a1i}$ and $\mathbf{r}_{a2i}$. Using equations (2.2.6), (2.3.5), (2.3.6) and the Fresnel approximation

$$k_0(\mathbf{r}_{sa1} - \mathbf{r}_{sa2}) \approx \mathbf{k} \cdot (\boldsymbol{\rho}_1 - \boldsymbol{\rho}_2) \quad \text{(A1.2)}$$

(A.1.1) can be written

$$\Gamma(\mathbf{R}_2, Z, \boldsymbol{\rho}_2) = \frac{1}{4\pi^2} \iint_{-\infty}^{\infty} \Gamma\left(\mathbf{R}_1 = \mathbf{R}_2 + \frac{Z}{k_0}\mathbf{k}, Z = 0, \boldsymbol{\rho}_1\right)$$

$$\times e^{-i\mathbf{k}\cdot(\boldsymbol{\rho}_1 - \boldsymbol{\rho}_2)}\, d^2\mathbf{k}\, d^2\boldsymbol{\rho}_1 \quad \text{(A.1.3)}$$

The propagation law (A.1.3) can be expressed in the form

$$\Gamma(\mathbf{R}_2, Z, \boldsymbol{\rho}_2) = \frac{1}{k_0^2} \int_{-\infty}^{\infty} N_G\left(\mathbf{R}_1 = \mathbf{R}_2 + \frac{Z}{k_0}\mathbf{k}, Z = 0, \mathbf{k}\right)$$

$$\times e^{i\mathbf{k}\cdot\boldsymbol{\rho}_2}\, d^2\mathbf{k} \quad \text{(A.1.4)}$$

where we have defined the generalized source radiance function in the plane σ_0 through the Fourier transform pair

$$\Gamma(\mathbf{R}_1, Z = 0, \boldsymbol{\rho}_1) = \frac{1}{k_0^2} \int_{-\infty}^{\infty} N_G(\mathbf{R}_1, Z = 0, \mathbf{k})\, e^{i\mathbf{k}\cdot\boldsymbol{\rho}_1}\, d^2\mathbf{k} \quad \text{(A.1.5)}$$

$$N_G(\mathbf{R}_1, Z = 0, \mathbf{k}) = \left(\frac{k_0}{2\pi}\right)^2$$

$$\times \int_{-\infty}^{\infty} \Gamma(\mathbf{R}_1, Z = 0, \boldsymbol{\rho}_1)\, e^{-i\mathbf{k}\cdot\boldsymbol{\rho}_1}\, d^2\boldsymbol{\rho}_1 \quad \text{(A.1.6)}$$

In analogy with (A.1.5) we define the generalized field radiance function in the plane σ_z through the Fourier transform pair

$$\Gamma(\mathbf{R}_2, Z, \boldsymbol{\rho}_2) = \frac{1}{k_0^2}\int_{-\infty}^{\infty} N_G(\mathbf{R}_2, Z, \mathbf{k})\, e^{i\mathbf{k}\cdot\boldsymbol{\rho}_2}\, d^2\mathbf{k} \qquad \text{(A.1.7)}$$

$$N_G(\mathbf{R}_2, Z, \mathbf{k}) = \left(\frac{k_0}{2\pi}\right)^2 \int_{-\infty}^{\infty} \Gamma(\mathbf{R}_2, Z, \boldsymbol{\rho}_2)\, e^{-i\mathbf{k}\cdot\boldsymbol{\rho}_2}\, d^2\boldsymbol{\rho}_2 \qquad \text{(A.1.8)}$$

Comparing (A.1.8) and (A.1.4), we find

$$N_G(\mathbf{R}_2, Z, \mathbf{k}) = N_G\left(\mathbf{R}_1 = \mathbf{R}_2 + \frac{Z}{k_0}\mathbf{k}, Z = 0, \mathbf{k}\right) \qquad \text{(A.1.9)}$$

which establishes an invariance property for the generalized radiance that is entirely analogous to the familiar traditional radiance invariance property (Gershun[6]; Nicodemus[7]). In particular, Figure 2.1 and (A.1.19) show that in the Fresnel diffraction zone the generalized radiance is invariant along the path $\hat{\boldsymbol{\eta}}$ defined by the vector $\mathbf{k}$ through (12). It is also clear from the development leading to (A.1.3) that (A.1.5) through (A.1.9) are consistent with the Rayleigh–Sommerfeld scalar diffraction theory only in the Fresnel diffraction zone. We wish to emphasize the fact that, in contrast to the results derived by Collett *et al.*[8], this consistency is achieved without requiring statistical homogeneity of the radiation field.

The function that we have called the generalized radiance is directly related to the generalized spectral radiance $N_G(\mathbf{R}, Z, \mathbf{k}, \omega)$, which was first introduced by Walther[1] and later used by Marchand and Wolf[2] as the cornerstone for their development of the mathematical structure underlying the generalized radiometry. The generalized spectral radiance is defined by the equation

$$N_G(\mathbf{R}, Z, \mathbf{k}, \omega) = \left(\frac{k_0}{2\pi}\right)^2 \cos(\theta) \int_{-\infty}^{\infty} \hat{\Gamma}(\mathbf{R}, Z, \boldsymbol{\rho}, \omega)\, e^{-i\mathbf{k}\cdot\boldsymbol{\rho}}\, d^2\boldsymbol{\rho} \qquad \text{(A.1.10)}$$

where $\hat{\Gamma}(\mathbf{R}, Z, \boldsymbol{\rho}, \omega)$ is the mutual spectral density function. For quasi-monochromatic fields in the small angle approximation the generalized radiance is given by

$$\begin{aligned} N_G(\mathbf{R}, Z, \mathbf{k}) &= \int_{-\infty}^{\infty} N_G(\mathbf{R}, Z, \mathbf{k}, \omega)\, d\omega \\ &= \left(\frac{k_0}{2\pi}\right)^2 \int_{-\infty}^{\infty} \Gamma(\mathbf{R}, Z, \boldsymbol{\rho})\, e^{-i\mathbf{k}\cdot\boldsymbol{\rho}}\, d^2\boldsymbol{\rho} \end{aligned} \qquad \text{(A.1.11)}$$

which is equivalent to our (A.1.8). Marchand and Wolf[(2)] have investigated some of the properties of the generalized radiance and have found that it exhibits several features not normally attributable to radiance. In particular, the generalized radiance must be interpreted with care, since it may be negative for some values of its argument, it is not in general measurable, and it does not in general vanish for all points in the object plane that lie outside the region occupied by the source. The generalized radiance is in fact a pseudoradiance that incorporates certain "unphysical" features that, as we will presently show, are necessary to account for spatial coherence effects correctly.

A.2. The Mutual Radiance

We now write the mutual coherence function in the source plane, in the form

$$\Gamma(\mathbf{R}_1, Z=0, \boldsymbol{\rho}_1) = \frac{4\pi}{k_0^2} I(\mathbf{R}_1, Z=0)\delta(\boldsymbol{\rho}_1) + M(\mathbf{R}_1, Z=0, \boldsymbol{\rho}_1) \quad \text{(A.2.1)}$$

where $\delta(\boldsymbol{\rho}_1)$ is the Dirac delta function. Equation (A.2.1) is a convenient artifice that allows separation of those effects that are attributable to spatial incoherence from those resulting from spatial partial coherence. The first term is the mutual coherence function associated with a spatially incoherent source, while the second term accounts for spatial coherence when $\boldsymbol{\rho}_1 \neq 0$ and is defined by the equation

$$M(\mathbf{R}_1, Z=0, \boldsymbol{\rho}_1) = \Gamma(\mathbf{R}_1, Z=0, \boldsymbol{\rho}_1) - \frac{4\pi}{k_0^2} I(\mathbf{R}_1, Z=0)\delta(\boldsymbol{\rho}_1) \quad \text{(A.2.2)}$$

The source intensity $I(\mathbf{R}_1, Z = 0)$, which has units of watts per square meter, is defined by the relationship

$$I(\mathbf{R}_1, Z = 0) = dP(\mathbf{R}_1, Z = 0)/d\sigma \quad \text{(A.2.3)}$$

with $dP(\mathbf{R}_1, Z = 0)$ the power transported across a source area element $d\sigma$ surrounding the point $\mathbf{R}_1$. As is customary in physical optics, the intensity is taken to be the average of the square of the complex scalar field variable and is usually assumed to be equal to the absolute value of the average Poynting vector in appropriate units.

For strictly spatially incoherent sources, (22) reduces to:

$$\Gamma(\mathbf{R}_1, Z = 0, \boldsymbol{\rho}_1) = \frac{4\pi}{k_0^2} I(\mathbf{R}_1, Z = 0)\delta(\boldsymbol{\rho}_1) \quad \text{(A.2.4)}$$

Equation (A.2.4) is a mathematical idealization that is usually assumed to be valid for substantially spatially incoherent fields such as those generated at the source plane of blackbody radiators, even though such fields can be spatially coherent over distances of the order of the mean wavelength of the radiation. For the purposes of the present analysis, an optical field will be said to be wide sense spatially incoherent if, in calculations involving the mutual intensity, the relationship (A.2.4) yields results that are, to a good approximation, equivalent to those obtained by using the true source mutual intensity function.

Substituting (A.2.1) in (A.1.6) gives

$$N_G(\mathbf{R}_1, Z=0, \mathbf{k}) = [I(\mathbf{R}_1, Z=0)/\pi] + N_M(\mathbf{R}_1, Z=0, \mathbf{k}) \quad \text{(A.2.5)}$$

where we have defined the mutual radiance in the plane σ_0:

$$N_M(\mathbf{R}_1, Z=0, \mathbf{k}) = \left(\frac{k_0}{2\pi}\right)^2 \int_{-\infty}^{\infty} M(\mathbf{R}_1, Z=0, \boldsymbol{\rho}_1)\, e^{-i\mathbf{k}\cdot\boldsymbol{\rho}_1} d^2\boldsymbol{\rho}_1 \quad \text{(A.2.6)}$$

The source intensity function $I(\mathbf{R}_1, Z = 0)$ is related to the traditional source radiance by the following equation:

$$I(\mathbf{R}_1, Z = 0) = \int_{2\pi} N_T(\mathbf{R}_1, Z = 0, \mathbf{S}) \cos\theta\, d^2\Omega \quad \text{(A.2.7)}$$

where $N_t(\mathbf{R}_1, Z=0, \mathbf{S})$ has units of $\text{W s m}^{-2}\,\text{sr}^{-1}$ and is defined by the relationship

$$N_T(\mathbf{R}_1, Z = 0, \mathbf{S}) = dP(\mathbf{R}_1, Z = 0)/d\sigma \cos\theta\, d\Omega \quad \text{(A.2.8)}$$

where $dP(\mathbf{R}_1, Z = 0)$ is the radiant power that is transported across a source area element $d\sigma$ surrounding the point $\mathbf{R}_1$ into a solid angle $d\Omega$ about the direction specified by the vector $\mathbf{S}$ and θ is the angle between $\mathbf{S}$ and the normal to the surface. It appears worthwhile to note that the basic postulates of the phenomenological theory of geometrical radiometry, including the definition of transitional radiance, have recently become suspect because they lack a firm foundation in the classical theory of the electromagnetic field. While it appears that this issue is not completely resolved, Wolf(9) has shown that the relationship (A.2.8) is a rigorous consequence of the basic electromagnetic field-equations for statistically stationary and homogeneous fields. His analysis also showed that (A.2.8) is approximately correct for nonhomogeneous fields in which "angular correlations" extend over a sufficiently small solid angle about each direction specified by $\mathbf{S}$.

Substituting (A.2.7) in (A.2.5), we have

$$N_G(\mathbf{R}_1, Z=0, \mathbf{k}) = \frac{1}{\pi}\int_{2\pi} N_T(\mathbf{R}_1, Z=0, \mathbf{S}) \times \cos(\theta)\, d^2\Omega + N_M(\mathbf{R}_1, Z=0, \mathbf{k}) \quad \text{(A.2.9)}$$

which establishes the relationship between the generalized, traditional, and mutual radiance functions in the plane σ_0.

We now recall that the source radiance function is independent of the direction of propagation **S** if and only if the source is Lambertian:

$$N_T(\mathbf{R}_1, Z = 0, \mathbf{S}) = N_T(\mathbf{R}_1, Z = 0) \quad \text{(A.2.10)}$$

Substituting (A.2.10) in (A.2.9) gives

$$N_G(\mathbf{R}_1, Z = 0, \mathbf{k}) = N_T(\mathbf{R}_1, Z = 0) + N_M(\mathbf{R}_1, Z = 0, \mathbf{k}) \quad \text{(A.2.11)}$$

Using the generalized radiance invariance property, we have

$$\begin{aligned} N_G(\mathbf{R}_2, Z, \mathbf{k}) &= N_T\left(\mathbf{R}_1 = \mathbf{R}_2 + \frac{z}{k_0}\mathbf{k}, Z = 0, \mathbf{k}\right) \\ &\quad + N_M\left(\mathbf{R}_1 = \mathbf{R}_2 \frac{Z}{k_0}\mathbf{k}, Z = 0, \mathbf{k}\right) \\ &= N_T(\mathbf{R}_2, Z, \mathbf{k}) + N_M(\mathbf{R}_2, Z, \mathbf{k}) \end{aligned} \quad \text{(A.2.12)}$$

Note that in the Fresnel diffraction zone the mutual radiance is also invariant along paths defined by the vector **k**, that is,

$$N_M(\mathbf{R}_2, Z, \mathbf{k}) = N_M\left(\mathbf{R}_1 = \mathbf{R}_2 + \frac{Z}{k_0}\mathbf{k}, Z = 0, \mathbf{k}\right) \quad \text{(A.2.13)}$$

Equations (A.2.11) and (A.2.12) show that the generalized radiance can be expressed as the sum of the traditional and mutual radiance functions in the source plane if and only if the source is quasi-monochromatic and Lambertian and in the radiation field if and only if the region of interest lies in the Fresnel diffraction zone of a quasi-monochromatic Lambertian source.

We wish to emphasize the fact that, by the construction of (A.2.1), the mutual radiance is a manifestation of source spatial coherence. In particular, the mutual radiance in a given direction specified by the vector **k** may be regarded as arising from the partially coherent interaction of all possible pairings of source points in the plane σ_0. Herein lies the departure from the traditional radiometry, since it is clear from (A.2.8) that there is no counterpart to this term in the definition of traditional radiance. Referring to the right-hand side of (A.2.11), the first term, by definition, is nonnegative and vanishes identically in regions of the source plane not occupied by the source. It is therefore clear that the peculiar unphysical behavior associated with the generalized radiance resides in the mutual radiance function. Thus the generalized radiance will always possess the properties associated with traditional radiance if and only if the mutual radiance function is identically equal to zero:

$$\int_{-\infty}^{\infty}\left[\Gamma(\mathbf{R}_1, Z=0, \boldsymbol{\rho}_1) - \frac{4\pi}{k_0^2} I(\mathbf{R}_1, Z=0)\delta(\boldsymbol{\rho}_1)\right] e^{i\mathbf{k}\cdot\boldsymbol{\rho}_1}\, d^2\boldsymbol{\rho}_1 = 0 \qquad \text{(A.2.14)}$$

Equation (A.2.14) will be valid for all values of $\mathbf{k}$ if and only if the term in brackets under the integral is identically equal to zero:

$$\Gamma(\mathbf{R}_1, Z=0, \boldsymbol{\rho}_1) = \frac{4\pi}{k_0^2} I(\mathbf{R}_1, Z=0)\delta(\boldsymbol{\rho}_1) \qquad \text{(A.2.15)}$$

Thus the mutual radiance is identically equal to zero if and only if the source is wide sense spatially incoherent. Equations (A.2.11) and (A.2.12) can then be written as

$$N_G(\mathbf{R}_1, Z=0, \mathbf{k}) = N_T(\mathbf{R}_1, Z=0, \mathbf{k}) \qquad \text{(A.2.16a)}$$

$$N_G(\mathbf{R}_2, Z, \mathbf{k}) = N_T(\mathbf{R}_2, Z, \mathbf{k}) \qquad \text{(A.2.16b)}$$

A.3. The Generalized Radiant Emittance

The generalized radiant emittance, or intensity, of the field in the plane σ_z is defined in direct analogy with Equation (1.2.14)

$$E_G(\mathbf{R}, Z) = \frac{1}{k_0^2}\int N_G(\mathbf{R}, Z, \mathbf{k}) \cos\theta\, d^2\mathbf{k} \qquad \text{(A.3.1)}$$

Like the generalized radiance, the generalized radiant emittance can take on negative values for certain values of its arguments. For $Z = 0$, $N_G(\mathbf{R}, Z = 0, \mathbf{k})$ gives the source intensity. Like the generalized source radiance, the generalized source intensity does not necessarily vanish in regions of the source plane not occupied by the source.

A.4. The Generalized Radiant Intensity

The generalized radiant intensity is given by the expression

$$J_G(Z, \mathbf{k}) = \frac{\cos\theta}{k_0^2} \int N_G(\mathbf{R}, Z, \mathbf{k})\, d^2\mathbf{R} \tag{A.4.1}$$

It can be shown that the generalized radiant intensity represents the flux per unit solid angle around the direction $\hat{\mathbf{n}}$ emitted by the entire field in the plane σ_Z. It therefore has the same physical significance as the traditional radiant intensity, even though the generalized radiance can be negative. When $Z = 0$, $J(Z = 0, R)$ gives the source radiant intensity and is identically equal to zero in regions of the source plane not occupied by the source.

References

1. A. J. Walther, Radiometry and coherence, *J. Opt. Soc. Am.* **59**, 1256-1259 (1968).
2. E. W. Marchand and E. Wolf, Radiometry with sources of any state of coherence, *J. Opt. Soc. Am.* **64**, 1219-1226 (1979).
3. Steven E. Moran, On the transport theory of mutual intensity propagation in a random medium, *Radio Science* **15**, 1195-1205 (1980).
4. F. Zernike, The concept of degree of coherence and its application to optical problems, *Physica* **5**(8), 725-795 (1938).
5. M. Born and E. Wolf, *Principles of Optics*, Macmillan, New York (1964), 806 pp.
6. A. Gershun, The light field, *J. Math. Phys.* **18**, 51-151 (1939).
7. F. Nicodemus, Radiance, *Amer. J. Phys.* **31**, 368-377 (1963).
8. E. Collett, J. T. Foley, and E. Wolf, On an investigation of Tatarski into the relationship between coherence theory and the theory of radiative transfer, *J. Opt. Soc. Am.* **67**(4), 465-467 (1977).
9. E. Wolf, New theory of radiative transfer in free electromagnetic fields, *Phys. Rev. D* **13**, 869-886 (1976).

Appendix B

Transport Theory

B.1. Heuristic Derivation of the Radiative Transport Equation

We consider the radiance, $N(\mathbf{R}, Z, \mathbf{S})$, which is incident at the point $(\mathbf{R}, Z)$ on a unit cross section cylindrical differential volume element which has its axis aligned with the unit vector $\mathbf{S}$ (Figure B.1). The differential volume has length dS and contains $dN = \rho dS$ particles. The vector $\mathbf{S}$ is given in terms of the direction cosines (l, m, n) by the equation:

$$\mathbf{S} = l\hat{\mathbf{c}} + m\hat{\jmath}_n + n\hat{\mathbf{k}} = n_1 + n\hat{\mathbf{k}} \tag{B.1.1}$$

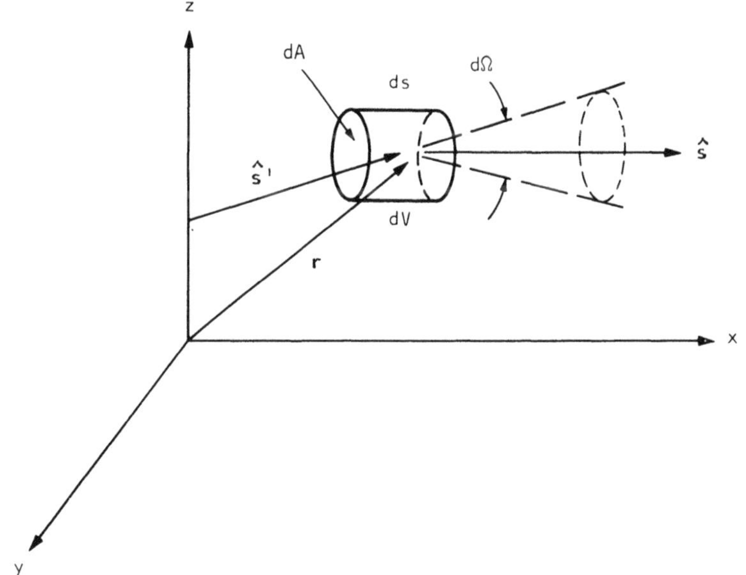

Figure B.1. Geometry for derivation of radiative transport equation.

with (l, m, n) defined in spherical coordinates as

$$\begin{aligned} l &= \sin\theta\cos\phi \\ m &= \sin\theta\sin\phi \\ n &= \cos\theta \end{aligned} \tag{B.1.2}$$

Let σ_A and σ_S represent, respectively, the absorption and scattering cross section per particle. The decrease in radiance in the direction **S** attributable to scattering and absorption from the differential volume element is given by

$$\begin{aligned} dN^-(\mathbf{R}, Z, \mathbf{S}) &= -N(\mathbf{R}, Z, \mathbf{S})(\sigma_A + \sigma_S)\rho\, dS\, dA \\ &= -N(\mathbf{R}, Z, S)c\, dS\, dA \end{aligned} \tag{B.1.3}$$

where we have defined the volume attenuation coefficient

$$c = \rho(\sigma_A + \sigma_S) = \rho\sigma_T \tag{B.1.4}$$

We must also include the increase in radiance in the direction **S** due to scattering from radiance incident on the differential volume element from other directions **S**′. This can be expressed as

$$dN^+(\mathbf{R}, Z, \mathbf{S}) = dA\, dS \int_{4\pi} \rho f(\mathbf{S}, \mathbf{S}')N(\mathbf{R}, Z, \mathbf{S}')\, d^2\mathbf{\Omega}' \tag{B.1.5}$$

where $f(\mathbf{S}, \mathbf{S}')$ gives the power scattered in the Fraunhoffer diffraction zone in the direction **S** when the particle is illuminated with a unit amplitude plane wave propagating in the direction **S**′. We define the scattering cross section per unit volume:

$$p(\mathbf{S}, \mathbf{S}') = \rho f(\mathbf{S}, \mathbf{S}') \tag{B.1.6}$$

Equation (B.1.5) can then be written

$$dN^+(\mathbf{R}, \mathbf{Z}, \mathbf{S}) = dA\, dS \int_{4\pi} p(\mathbf{S}, \mathbf{S}')N(\mathbf{R}, Z, \mathbf{S}')\, d^2\mathbf{\Omega}' \tag{B.1.7}$$

The total change in the radiance is obtained by adding Equations (B.1.3) and (B.1.5):

$$\begin{aligned} dN(\mathbf{R}, Z, \mathbf{S}) &= dN^-(\mathbf{R}, Z, \mathbf{S}) + dN^+(\mathbf{R}, Z, \mathbf{S}) \\ &= -cN(\mathbf{R}, Z, \mathbf{S})dS + \int_{4\pi} p(\mathbf{S}, \mathbf{S}')N(\mathbf{R}, Z, \mathbf{S}')\, dS\, d^2\mathbf{\Omega}' \end{aligned} \tag{B.1.8}$$

Dividing both sides of Equation (B.1.8) by dS we obtain the radiative transport equation:

$$\frac{d(\boldsymbol{R}, Z, \mathbf{S})}{dS} = -cN(\mathbf{R}, Z, \mathbf{S}) + \int_{4\pi} p(\mathbf{S}, \mathbf{S}')N(\mathbf{R}, Z, \mathbf{S}')\, \mathrm{d}^2\boldsymbol{\Omega}' \qquad \text{(B.1.9)}$$

We can write Equation (B.1.9) in terms of the gradient operator by observing that the left-hand side is just the directional derivative of the radiance function in the direction **S**:

$$\frac{dN(\boldsymbol{R}, Z, \mathbf{S})}{dS} = \mathbf{S} \cdot \nabla N(\mathbf{R}, Z, \mathbf{S}) \qquad \text{(B.1.10)}$$

Substituting Equation (B.1.10) in Equation (B.1.9) we obtain

$$\mathbf{S} \cdot \nabla N(\mathbf{R}, Z, \mathbf{S}) = -cN(\mathbf{R}, Z, \mathbf{S}) + \int_{4\pi} p(\mathbf{S}, \mathbf{S}')N(\mathbf{R}, Z, \mathbf{S}')\, d^2\boldsymbol{\Omega}' \qquad \text{(B.1.11)}$$

We wish to emphasize the fact that the above heuristic derivation of the radiative transport equation is based only on energy balance considerations and assumes that the scattering process is linear in power, not complex field amplitude. This incoherent scattering assumption consists of two parts. First, Equation (B.1.5) implies that the scattered fields from different scatterers are mutually incoherent. Second, Equation (B.1.8) implies that the scattered and unscattered fields are mutually incoherent. It is usually assumed that incoherent scattering is a sufficient condition for the validity of the radiative transport equation. However, the Van Cittert–Zernike theorem of optical coherence theory shows that even spatially incoherent sources can generate fields which are highly coherent over large regions of space, a fact that raises serious questions concerning both the conditions required for the validity of the traditional radiance transport equation and its relationship with electromagnetic theory. In the following sections we will establish the conditions under which the traditional small-angle transport equation can be derived from the parabolic wave optics theory.

B.2. Radiative Transfer Theory in the Small-Angle Scattering Limit

We begin our development with the equation of radiative transfer for a plane-parallel scattering atmosphere in the presence of a source. In particular, for time-independent illumination we have

$$\left(\cos\phi\frac{d}{d\tau}+1\right)N(\tau,\mathbf{r},\theta,\varphi)=N_0(\tau,\mathbf{r},\theta,\varphi)$$
$$+\,\omega_0\int_0^{2\pi}\int_0^{\pi}p(\theta,\phi;\theta',\phi')N(\tau,\mathbf{r},\mu',\phi')\,d\mu',d\varphi' \tag{B.2.1}$$

where $p = P/4\pi$, $\mu = \cos\theta$, $N_0 \equiv$ radiance due to a distributed source within the scattering medium, $z \equiv$ physical thickness of the medium, and $\mathbf{r} \equiv$ planar spatial coordinates. Figure B.2 depicts the spatial and angular coordinate system pictorially. Following Heggestad,[1] we neglect backscatter and apply a bilinear transformation to the above angular coordinates such that the point $y = (\theta, \phi)$ on the $\{\theta, \phi; 0 \leqslant \theta \leqslant \pi/2,\ 0 \leqslant \phi \leqq 2\pi\}$ hemisphere maps to a unique point on a planar surface. Specifically, we define the transformation variables to be

$$\alpha = \theta\cos\phi \text{ radians}$$
$$\beta = \theta\sin\phi \text{ radians}$$

This mapping is illustrated in Figure B.3. This implies that

$$\theta = (\alpha^2+\beta^2)^{1/2} - |\gamma|^2 = \gamma^2$$
$$\phi = \tan^{-1}(\beta/\alpha)$$

The Jacobian for this transformation is equal to θ^{-1}. Recall that the scalar phase functions for the marine and atmospheric channels are generally:

1. Isoplanatic
2. Circularly symmetric
3. Highly forward-biased in scattering.

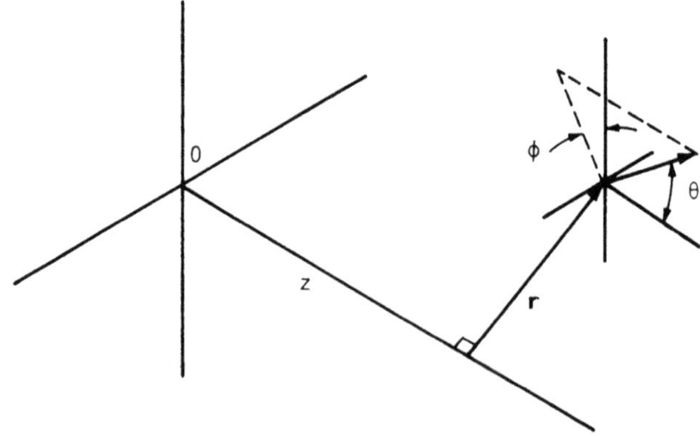

Figure B.2. Spatial and angular coordinate system used in this problem.

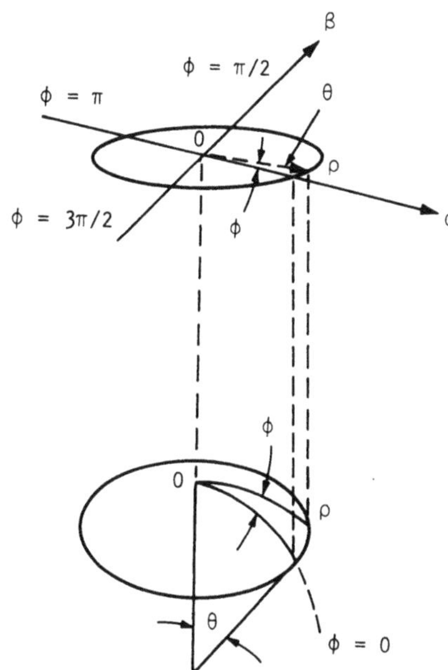

Figure B.3. Mapping a unit sphere onto the (r, θ)-plane.

In light of property (3), we expect most of the scattering events which occur to be of a small-angle scattering nature. Hence one should be able to apply the small-angle scattering approximation to Equation (B.1.1) without significant errors being generated. From Figure B.3, it is apparent that a distance on the plane differs from the equivalent great-circle distance by a factor roughly upper-bounded by

$$\frac{\theta_m}{\sin \theta_m} \tag{B.2.2}$$

where

$$\theta_m = \max [\theta_1, \theta_2]$$

and θ_1 and θ_2 are the two polar angles involved in defining said distances.[1] Hence a 10% error in Equation (B.1.2) occurs when $\theta_m \sim 43° = 0.75$ radius and a 20% error when $\theta_m \sim 50° = 1.03$ radians. Thus, through the use of the chain rule, we can rewrite Equation (B.2.1) in the small-angle scattering limit as

$$\left[\frac{\partial}{\partial Z} + \boldsymbol{\gamma} \cdot \frac{\partial}{\partial \mathbf{r}} + c\right] N(z, \mathbf{r}, \boldsymbol{\gamma}) \approx \mathrm{cN_0}(z, \mathbf{r}, \boldsymbol{\gamma}) + \omega_0 \iint_{|\boldsymbol{\gamma}'| \leq \pi/2} p(|\hat{\boldsymbol{\gamma}} - \boldsymbol{\gamma}'|) N(z, \mathbf{r}, \boldsymbol{\gamma}') \, d^2\boldsymbol{\gamma}' \tag{B.2.3}$$

where $\boldsymbol{\gamma} = (\alpha, \beta)$. In addition, because of the small-angle scattering assumption, we should be able to extend the $|\gamma'| \leqslant \pi/2$ circle to infinity without too much error.[1] In other words, we can write Equation (B.2.3) as

$$\left[\frac{\partial}{\partial z} + \boldsymbol{\gamma} \cdot \frac{\partial}{\partial \mathbf{r}} + c\right] N(z, \mathbf{r}, \boldsymbol{\gamma}) \approx c N_0(z, \mathbf{r}, \boldsymbol{\gamma}) + \omega_0 c \iint_{-\infty}^{\infty} p(|\boldsymbol{\gamma} - \boldsymbol{\gamma}'|) N(z, \mathbf{r}, \boldsymbol{\gamma}')\, d^2\boldsymbol{\gamma}' \tag{B.2.4}$$

Equation (B.2.4) is the equation of radiative transfer under the small-angle scattering approximation.

B.3. Derivation of the Small-Angle Transport Equation from Parabolic Wave Optics Theory

We begin with the parabolic wave optics equation for the propagation of the mutual intensity function which was derived in Section 5.2:

$$2ik_0 \frac{\partial \Gamma(\mathbf{R}, Z, \boldsymbol{\rho})}{\partial z} + 2\boldsymbol{\nabla}_{\boldsymbol{\rho}} \cdot \boldsymbol{\nabla}_{\mathbf{R}} J(\mathbf{R}, Z, \boldsymbol{\rho}) + \frac{ik_0^3}{2} H(Z, \boldsymbol{\rho}) \Gamma(\boldsymbol{R}, Z, \boldsymbol{\rho}) = 0 \tag{B.3.1}$$

Dividing both sides by $2ik_0$ and using the definitions given in Section 5.2

$$H(Z, \boldsymbol{\rho}) = A(Z, 0) - A(Z, \boldsymbol{\rho}) \tag{B.3.2}$$

$$b = k_0^2 A(0)/4 \tag{B.3.3}$$

Equation (B.3.1) can be written:

$$\frac{\partial \Gamma(\mathbf{R}, Z, \boldsymbol{\rho})}{\partial z} + \frac{1}{ik_0} \boldsymbol{\nabla}_{\boldsymbol{\rho}} \cdot \boldsymbol{\nabla}_{\mathbf{R}} \Gamma(\mathbf{R}, Z, \boldsymbol{\rho}) + SJ(\mathbf{R}, Z, \boldsymbol{\rho}) - \frac{k_0^2}{4} A(Z, \boldsymbol{\rho}) \Gamma(\mathbf{R}, Z, \boldsymbol{\rho}) = 0 \tag{B.3.4}$$

Fourier transforming both sides of Equation (B.3.4) with respect to the variable $\boldsymbol{\rho}$, we obtain:

$$\frac{1}{k_0^2} \frac{\partial N_G(\mathbf{R}, Z, \mathbf{k})}{\partial Z} + \frac{1}{k_0^3} \mathbf{k} \cdot \boldsymbol{\nabla}_{\mathbf{R}} N_G(\mathbf{R}, Z, \mathbf{k}) + \frac{b}{k_0^2} N_G(\mathbf{R}, Z, \mathbf{k}) = \frac{k_0^2}{4} \left\{ \frac{1}{(2\pi)^2} \iint_{-\infty}^{\infty} A(Z, \mathbf{P}) \Gamma(\mathbf{R}, Z, \mathbf{P})\, e^{-i\mathbf{k}\cdot\boldsymbol{\rho}} d^2\boldsymbol{\rho} \right\} \tag{B.3.5}$$

where

$$N_G(\mathbf{R}, Z, \mathbf{k}) = \left(\frac{k_0}{2\pi}\right)^2 \int \Gamma(\mathbf{R}, Z, \boldsymbol{\rho})\, e^{-i\mathbf{k}\cdot\boldsymbol{\rho}\, d^2\boldsymbol{\rho}} \tag{B.3.6}$$

Using the identities of Section 5.2:

$$8\pi^3\phi_n(Z, \mathbf{k}) = \int_{-\infty}^{\infty} A(Z, \boldsymbol{\rho})\, e^{-i\mathbf{k}\cdot\boldsymbol{\rho}}\, d^2\boldsymbol{\rho} \tag{B.3.7}$$

the convolution theorem result:

$$\frac{1}{(2\pi)^2}\iint_{-\infty}^{\infty} A(Z, \boldsymbol{\rho})\Gamma(\mathbf{R}, Z, \boldsymbol{\rho})\, e^{-i\mathbf{k}\cdot\boldsymbol{\rho}}\, d^2\boldsymbol{\rho} = \frac{2\pi}{k_0^2}\iint_{-\infty}^{\infty} \Phi_\varepsilon(\mathbf{k}')N_G(\mathbf{R}, t, \mathbf{k} - \mathbf{k}')\, d^2\mathbf{k}' \tag{B.3.8}$$

and the fact that $d^2\Omega \approx e^2\mathbf{k}/\mathbf{k}_0^2$, we have:

$$\frac{1}{k_0^2}\frac{\partial N_G(\mathbf{R}, Z, \mathbf{K})}{\partial Z} + \frac{1}{k_0^3}\mathbf{k}\cdot\nabla_{\mathbf{R}}N_G(\mathbf{R}, Z, \mathbf{k}) + \frac{b}{k_0^2}N_G(\mathbf{R}, Z, \mathbf{k})$$

$$= 2\pi k_0^2 \iint_{2\pi} \phi_n(\mathbf{k}')N_G(\mathbf{R}, Z, \mathbf{k} - \mathbf{k})\, d^2\boldsymbol{\Omega} \tag{B.3.9}$$

Multiplying both sides of Equation (B.3.9) by k_0^2 and using the relationship

$$\beta(\mathbf{k}) = 2\pi k_0^4\Phi_n(\mathbf{k}) \tag{B.3.10}$$

derived in Section 5.2, Equation (B.3.9) can be written:

$$\frac{\partial N_G(\mathbf{R}, Z, \mathbf{k})}{\partial Z} + \frac{1}{k_0^3}k\cdot\nabla_{\mathbf{R}}N_G(\mathbf{R}, Z, \mathbf{k}) + bN_G(\mathbf{R}, Z, \mathbf{k})$$

$$= \int_{-\infty}^{\infty} \beta(\mathbf{k}')N_G(\mathbf{R}, Z, \mathbf{k} - \mathbf{k}')\, d^2\boldsymbol{\Omega}' \tag{B.3.11}$$

Using Equation (B.3.10), Equation (B.3.11) becomes:

$$\frac{\partial N_G(\mathbf{R}, Z, \mathbf{n})}{\partial Z} + \mathbf{n}_1\cdot\nabla_{\mathbf{R}}N_G(\mathbf{R}, Z, \mathbf{n}_1) + bN_G(\mathbf{R}, Z, \mathbf{n}_1)$$

$$= \int_{4\pi} \rho(\mathbf{n}_1')N_G(\mathbf{R}, Z, \mathbf{n}_1 - \mathbf{n}_1')\, d^2\Omega' \tag{B.3.12}$$

From the results of Appendix A it is clear that the small angle transport equation (B.3.11) can be used to describe the propagation of the traditional radiance if and only if the generalized and traditional radiance functions are identical throughout the scattering volume. It can be shown that incoherent scattering is not a sufficient condition for this to occur. Equation (B.3.12) shows that for sources of arbitrary degree of spatial coherence, the small-angle transport equation correctly describes the propagation of the generalized radiance. This equation reduces to a transport equation for the traditional radiance if and only if the generalized and traditional radiance functions are identical throughout the scattering volume.

Appendix C

Solution of the Small-Angle Transport Equation for the Generalized Radiance and the Mutual Coherence Function

The general solution of the small-angle transport equation can be derived by first Fourier transforming Equation (B.3.12) with respect to the variable **R**:

$$\frac{\partial f'(\mathbf{q}, Z, \mathbf{n}_1)}{\partial Z} + i\mathbf{q} \cdot \mathbf{n}_1 f'(\mathbf{q}, Z, \mathbf{n}_1) + if'(\mathbf{q}, Z, \mathbf{n}_1) - \int \beta(\mathbf{n}_1 - \mathbf{n}_1')f'(\mathbf{q}, Z, \mathbf{n}_1')\, d^2\mathbf{n}_1' = 0 \qquad \text{(C.1)}$$

where

$$f'(\mathbf{g}, Z, n_1) = \int_{-\infty}^{\infty} N_G(\mathbf{R}, Z, \mathbf{n}_1)\, e^{-i\mathbf{q}\cdot\mathbf{R}} d^2\mathbf{r} \qquad \text{(C.2)}$$

Equation (C.1) can be further simplified by defining the function

$$f'(\mathbf{q}, Z, \mathbf{n}_1) = f'(\mathbf{g}, Z, \mathbf{n}_1)\, e^{-(i\mathbf{q}\cdot n_1 - \alpha)Z} \qquad \text{(C.3)}$$

Substituting Equation (C.3) in Equation (C.1) we have

$$\frac{\partial f(\mathbf{q}, Z, \mathbf{n}_1)}{\partial Z} - \int \beta(\mathbf{n}_1 - \mathbf{n}_1')f(\mathbf{q}, Z, \mathbf{n}_1')\, e^{-i\mathbf{q}\cdot(\mathbf{n}_1 - \mathbf{n}_1')z} d^2\mathbf{n}_1' = 0 \qquad \text{(C.4)}$$

Fourier transforming Equation (C.4) with respect to the variable n_1 and using the convolution and shift theorems, one obtains the linear first-order differential equation

$$\frac{\partial F(\mathbf{q}, Z, \boldsymbol{\rho})}{\partial Z} - \hat{\beta}(\boldsymbol{\rho} - \mathbf{q}Z)F(\mathbf{q}, Z, \boldsymbol{\rho}) = 0 \tag{C.5}$$

where

$$F(\mathbf{q}, Z, \boldsymbol{\rho}) = \int_{-\infty}^{\infty} f(\mathbf{q}, Z, \mathbf{n}_1)\, e^{i\mathbf{n}_1 \cdot \boldsymbol{\rho}}\, d^2\mathbf{n}_1$$
$$\hat{\beta}(\boldsymbol{\rho}) = \int_{-\infty}^{\infty} \sigma(\mathbf{n}_1)\, e^{i\mathbf{n}_1 \cdot \boldsymbol{\rho}}\, d^2\mathbf{n}_1 \tag{C.6}$$

Equation (C.5) can be written in the form

$$\frac{dF(\mathbf{q}, Z, \boldsymbol{\rho})}{F(\mathbf{q}, Z, \boldsymbol{\rho})} = \hat{\sigma}(\boldsymbol{\rho} - \mathbf{q}Z)\, dZ \tag{C.7}$$

Integrating both sides of Equation (C.7) gives

$$\ln\{F(\mathbf{q}, Z, \boldsymbol{\rho})\} - \ln\{F(\mathbf{q}, 0, \boldsymbol{\rho})\} = \int_0^Z \hat{\sigma}(\boldsymbol{\rho} - \mathbf{q}Z')\, dZ' \tag{C.8}$$

which leads to the result:

$$F(\mathbf{q}, Z, \boldsymbol{\rho}) = F(\mathbf{q}, Z = 0, \boldsymbol{\rho}) \exp\left\{ \int_0^Z \hat{\beta}(\boldsymbol{\rho} - \mathbf{q}Z')\, dZ' \right\} \tag{C.9}$$

Using Equations (C.9), C.6), (C.3), and (C.2), and introducing the change in variables

$$\boldsymbol{\rho}' = (\boldsymbol{\rho} - \mathbf{q}Z)$$
$$\boldsymbol{\eta} = Z - Z' \tag{C.10}$$

and then removing the prime from $\boldsymbol{\rho}'$, we obtain the solution for the generalized radiance:

$$N_G(\mathbf{R}, Z, \mathbf{n}_1) = \left(\frac{1}{2\pi}\right)^4 e^{-\alpha Z} \iint_{-\infty}^{\infty} F(\mathbf{q}, \mathbf{p} + \mathbf{q}Z) \times \exp\left\{ -i(\mathbf{q} \cdot \mathbf{R} - \boldsymbol{\rho} \cdot \mathbf{n}_1) + \int_0^Z \hat{\beta}(\boldsymbol{\rho} - \mathbf{q}Z')\, dz' \right\} d^2\mathbf{q}\, d^2\boldsymbol{\rho} \tag{C.11}$$

where

$$F(\mathbf{q}, \boldsymbol{\rho}) = \iint_{-\infty}^{\infty} N_G(\mathbf{r}, Z = 0, \mathbf{n}_1)\, e^{i(\mathbf{q}\cdot\mathbf{R}+\boldsymbol{\rho}\cdot\mathbf{n}_1}\, \mathrm{d}^2\mathbf{r}\, d^2\mathbf{n}_1 \tag{C.12}$$

We now wish to find the solution for the mutual intensity function, $J(\mathbf{R}, Z, \boldsymbol{\rho})$. We begin by introducing the change of variables

$$\boldsymbol{\rho}' = \boldsymbol{\rho}/k_0 \tag{C.13}$$

in Equation (C.11) and then remove the prime from $\boldsymbol{\rho}'$. The result is

$$N_G(\mathbf{R}, Z, \mathbf{k}) = \left[\frac{k_0^2}{(2\pi)^4}\right] e^{-\alpha Z} \iint_{-\infty} - \infty\, F(\mathbf{q}, k_0\boldsymbol{\rho} + \mathbf{q}Z)$$

$$\times \exp\left\{-i(\mathbf{q}\cdot\mathbf{R} - \boldsymbol{\rho}\cdot\mathbf{k}) + \int_0^Z \hat{\beta}(\mathbf{k}_0\boldsymbol{\rho} + \mathbf{q}\eta)\, d\eta\right\} d^2\mathbf{q}\, d^2\boldsymbol{\rho} \tag{C.14}$$

with

$$\mathrm{F}(\mathbf{q}, \mathbf{r}) = \frac{1}{k_0^2} \int N_G(\mathbf{R}, Z = 0, \boldsymbol{\rho})\, e^{i(\mathbf{q}\cdot R+\boldsymbol{\rho}\cdot\mathbf{k}/\mathbf{k}_0)}\, d^2\mathbf{k} \tag{C.15}$$

Using the definitions of generalized radiance, Equations (A.1.1) and (A.1.3), we find:

$$\Gamma(\mathbf{R}, Z, \boldsymbol{\rho}) = \left(\frac{1}{2\pi}\right)^2 e^{-cZ} \int_{-\infty}^{\infty} F(\mathbf{q}, \mathbf{k}_0\boldsymbol{\rho} + qZ)$$

$$\times \exp\left\{-i\mathbf{q}\cdot\mathbf{R} + \int_0^Z \hat{\beta}(k_0\boldsymbol{\rho} + \mathbf{q}\eta)\, d\eta\right\} d^2\mathbf{q} \tag{C.16}$$

where

$$F(\mathbf{g}, \boldsymbol{\rho}) = \int_{-\infty}^{\infty} \Gamma(\mathbf{R}, Z = 0, \boldsymbol{\rho}/k_0)\, e^{i\mathbf{q}\cdot\mathbf{R}}\, d^2\mathbf{R} \tag{C.17}$$

A series of straightforward manipulations allows Equation (C.16) to be expressed in the following form which displays explicitly the dependence of our solution on the wavenumber spectrum of index of refraction fluctuations:

$$\Gamma(\mathbf{R}, Z, \boldsymbol{\rho}) = \left(\frac{k}{2\pi Z}\right)^2 \iint_{-\infty}^{\infty} \Gamma(\mathbf{R}, Z = 0, \rho)\, e^{i(k/x)(\boldsymbol{\rho}-\boldsymbol{\rho}')\cdot(\mathbf{R}-\mathbf{R}')-H}\, d^2\mathbf{R}'\, d^2\boldsymbol{\rho}' \tag{C.18}$$

where

$$H = \frac{k^2}{4}\int_0^Z \left\{A(0) - A\left[\boldsymbol{\rho}' + (\boldsymbol{\rho} - \boldsymbol{\rho}')\frac{z}{Z}\right]\right\} dz$$

$$= 4\pi^2 k^2 \int_0^Z \int_0^\infty [1 - J_0(\kappa p)]\Phi_n(\kappa)\kappa \, d\kappa$$

$$p = \left|\boldsymbol{\rho}' + (\boldsymbol{\rho} - \boldsymbol{\rho}')\frac{z}{Z}\right| \tag{C.19}$$

When $H = 0$, the effects of the random medium are removed from our solution and Equation (C.17) reduces to the parabolic or Fresnel free-space propagation equation for the mutual coherence function.

Appendix D

Power Flow in Fibers

In this section we will derive a power rate equation for electromagnetic propagation in fibers. The development closely parallels that given by Gloge[1] and begins by first considering propagation in a slab where m electromagnetic modes propagate at angles

$$\theta_m = m\lambda/4an$$

with a the radius of the slab, n its index of refraction, and λ the wavelength of the propagating electromagnetic radiation.

The variation in power, dP_m, of the mth mode is attributable to (1) scattering loss to the outside of the fiber, which we account for with the attenuation coefficient, α_m, for the mth mode, and (2) coupling to other modes through the coupling coefficient d_m between modes of order $m + 1$ and m. Taking these two factors into account we obtain the equation

$$\frac{dP_m}{dz} = -\alpha_m P_m + d_m(P_{m+1} - P_m) + d_{m-1}(P_{m-1} - P_m) \qquad \text{(D.1)}$$

Considering a continuum of modes we set

$$\frac{P_{m+1} - P_m}{\theta_{m+1} - \theta_m} = \frac{dP_m}{d\theta}$$

$$\theta_m - \theta_{m-1} = \Delta\theta \qquad \text{(D.2)}$$

and Equation (D.1) becomes

$$\frac{dP_m}{dz} = -\alpha_m P_m + \Delta\theta\left(d_m\frac{dP_m}{d\theta} - d_{m-1}\frac{dP_{m-1}}{d\theta}\right) \qquad \text{(D.3)}$$

Replacing the index m by a functional dependence on θ we obtain the equation for power flow

$$\frac{\partial P}{\partial z} = -\alpha(\theta)P + (\Delta\theta)^2 \frac{\partial}{\partial\theta}\left[d(\theta)\frac{\partial P}{\partial\theta}\right] \tag{D.4}$$

In the case of a cylindrical fiber, the index m signifies a group of m modes. Summing Equation (D.3) over all m modes in the mth group, we obtain

$$m\frac{dP_m}{dz} = -m\alpha_m P_m + md_m(P_{m+1} - P_m) + (m-1)\,d_{m-1}(P_{m-1} - P_m) \tag{D.5}$$

where α_m and d_m are the same for all modes in a group and we have used the fact that coupling to the group $(m-1)$ occurs only between $(m-1)$ members. Using steps similar to Equations (D.2) and (D.3) and replacing the dependence on m with a functional dependence on θ, we have

$$\frac{\partial P}{\partial z} = -\alpha(\theta)P + (\Delta\theta)^2 \frac{1}{\theta}\frac{\partial}{\partial\theta}\left[\theta d(\theta)\frac{\partial P}{\partial\theta}\right] \tag{D.6}$$

Next, using symmetry, we expand $\alpha(\theta)$ in the form

$$\alpha(\theta) = \alpha_0 + A\theta^2 + \cdots \tag{D.7}$$

and retain only the second-order term in θ. The term α_0 accounts for a loss common to all modes and can be accounted for by multiplying the final solution by $\exp(-\alpha_0 z)$. Since the power density at the core-cladding interface increases quadratically with θ, the second term accounts for losses at the interface. The coupling coefficient is also expanded in a power series in θ and the first term only is retained:

$$d(\theta) = d_0 \tag{D.8}$$

If we define

$$D = (\Delta\theta)^2\, d_0 = \left(\frac{\lambda}{4an}\right)^2 d_0 \tag{D.9}$$

Equations (D.4) and (D.5) can be written

$$\frac{\partial P}{\partial z} = \begin{cases} -A\theta^2 P + D\dfrac{\partial^2 p}{\partial \theta^2} & \text{for the slab} \\ \\ \tau A\theta^2 P + \dfrac{D}{\theta}\dfrac{\partial}{\partial \theta}\left(\theta \dfrac{\partial P}{\partial \theta}\right) & \text{for the fiber} \end{cases} \tag{D.10}$$

If P is a function of time we also have

$$dP = \frac{\partial P}{\partial z}\,dz + \frac{\partial P}{\partial t}\,dt \tag{D.11}$$

Equating Equations (D.10) and (D.11) and dividing by dz gives the equation

$$\frac{\partial P}{\partial z} + \frac{dt}{dz}\frac{\partial P}{\partial t} = -A\theta^2 P + \frac{1}{\theta}\frac{\partial}{\partial \theta}\left(\theta D \frac{\partial P}{\partial \theta}\right) \tag{D.12}$$

The derivative dz/dt is the group velocity at angle θ. Relative to the center mode, this is approximately $dz/dt \approx 2c/n\theta^2$. Substituting in (D.12), and inserting the Laplace transform

$$\hat{P}(\theta, z, s) = \int_{-\infty}^{\infty} e^{-st} P(\theta, z, t)\,dt \tag{D.13}$$

allows us to write (D.12) in the transform domain as

$$\frac{\partial \hat{P}}{\partial z} = -A\sigma^2\theta^2\hat{P} + \frac{1}{\theta}\frac{\partial}{\partial \theta}\left[\theta D \frac{\partial \hat{P}}{\partial \theta}\right] \tag{D.14}$$

where $\sigma^2 = [1 + (ns/2cA)]$. For an initial insertion function $\hat{P}_0 = f(0, s)\,e^{-\theta^2/\theta_0^2}$, (D.14) has the transform solution

$$\hat{P}(\theta, z, s) = f(z, s)\,e^{-\theta^2/\phi^2(z,s)} \tag{D.15}$$

where

$$\phi^2(z, s) = \frac{\theta_\infty[\sigma\theta_0^2 + \theta_\infty \tanh(\sigma\gamma_\infty z)]}{\sigma[\theta_\infty^2 + \sigma\theta_0^2 \tanh(\sigma\gamma_\infty z)]}$$

$$f(z, s) = \frac{f(0, s)\sigma\theta_0^2}{\theta_\infty^2 \sinh(\sigma\theta_\infty z) + \sigma\theta_0^2 \cosh(\sigma\gamma_\infty z)}$$

$$\theta_\infty = (4D/A)^{1/4}$$

$$\gamma_\infty = (4D/A)^{1/2}$$

Limiting forms of (D.15) are used in Section 4.3 of Chapter 4.

Reference

1. D. Gloge, Impulse response of clad optical multimode fibers, *Bell System Tech. J.* **52**, 801-816 (1973).

Appendix E

Optical and Physical Thickness Relations for Clouds at Various Locations on the Earth

After clear atmospheric conditions, clouds are the most common constituents of the optical scatter channel. Unfortunately, they are also the most stressing to optimum communication system performance and need to be considered from a statistical occurrence point of view. This is because time availability of the prospective optical link is a critical requirement of any system design, and cloud-induced propagation losses will set the necessary technology to meet the specification. For plane-wave source illumination scenarios, optical communication system designers need only take into account two cloud parameters in their link budget calculation (see Section 7.4). One is cloud optical thickness and the other is cloud physical thickness. Ciany *et al.* used empirically derived mean free path estimates and the Air Force Global Weather Central (AFGWC) 3DNEPH data base

Table E.1. Extinction Length versus Optical Thickness Used in Navy Laser Communications Simulation Studies

Optical thickness	Extinction length, b (m)
$0 \leq \tau < 11.2$	350
$11.2 \leq \tau < 18.5$	65
$18.5 \leq \tau < 27.5$	50
$27.5 \leq \tau < 36.5$	45
$36.5 \leq \tau < 44.0$	20
$44.0 \leq \tau < 63.0$	55
$63.0 \leq \tau < 97.0$	15
$97.0 \geq \tau$	25

Table E.2. Least-Squares Coefficients to 3DNEPH Thickness Distribution Curve Fits for Locations in Climatology Study

(°Lat., °long.)	N	f_0	a_0	a_1	a_2	a_3	a_4	a_5
(15 N, 130 E)	3	0.5657	0.49021E + 03	−0.18676E + 04	0.20455E + 04	−0.45536E + 03		
(25 N, 135 E)	3	0.6373	−0.3711E + 04	0.14371E + 05	−0.18511E + 05	0.80773E + 04		
(30 N, 150 E)	3	0.5356	−0.14118E + 04	0.60584E + 04	−0.86365E + 04	0.41943E + 04		
(40 N, 150 E)	3	0.3642	−0.27997E + 03	0.13528E + 04	−0.20256E + 04	0.11585E + 04		
(40 N, 165 E)	3	0.3175	−0.23589E + 03	0.12917E + 04	−0.21389E + 04	0.12939E + 04		
(40 N, 180 W)	3	0.4056	−0.53818E + 03	0.25015E + 04	−0.36658E + 04	0.18976E + 04		
(50 N, 165 E)	3	0.3579	−0.26036E + 03	0.12862E + 04	−0.19747E + 04	0.11554E + 04		
(50 N, 180 W)	3	0.2880	0.14918E + 03	0.83092E + 03	−0.13347E + 04	0.86181E + 03		
(60 N, 180 W)	3	0.2555	−0.14056E + 03	0.88687E + 03	−0.15805E + 04	0.10275E + 04		
(50 N, 165 W)	3	0.3307	−0.22535E + 03	0.11649E + 04	−0.18307E + 04	0.11154E + 04		
(20 N, 180 W)	4	0.6824	0.68687E + 05	−0.34118E + 06	0.63241E + 06	−0.51867E + 06	0.1589E + 06	
(30 N, 165 W)	3	0.5977	−0.16514E + 04	0.64740E + 04	−0.85105E + 04	0.38507E + 04		
(20 N, 165 W)	3	0.5524	−0.19181E + 04	0.79777E + 04	−0.11011E + 05	0.51683E + 04		
(20 N, 150 W)	3	0.6466	−0.47331E + 04	0.18194E + 05	−0.23331E + 05	0.10074E + 05		
(30 N, 150 W)	3	0.6086	−0.27808E + 04	0.11047E + 05	−0.14629E + 05	0.65481E + 04		
(30 N, 135 W)	3	0.5132	−0.13905E + 04	0.58560E + 04	−0.81064E + 04	0.38489E + 04		
(20 N, 120 W)	3	0.6044	−0.22301E + 04	0.86423E + 04	−0.11138E + 05	0.48707E + 04		
(20 N, 105 W)	3	0.6225	−0.18891E + 04	0.76333E + 04	−0.10344E + 05	0.47497E + 04		
(30 N, 120 W)	3	0.6923	−0.45813E + 04	0.17336E + 05	−0.21975E + 05	0.93783E + 04		
(30 N, 180 W)	3	0.6105	−0.16028E + 04	0.65780E + 04	−0.90887E + 04	0.42822E + 04		
(30 N, 165 E)	3	0.6302	−0.41800E + 04	0.16364E + 05	−0.21300E + 05	0.92964E + 04		
(40 N, 165 W)	3	0.4377	−0.78285E + 03	0.35953E + 04	−0.53078E + 04	0.26959E + 04		
(40 N, 150 W)	3	0.3807	−0.26998E + 03	0.12451E + 04	−0.17960E + 04	0.10197E + 04		

(40 N, 135 W)	3	0.4961	−0.97806E + 03	0.41322E + 04	−0.57186E + 04	0.27478E + 04		
(50 N, 150 W)	3	0.3516	−0.28646E + 03	0.14238E + 04	−0.21651E + 04	0.12309E + 04		
(50 N, 135 W)	3	0.3742	−0.22408E + 03	0.10294E + 04	−0.14769E + 04	0.87207E + 03		
(30 N, 75 W)	3	0.6327	−0.27112E + 04	0.10907E + 04	−0.14727E + 05	0.67346E + 04		
(30 N, 60 W)	3	0.6825	−0.41823E + 04	0.15559E + 05	−0.19396E + 05	0.81723E + 04		
(30 N, 45 W)	3	0.6877	−0.86521E + 04	0.32552E + 05	−0.40777E + 05	0.17067E + 05		
(30 N, 30 W)	4	0.7188	0.51088E + 05	−0.2558E + 06	0.47749E + 06	−0.39378E + 06	0.12120E + 06	
(36 N, 15 E)	3	0.6168	−0.71508E + 04	0.28702E + 05	−0.37991E + 05	0.16625E + 05		
(39 N, 74 W)	3	0.4406	−0.76731E + 03	0.38442E + 04	−0.63146E + 04	0.35003E + 04		
(50 N, 60 W)	3	0.3723	−0.38159E + 03	0.19986E + 04	−0.33274E + 04	0.1932E + 04		
(40 N, 60 W)	3	0.3522	−0.94651E + 02	0.39688E + 03	−0.51067E + 03	0.41702E + 03		
(40 N, 45 W)	3	0.4710	−0.56176E + 03	0.25021E + 04	−0.37045E + 04	0.19627E + 04		
(50 N, 45 W)	3	0.2679	−0.12116E + 03	0.70420E + 03	−0.11462E + 04	0.76809E + 03		
(50 N, 30 W)	3	0.3257	−0.14502E + 03	0.73016E + 03	−0.11120E + 04	0.72873E + 03		
(40 N, 30 W)	3	0.4986	−0.11924E + 04	0.53915E + 04	−0.80030E + 04	0.39833E + 04		
(40 N, 15 W)	3	0.5611	−0.16022E + 04	0.66000E + 04	−0.90262E + 04	0.41929E + 04		
(50 N, 15 W)	3	0.3277	−0.15279E + 03	0.75818E + 03	−0.11238E + 04	0.71074E + 03		
(35 N, 30 E)	3	0.7709	−0.36640E + 05	0.12809E + 06	−0.14909E + 06	0.57838E + 05		
(60 N, 46 W)	3	0.3756	−0.38695E + 03	0.20296E + 04	−0.34031E + 04	0.19765E + 04		
(67 N, 26 W)	3	0.2339	−0.99433E + 02	0.65444E + 03	−0.11787E + 04	0.84662E + 03		
(60 N, 30 W)	3	0.2852	−0.11326E + 03	0.6710E + 03	−0.10002E + 04	0.67947E + 03		
(60 N, 15 W)	3	0.3206	−0.19413E + 03	0.10649E + 04	−0.17907E + 04	0.1163E + 04		
(70 N, 15 W)	3	0.4621	−0.79140E + 03	0.36455E + 04	−0.54635E + 04	0.27713E + 04		
(70 N, 0 W)	3	0.3549	−0.20658E + 03	0.10509E + 04	−0.16587E + 04	0.95151E + 03		
(60 N, 0 W)	3	0.2404	−0.11791E + 03	0.75450E + 03	−0.1391E + 04	0.876694 + 03		
(50 N, 0 W)	3	0.2856	−0.13826E + 03	0.78299E + 03	−0.12858E + 04	0.83644E + 03		
(40 N, 0 W)	5	0.5106	−0.23734E + 05	0.17225E + 06	−0.49405E + 06	0.69989E + 06	−0.48964E + 06	0.13543E + 06

to develop least-mean-square polynomial relations for optical and physical thickness statistics.[1] In particular, they derived polynomial expressions for the cumulative probability distributions of optical and physical thickness at several locations on the earth. Their intent was to give a feel for the type of cloud conditions one might encounter for Navy Laser Communications. This appendix presents those relations in a form useful to the optical communication system designer concerned with that environment.

In their study, Ciany *et al.* showed that analytical expressions for the cumulative probability distribution for cloud optical thickness could be derived from the AFGWC 3DNEPH data base in the form

$$\geqq \; = \begin{cases} \sum_{i=0}^{N} a_i f^i, & f_0 < f \leqslant 1 \\ 0, & 0 \leqslant f \leqslant f_0 \end{cases} \tag{E.1}$$

where f is the probability that the optical thickness $< \tau$, f_0 is the probability that the optical thickness $= 0$, the a_i are least-squares coefficients, and N is the order of the polynomial fit. In a link budget calculation, f would be the desired time availability of the prospective optical communication link and the corresponding optical thickness would be computed using Equation (E.1). Precision of this estimate was shown to be accurate to within a few percent.[1] By using the characteristic optical thickness data shown in Table E.1, an extinction length for the computed cloud optical thickness can be identified and a corresponding physical thickness for this cloud can be calculated. Recent experimental work indicates that the values in Table E.1 may be conservative estimates of extinction length, but more measurements will be necessary to confirm this.[2]

Table E.2 is a summary of the least-squares coefficients for the 50 ocean locations processed in the study.[1] These coefficients, when used in Equation (E.1), indicate that the range of cloud optical and physical thicknesses one must overcome for high link availability are worse in the upper North Atlantic latitudes than anywhere else. For example, in similar water conditions, a 90% time availability system requirement implies that enough system margin must exist to compensate for $\tau = 145$ cloud conditions in the North Atlantic (67° N, 25° W) versus $\tau = 50$ cloud conditions for the Central Pacific area (30° N, 120° W). In addition, the analytical results show that the upper-latitude cloud conditions in both oceans are worse than in the lower latitudes near the equator.

References

1. C. Ciany, R. Farrel, and F. Saum, Laser Communications Climatology Study, McDonnell-Douglas Astronautics, Final Report, Contract No. N00014-78-C-0539 (August 31, 1979).
2. G. M. Lee, Titan Systems Inc., private communication.

Appendix F

Atmospheric Optical Loss Model

Figure F.1 depicts the assumed source/receiver geometry. Let

$$R \equiv \text{radius of the "4/3" earth}$$
$$= 8393 \text{ km}$$
$$R + h_0 \equiv \text{distance from the center of the earth to the source at } \mathbf{r}_0$$
$$R + h_R \equiv \text{distance from the center of the earth to the receiver}$$
$$R + h \equiv \text{distance from the center of the earth to the point } \mathbf{r}$$

Referring to the figure, let the coordinate origin be located at the transmitter such that the $\hat{z}$ axis points to the right, $\hat{x}$ points upwards, and $\hat{y}$ points out of the page. Thus, the position of the center of the earth would be given by

$$\mathbf{r}_e = (R + h_{R}, \pi/2, \pi)$$

Hence, the angle, α, between $\mathbf{r}_0$ and r_e is equal to

$$\cos^{-1}(-\sin\theta_0 \cos\varphi_0)$$

where $\mathbf{r}_0 = (r_0, \theta_0, \varphi_0)$. This implies that the distance from the surface of the earth to the point $\mathbf{p}$ along the $(R + h)$-projection is given by

$$h = (R' + r^2 + 2R'r \sin\theta_0 \cos\varphi_0)^{1/2} - R$$

where $|\mathbf{p}| = r$ and $R' = R + h_R$.

Using classical radiative transfer theory, i.e., the Lambert-Bouger law, it can be shown that the atmospheric attenuation term, or transmissivity,

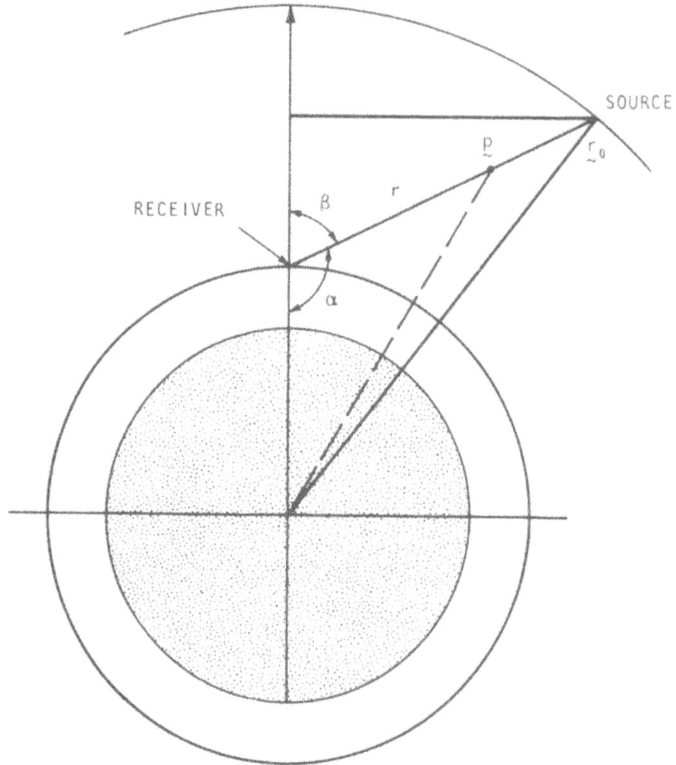

Figure F.1. Source/receiver geometry.

has the form

$$\tau(\mathbf{r}_0) = \exp\left\{-\int_0^{r_0} \alpha[h(r)]\,dr\right\}$$

where α is the total atmospheric attenuation coefficient. In general, one finds that this coefficient falls off exponentially. That is,

$$\alpha = A_0 \exp\{-A_1[(R'^2 + r^2 + 2R'r\sin\theta_0\cos\varphi_0)^{1/2} - R]\}$$

where A_0, A_1 denote coefficients proportional to inverse meters. Unfortunately, the integration of α over the range does not yield a closed-form solution. However, the fact that $R' \gg r_0$ within our range of interest does allow us to approximate α in such a way as to give closed-form representations for τ.

$$\text{CASE I:}\quad R'^2 \gg r_0^2 \gg /2r_0R' \sin\theta_0 \cos\varphi_0$$

Under this condition, we find that

$$(R'^2 + r^2 + 2rR \sin\theta_0 \cos\varphi_0)^{1/2} - R$$
$$\cong h_R + \frac{r^2}{2R'}[1 - \sin^2\theta_0 \cos^2\varphi_0] + r\sin\theta_0\cos\varphi_0$$

This implies that

$$\int_0^{r_0} \alpha(r)\,dr \approx A_0\sqrt{\frac{\pi R'}{A_1(1-\sin^2\theta_0\cos^2\varphi_0)}}$$
$$\times \exp\left\{-\frac{2A_1R'\cos^2\varphi_0\sin^2\theta_0}{(1-\sin^2\theta_0\cos^2\varphi_0} - A_1h_R\right\}$$
$$\times\left\{\operatorname{erf}\left(\frac{A_1(1-\sin^2\theta_0\cos^2\varphi_0}{2R'}\right)^{1/2} r_0\right.$$
$$+\frac{(2R'A_1)^{1/2}\sin\theta_0\cos\varphi_0}{(1-\sin^2\theta_0\cos^2\varphi_0)^{1/2}}$$
$$\left.-\operatorname{erf}\left[\frac{(2R'A_1\sin\theta_0\cos\varphi_0)^{1/2}}{(1-\sin^2\theta_0\cos^2\varphi_0)^{1/2}}\right]\right\}$$

where erf(x) is the error function.

$$\text{Case II:}\quad R'^2 \gg |2r_0R'\sin\theta_0\cos\varphi_0| \gg r_0^2$$

Under this condition,

$$(R'^2 + r^2 + 2rR'\sin\theta_0\cos\varphi_0)^{1/2} - R \cong h_R + r\cos\varphi_0\sin\theta_0$$

Hence,

$$\int_0^{r_0} \alpha(r)\,dr \cong \frac{A_0\sec\varphi_0\csc}{A_1}\theta_0\, e^{-A_1h_R}(1 - e^{-A_1r_0\sin\theta_0\cos\varphi_0})$$

Reference

1. W. G. Wells, G. Gal, and M. W. Munn, Aerosol distributions in maritime air and predicted scattering coefficients in the infrared, *Appl. Opt.* **16**, 654–659 (1977).

Index

GPSR Compliance
The European Union's (EU) General Product Safety Regulation (GPSR) is a set of rules that requires consumer products to be safe and our obligations to ensure this.

If you have any concerns about our products, you can contact us on

ProductSafety@springernature.com

In case Publisher is established outside the EU, the EU authorized representative is:

Springer Nature Customer Service Center GmbH
Europaplatz 3
69115 Heidelberg, Germany

www.ingramcontent.com/pod-product-compliance
Ingram Content Group UK Ltd.
Pitfield, Milton Keynes, MK11 3LW, UK
UKHW012148240726
13966UKWH00001B/207
9781489908070